S0-AWV-367

Surface Engineering
Volume III: Process Technology and Surface Analysis

Surface Engineering
Volume III: Process Technology and Surface Analysis

Edited by

P.K. Datta and J.S. Gray

University of Northumbria at Newcastle, Newcastle upon Tyne

Based on the Proceedings of the Third International Conference on Advances in Coatings and Surface Engineering for Corrosion and Wear Resistance, and the First European Workshop on Surface Engineering Technologies and Applications for SMEs. Both events were held at the University of Northumbria at Newcastle, Newcastle upon Tyne, UK, on 11–15 May 1992.

The front cover illustration was kindly provided by Metco Ltd., UK.

Special Publication No. 128

ISBN 0-85186-685-9

A catalogue record for this book is available from the British Library

Published by The Royal Society of Chemistry, Thomas Graham House, Science Park, Cambridge CB4 4WF

Printed in Great Britain by Redwood Books Ltd., Trowbridge, Wiltshire

Preface

Surface Engineering is an enabling technology widely applicable to a range of industrial sector activities. It encompasses techniques and processes capable of creating and/or modifying surfaces to provide enhanced performance such as wear, corrosion and fatigue resistance and bio-compatibility. Surface engineering processes can now be used to produce multilayer and multicomponent surfaces, graded surfaces with novel properties and surfaces with highly non-equilibrium structures. In a broad sense surface engineering covers three interrelated activities:-

1. *Optimization of surface/substrate properties and performance* in terms of corrosion, adhesion, wear and other physical and mechanical properties.

2. *Coatings technology* including the more traditional techniques of painting, electroplating, weld surfacing, plasma and hypervelocity spraying, various thermal and thermochemical treatments such as nitriding and carburizing, as well as the newer processes of laser surfacing, physical and chemical vapour deposition, ion implantation and ion mixing.

3. *Characterization and evaluation of surfaces and interfaces* in terms of composition and morphology, and mechanical, electrical and optical properties.

It is now widely recognized that the successful exploitation of these processes and coatings may enable the use of simpler, cheaper and more easily available substrate or base materials, with substantial reduction in costs, minimization of demands for strategic materials and improvement in fabricability and performance. In demanding situations where the technology becomes constrained by surface-related requirements, the use of specially developed coating systems may represent the only real possibility for exploitation.

The three volumes of ***"Surface Engineering"*** have been prepared focusing attention on these comments and have been principally based on papers presented at the 3rd International Conference on Advances in Coatings and Surface Engineering for Corrosion and Wear Resistance. The structure and contents of the volumes in this series have been conceived to provide a number of interrelated themes and a coherent philosophy. Additional material has been incorporated to complement the information delivered at the Conference. As such, the text provides a useful blend of Keynote, review, scientific and state-of-the-art type papers by international authorities and experts in surface engineering.

"Surface Engineering" is structured in three volumes. The first volume, *Fundamentals of Coatings* considers principles of coating/substrate design in high temperature and aqueous corrosion and wear environments, and scans the coatings' spectrum from organic, through metallic to inorganic. Here there is a general emphasis on the science and design of coating/substrate systems rather than technology. The second volume, *Engineering Applications*, is dedicated to topics concerning the performance of coatings and surface treatments embracing four main areas - the inhibition of wear and fatigue, corrosion control, application of coatings in heat engines and machining, and quality and properties of coatings. Finally, the third volume has two main thrusts: *Process Technology and Surface Analysis*. Both areas are clearly central to surface engineering, and each holds particular promise, not only for improvements in existing types of coatings' performance, but also in the design, development and evaluation of totally new - for example hybrid - coating/substrate systems.

The editors wish to pay tribute to Dr. Tom Rhys-Jones who recently died.

Contents

PART 3: PROCESS TECHNOLOGY

Section 3.4 PVD, Sputtering, Plasma Nitriding, and Ion Implantation

PART 4: SURFACE ANALYSIS

Section 4.1 Corrosion Studies

Section 4.2 Wear Studies

Section 4.3 Surface Integrity and Properties

Contents
Volume I: Fundamentals of Coatings

Section 1.5 Metallic Coatings

Section 1.6 Ceramic and Glass Ceramic Coatings

Contents
Volume II: Engineering Applications

PART 2: ENGINEERING APPLICATIONS

Section 2.1 Wear and Fatigue Resistance

Section 2.2 Corrosion Control

Section 2.3 Heat Engines

Section 2.4 Machining

Acknowledgements

The editors wish to express their gratitude for the support extended by:

> The Commission of the European Communities, The Department of Trade and Industry, The Institute of Materials, The Institute of Corrosion, Northern Electric, METCO Ltd., Cobalt Development Institute, LECO Instruments and Severn Furnaces.

Special thanks are due to Dr N E W Hartley of the CEC for his continual work and support of surface engineering.

The support and encouragement of Dr C Armstrong, Head of the Department of Mechanical Engineering and Manufacturing Systems, is gratefully acknowledged. Thanks are also due to Prof J Rear for opening the Conference.

The human commitment to any conference or book is substantial and often not fully acknowledged. In this final regard the work of Kath Hynes, Pauline Bailey, the technicians from the Department of Mechanical Engineering and Manufacturing Systems and the members of the Surface Engineering Research Group should be fully recognized.

Finally special commendation is reserved for Lee Comstock who administered the 3rd International Conference on Advances in Coating and Surface Engineering for Corrosion and Wear Resistance and the 1st European Workshop on the Application of Surface Engineering for SMEs, and was also responsible for the retyping of many conference papers.

P.K. Datta
J.S. Gray
University of Northumbria at Newcastle

Part 3: Process Technology

Section 3.1 Thermal Spraying and Hardsurfacing

3.1.1
Development and Applications of Corrosion Resistant Thermal Sprayed Coatings

M. R. Dorfman, B. A. Kushner, A. J. Rotolico, and J. A. DeBarro

METCO DIVISION / PERKIN-ELMER, 1101 PROSPECT AVENUE, WESTBURY, NY 11590, USA

1 INTRODUCTION

In the past, the application of thermal sprayed coatings used to protect steel components for various types of corrosion have been limited due to their inherent heterogeneous nature. Overlays applied by conventional plasma, combustion and electric arc processes typically demonstrate porosity, oxides, unmelted particles and interparticle boundaries. Porosity and interparticle boundaries act as passage ways for the electrolyte to penetrate through to the substrate.[2] If a coating is more noble than the base material, which is the case for many ferrous, nickel and cobalt base alloys, the substrate is attacked and the coating will be undermined. Active materials, such as zinc and aluminum, are used to cathodically protect steel structures for industrial and marine environments, but do not have the wear resistance of iron, nickel and/or cobalt alloys. Due to the rapid solidification associated with these processes grain sizes are fine (usually between 1-10 microns). Comparable wrought materials are dense and more homogeneous with larger grain sizes which are dependent on the thermal and mechanical history of the alloy. Grain size and chemical homogeneity are important in understanding the electrochemistry of materials. Typically, the smaller the grain size the greater the chemical activity of a particular material. Materials with smaller grain sizes offer a higher area of attack sites. Reduced chemical homogeneity results in an increase in galvanic action between various phases.

In addition to microstructural concerns, thermal spray coatings are often deficient in critical elements needed to generate passive films. A material's ability to generate, maintain and reform dense adherent oxide films is necessary for the corrosion resistance of many alloys. Elements such as chromium are necessary in the design of stainless steel alloys.[2-4] Standard stainless steel alloys, however, experience oxidation and volatilization during the spraying process. The interaction of the environment, process and material influences coating chemistry. The specific loss of chromium is related to the amount of dwell time a particular wire or powder experiences in the flame. This results in a loss of corrosion protection. Powder or wire chemistries must be modified for anticipated elemental loss when using conventional thermal spray processes. An alternate approach to modifying traditional material chemistries in order to maintain coating chemistry is to use the HVOF process. The HVOF process minimizes decomposition loss of critical elements and achieves high density coatings.

The HVOF process represents the state-of-the-art for thermal sprayed metallic coatings. The technology uses extremely high kinetic energy and controlled thermal energy output to produce very low porosity coatings that exhibit high bond strength, fine as-sprayed surface finish, and low residual stresses.

The process operates with an oxygen-fuel mixture consisting of either propylene, propane, or hydrogen fuel gas depending on coating requirements. Fuel gases flow through a siphon system where they are thoroughly mixed with oxygen. In one design, the mixed gases are ejected from the gun nozzle and are ignited. The high velocity gases produce a unique characteristic of multiple shock diamond patterns which are visible in the flame. Combustion temperatures approach 2745°C (5000°F) and form a circular flame configuration. Powder is injected into the flame axially to provide uniform heating and powder particles are accelerated by the high velocity gases. The velocity typically approaches 1365 m/sec (4500 ft./sec.). The low residual coating stress produced by the HVOF process allows for significantly greater thickness capability than the plasma method while providing lower porosity, lower oxide content, and higher coating adhesion.[5]

Therefore, to better understand the behavior of thermal spray coatings, the electrochemical behavior of wrought 316 alloys has been compared to its thermal spray counterpart. In addition, two other commercially available austenitic state-of-the-art stainless steels and one proprietary material were also evaluated in this paper to determine if spraying "hi-tech" stainless steels offer any advantage to less exotic, inexpensive materials. The effect of spray process was investigated to determine if there is an advantage of using the HVOF process compared to traditional combustion and/or plasma processes.

2 EXPERIMENTAL

The objective in this paper may be met by evaluating the electrochemical behavior of four stainless steel coatings along with wrought Type 316 stainless steel and low carbon steel substrates. Chemistries for these powders can be seen in Table 1.

Table 1 Chemical Composition of Various Powders (Wt.%)

Sample	Fe	Ni	Cr	Mo	C	Others*
S1	Bal.	11	17	2.4	.03	2.5
S2	Bal.	18	20	6.2	.02	2.1
S3	Bal.	25	20	4.4	.01	3.5
S4	Bal.	21	20	← PROPRIETARY INFORMATION →		

Others Include P, Mn, N, W, Si, Cu

Type 316 Wrought stainless steel is comparable in chemistry to the S1 powder. State-of-the-art stainless steel powders are identified as samples S2 and S3. While Alloy S4, a proprietary powder, was developed for corrosive wear applications. Three process conditions were evaluated for the S4 material chemistry. Analytical techniques used to monitor the corrosion resistance of these materials included open circuit potential and polarization measurements (Figure 1). Polarization measures the current associated with metal corrosion as a function of a change in applied potentials. This test indicates how the coating responds to shifts in the oxidizing power of an electrolyte. In addition, polarization can be used as an engineering tool to determine corrosion current densities while also observing and understanding passivity for a specific environmental condition. The corrosion current is a direct measure of the degree of coating corrosion. Open circuit verses time studies show the susceptibility of a coating to the penetration of an electrolyte. As the electrolyte diffuses to the substrate, the mixed potential will approach the base material. No penetration results in a steady state potential suggesting that the coating is dense and impervious to the electrolyte. The electrochemical data is summarized in terms of the corrosion potential [mV referenced to the standard calomel electrode (SCE)] and corrosion current (Icorr) in micro amps/cm^2. The chosen corrodent was 0.1M hydrochloric acid deaerated with nitrogen. This was chosen because it readily indicates localized weaknesses in the corrosion resistance of a coating due to surface heterogeneities.

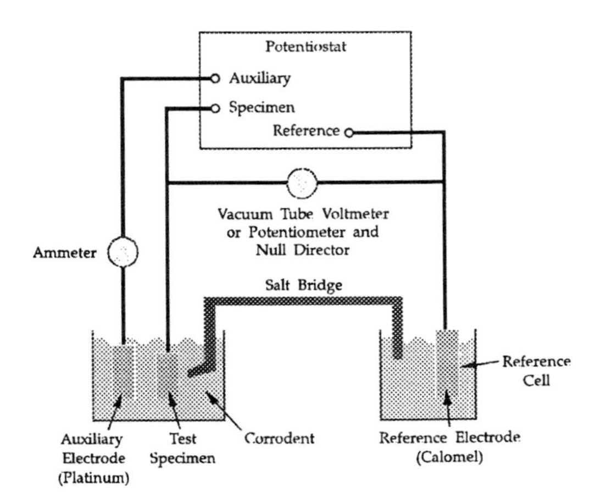

Figure 1 Experimental Set-up for Polarization Measurement

Immersion testing on specified coatings and substrates was also carried out to see if there was a correlation between corrosion rates and corrosion current density measurements. Hydrochloric acid (10 wt.%) was again used as the electrolyte. Additional studies centered on understanding the effect of coating microstructure and process on corrosion rate. Therefore, a number of coatings were metallurgically evaluated to determine through porosity using a gas permeameter.

The permeameter is used to determine the permeability of porous solids by forcing a gas such as air to flow through the sample. The steady state gas flow rate and the corresponding gas pressures into and out of the sample provide the necessary data for calculation of the permeability from Darcy's Law. To determine the permeability of different materials, a 25.4 mm (1 inch) diameter, 0.5 mm (0.020") thick free standing sample was used.

Besides understanding the various levels of porosity associated with the HVOF, Plasma and Combustion process, powders and coatings were analysed for chemistry. Specified levels of chromium and other proprietary elements were compared in alloy S4 in order to understand which process will be best suited for developing corrosion resistant coatings.

3 RESULTS AND DISCUSSION

The polarization behavior of Type 316 stainless steel and its thermal spray counterpart (S1) is seen in Figure 2. They both show substantially different electrochemical behavior when evaluated in 0.1M hydrochloric acid. The corrosion currents determined from the intersection of the Tafel slope[3] of the anodic and the cathodic curve measure 163 $\mu A/cm^2$ verses 9 $\mu A/cm^2$ for a High Velocity Oxygen Fuel (HVOF) coating compared to a Type 316 wrought alloy. The difference in corrosion rates was confirmed using immersion testing and metallography. Immersion testing reveals 12,000 µm/yr. (480 mpy) and 400 µm/yr. (16 mpy) for Type 316 coatings and substrates respectively. In addition to corrosion current differences, thermal spray coatings show limited passivity for various potential settings. Wrought alloys, however, show passivity between -250 to 100 mV.

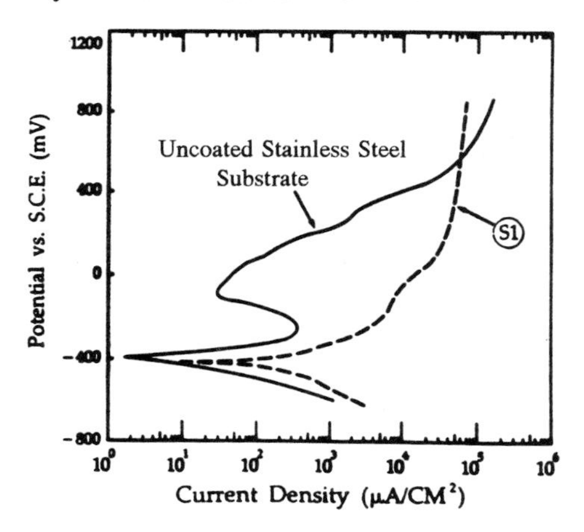

Figure 2 Polarization Curves showing the electrochemical behavior of S1 (Type 316 Stainless Steel HVOF sprayed coating) compared to an uncoated Type 316 Stainless Steel substrate in 0.1M Hydrochloric Acid

The reason why thermal spray coatings show limited passivity in non-oxidizing acids, such as hydrochloric, may be explained in Figures 3 and 4. Cross-sectional photomicrographs of wrought 316 stainless steel and an HVOF Type 316 stainless steel coating (S1) are seen in Figure 3 and show that the HVOF coating is a conglomerate of many particles with an average particle size under 44 microns. Oxides of iron, chrome and nickel are present at the interparticle boundaries. The grain size of wrought type 316 stainless steel is comparable to the particle size of a Type 316 stainless coating. The actual grain size within the particle of the S1 coating is 0.45 - 4.5 microns. Upon exposure to hydrochloric acid, microcrevices form, leading to localized corrosion. These microcrevices are a strong source of anodic dissolution. Embedded oxide particles are likely sites of enhanced anodic action leading to surface microcrevices or pits. In addition, the interparticle boundaries of a thermal spray coating are more active than the grain boundaries of a wrought alloy. This is clearly shown in Figure 4 which shows chloride acting at the interparticle boundaries leading to microcrevices which eventually undermine the coating.

A)

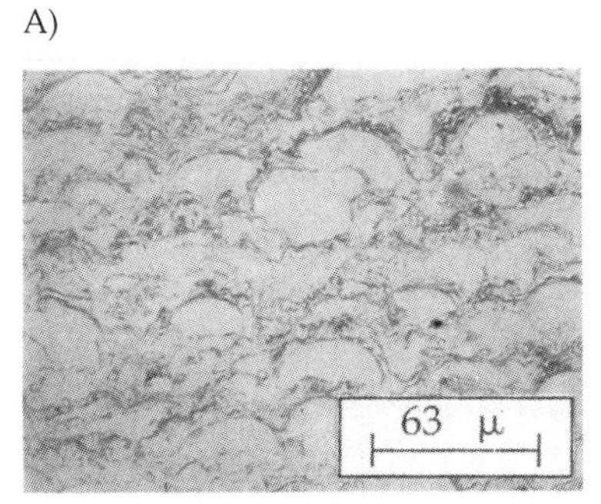

Type 316 Stainless Steel Coating
(Unetched)

B)

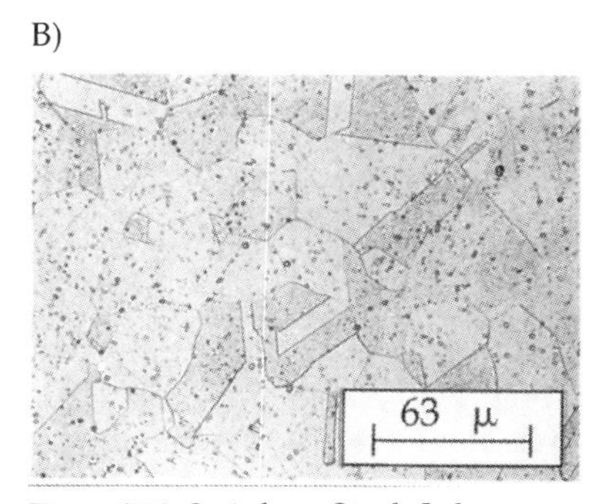

Type 316 Stainless Steel Substrate
(Etched 10% Oxalic)

Figure 3 **Cross-sectional Photomicrographs showing the fundamental difference in microstructure between a Type 316 Stainless Steel HVOF coating and a wrought material**

A) B)

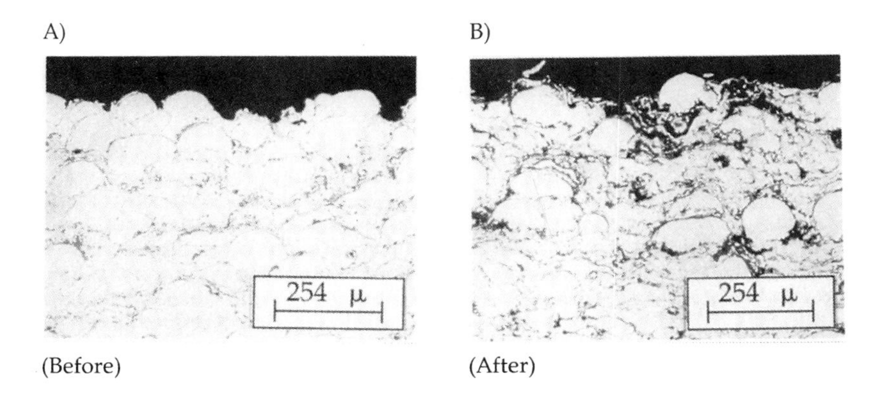

(Before) (After)

Figure 4 **Cross-sectional Photomicrographs of a Type 316 stainless steel coating: A) Before immersion testing (10% HCl); B) After immersion testing (10% HCl)**

Open circuit verses time studies show the attack of the electrolyte at the particle boundaries to be relatively slow. Potential measures approximate -.398mV for approximately 30 minutes suggesting that the 0.5 mm (0.020") thick HVOF coating was initially dense enough to retard the penetration of the electrolyte toward the substrate (Figure 5). Prolonged exposure, however, would result in the coating potential approaching the steel. At this point, the electrolyte penetrates through the coating, attacks the substrate and undermines the entire coating.

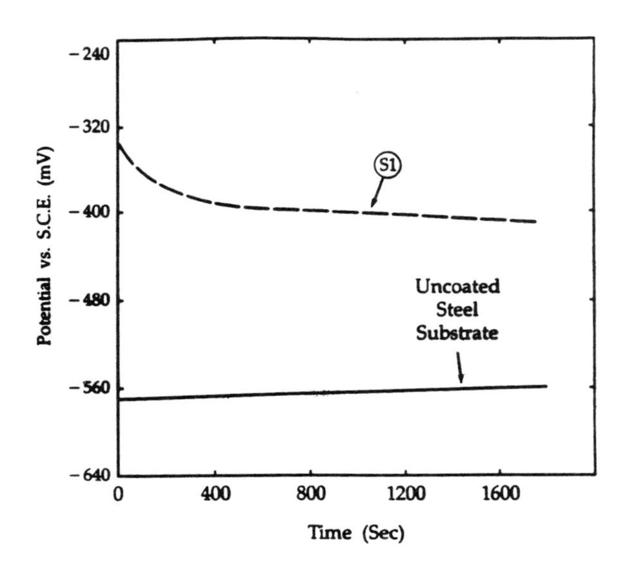

Figure 5 Open Circuit Potential Measurements E_{corr} versus Time for S1 (HVOF Sprayed) and uncoated steel substrate

Material Selection

A number of alternative alloy compositions were evaluated to improve the corrosion resistance of S1 coatings to levels approaching wrought alloys. Commercially available "state-of-the-art" austenitic stainless steels identified as S2 and S3 are known to have excellent corrosion resistance (as wrought alloys). Based on this information, these compositions were gas atomized, as spherical powders, with a -44 +5 micron particle size distribution. The fine size is ideally suited for producing dense coatings through the HVOF gun. In addition to these materials, a fourth proprietary composition (S4) was developed using deoxidizers. These deoxidizers would minimize interparticle oxidation during spraying. In addition to deoxidizers, Table 1 illustrates the higher nickel and chromium additions for alloys S2, 3 and 4, compared to S1. High chromium and nickel tend to improve the corrosion resistance of austenitic stainless steels.

Polarization data on these four materials is shown in Figure 6 and Table 2. Based on Table 2 and Figure 6, it is observed that the corrosion current density is substantially reduced to ranges between 9-18 $\mu A/cm^2$

while the open circuit potential drops from -.398 for S1 coating to -.270 mV for S4 coatings. Coating samples S2 and S3 have open circuit potential values of -.350mV and -.355mV respectively. The more positive potentials indicate that chemistry additions result in more noble overlays.

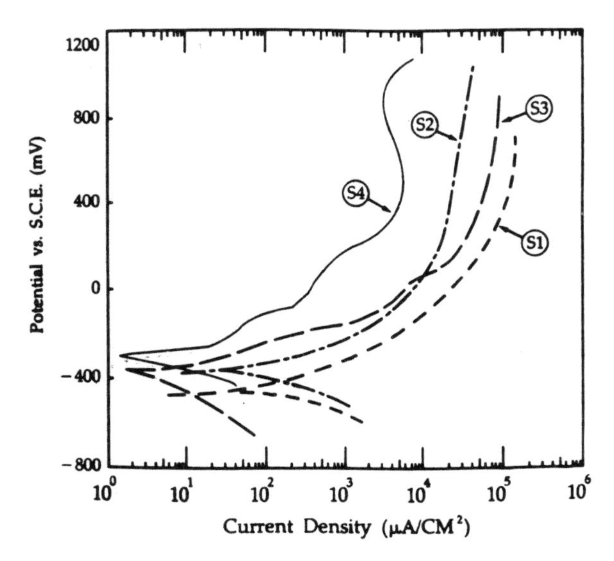

Figure 6 Polarization curves showing differences in performance based on material selection using HVOF technology

Table 2 Corrosion Data on Various HVOF Coatings and Substrates

Material	Polarization Studies		Immersion	
	E_{corr} (mV)	I_{corr} $\frac{\mu A}{cm^2}$	$\frac{\mu m}{yr}$	(MPY)
S-1	-.398	163	12,000	(480)
S-2	-.350	14	250	(10)
S-3	-.355	18	300	(12)
S-4	-.270	9	175	(7)
Type 316 Stainless Steel	-.403	9	400	(16)
Low Carbon Steel	-.557	266	8,875	(355)

Although these new alloys exhibit improved corrosion current densities and are less active, they still due not exhibit classical passivity. Results seen in cross-sectional photomicrographs (Figure 7) of alloy S4, reveal S4 coatings to be better melted with less oxide at interparticle boundaries. This, in combination with chemistry improvements results in less attack at interparticle boundaries. Immersion testing (Table 2) reveals only 175 µm/yr. (7 mpy) compared to 12,000 µm/yr. (480 mpy) for S1 coatings. S2 and S3 powders have corrosion rates of 250 µm/yr. (10 mpy) and 300 µm/yr. (12 mpy) when sprayed as HVOF coatings respectively.

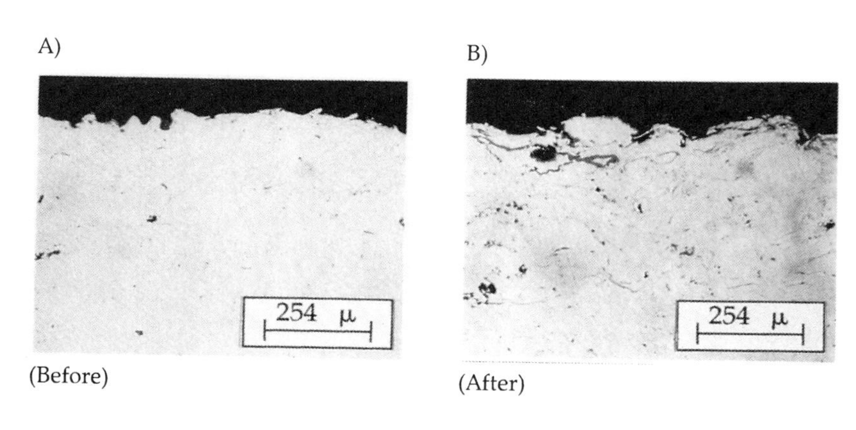

Figure 7 Cross-sectional Photomicrographs on an Iron Nickel Chromium Coating (S4) before and after Immersion Testing in 10 Wt.% Hydrochloric Acid

Spray Process Selection

The corrosion performance of a thermal spray coating changes as a function of the spray process. Figure 8 illustrates this variation for the S4 powder. Results show HVOF coatings to be more cathodic with reduced corrosion current densities than combustion or plasma coatings. This is confirmed in Table 3 which shows corrosion rates of 175 µm/yr for S4 HVOF coatings compared to 525 µm/yr and 300 µm/yr. for plasma and combustion coatings. The variation in coating microstructure and chemistry explains why plasma and combustion coatings are inferior to HVOF coatings. Metallurgical review of S4 coatings sprayed using combustion, plasma and HVOF processes reveal that combustion coatings have the highest porosity (2-3 vol.%) compared to plasma which has 1-2 vol.%. HVOF coatings have the least amount of porosity (<1%). Permeability studies confirm the trends seen in the S4 coating microstructures. In fact, Darcy's value correlates reasonably well with porosity. The most interesting result seen in Table 3 is the lower corrosion rate of combustion coatings compared to plasma. This may be explained by the fact that plasma is the hottest process evaluated. As a consequence, it is prone to a higher degree of elemental burnout. Chromium lost approximately 15% of its input amount after plasma spraying compared to less than 3% using the combustion or HVOF processes.

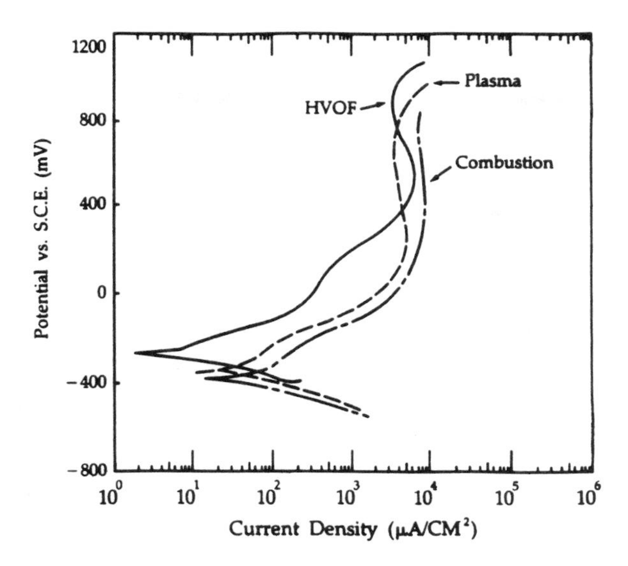

Figure 8 Polarization curves showing performance variations when S4 Chemistry is sprayed using different Thermal spray processes (0.1 M Hydrochloric Acid)

Table 3 Effect of Spray Process on Porosity, Permeability and Corrosion Rate

Coating	Process	Corrosion Rate* $\frac{\mu m}{yr}$ (MPY)	Permeabilty (Darcy's)	Porosity (Vol. %)
S-4	HVOF	175 (7)	2.05×10^{-6}	<1
S-4	Plasma	525 (21)	18.96×10^{-6}	1-2
S-4	Combustion	300 (12)	38.26×10^{-6}	2-3

*Immersion Testing in 0.1 M Hydrochloric Acid (Mils Per Year)

To obtain less elemental burnout of chromium during plasma spraying, parameter modifications need to be developed. Combustion coatings lose their corrosion resistance due to higher porosity levels. An explanation why porosity differences were not greater for the different processes reviewed can be understood by reviewing particle size distribution. In this program, a -44 micron powder was used for all three processes.

Application Selection

The HVOF process appears to be the most reliable process for spraying corrosion resistant coatings. The high velocity associated with this process will result in dense coating morphologies with minimal burnout of key elements. HVOF coatings may be designed and developed for various OEM applications as long as the edges are sealed. Coatings should be applied to a thickness of 0.5 mm (0.020") or greater. The intent of these coatings may be to improve the corrosive wear resistance of an uncoated carbon steel component. Although a coating life and versatility are lower than the life of a wrought material, coatings may be used if properly tested for the environment of choice. The coatings will eventually be replaced while protecting and prolonging the life of the base from eventual failure. Finally, Figure 9 is a polarization curve showing the potential benefit of using an HVOF S4 and S1 coating when applied onto low carbon steel. Corrosion rates can be reduced two orders of magnitude by using the proper HVOF coating and sealing the edges.

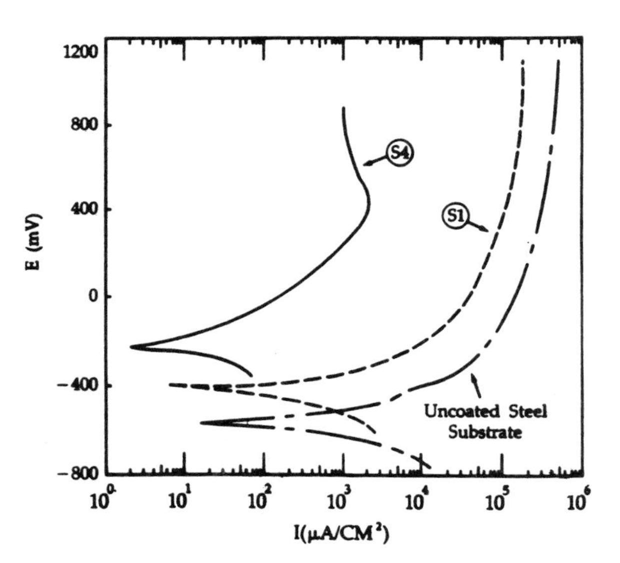

Figure 9 Polarization Curves Showing the Benefits of Depositing an HVOF Coating onto Steel

4 CONCLUSIONS

Polarization studies and immersion testing have demonstrated that Type 316 stainless steel coatings are inferior to wrought Type 316 alloys. The heterogeneous nature of a coating microstructure results in limited passivity in 0.1M hydrochloric acid. High Velocity Oxy-Fuel (HVOF) coatings, however, show superior corrosion resistance compared to other spray processes tested. HVOF coatings experience the least amount of elemental burnout (reduced oxidation) and have the highest density compared to traditional low velocity combustion and plasma processes. Further improvements in corrosion current densities were obtained by incorporating deoxidizers into the thermal spray powders. These deoxidizers minimize oxidation at interparticle boundaries. Particle boundaries are active sites for galvanic attack and eventual crevice corrosion.

5 FUTURE WORK

Alternate and/or additional processes should be evaluated in order to modify the surface structure of thermal spray coatings. Surface modification through heat treating (laser glazing, or fusing) techniques should eliminate interparticle boundaries and increase the overall density of the overlay. This, with the proper material section to minimize heterogeneous phases, should result in long range passivity of ferrous base alloys. Laser glazing may also lead to amorphous structures, which are beneficial to the corrosion behavior of iron base alloys.

REFERENCES

1. P.E. Arvidsson, "Plasma and HVOF Sprayed Coatings of Alloy 625 and 718", Thermal Spray Coatings: Properties, Processes and Applications ed. T.F. Bernecki ASM International, Ohio 1992. Page 295.

2. M. Magome, K. Kawarada and S. Ogawa, Computer Application on Anti-Corrosive Properties of Ni-Cr Plasma Coatings, Thermal Spray Coatings: Properties, Processes and Applications, ed. T.F. Bernecki ASM International, Ohio, 1992. Page 295.

3. R.W. Kirchner, "Alloy That Fights Corrosion", Machine Design, Dec. 6, 1991.

4. R.D. Willenburch, MS Thesis, SUNY at Stony Brook, 1989.

5. E.C. Hoxie, "Some Corrosion Considerations in the Selection of Stainless Steel for Pressure Vessels and Piping", Pressure Vessels and Piping: - Decade of Progress, Materials and Fabrication, ASME, NY, 1975, Vol.3.

ACKNOWLEDGEMENT

The authors of this paper would like to acknowledge the support of Dr. C.R. Clayton, Dept. of Material Science and Engineering, SUNY at Stony Brook, Stony Brook, NY.

3.1.2
Thermally Sprayed Al-Zn-In-Sn Coatings for the Protection of Welded 7000 Series Structures

P. D. Green,[1] S. J. Harris,[1] and R. C. Hobb[2]

[1] DEPARTMENT OF MATERIALS ENGINEERING AND MATERIALS DESIGN, UNIVERSITY OF NOTTINGHAM, NOTTINGHAM NG7 2RD, UK

[2] DEPARTMENT OF MANUFACTURING ENGINEERING AND OPERATIONS MANAGEMENT, UNIVERSITY OF NOTTINGHAM, NOTTINGHAM NG7 2RD, UK

1 INTRODUCTION

The combination of high strength and light weight has created a number of potential applications for 7000 series alloys besides those in aerospace, e.g. in portable bridge construction and in certain land-based transport equipment. These alloys are prone in damp humid conditions to exfoliation corrosion, and after welding stress corrosion problems can arise in the heat affected zone immediately adjacent to the weld which leads to weld toe cracking, see fig. 1. Four possible solutions have been considered in order to prevent these deleterious effects from occurring in service; these are:

(a) modification of composition of the alloy used in the parent plate or in the weld filler wire;

(b) reducing internal stresses within the weld region with a post-weld heat treatment;

(c) cathodic protection to reduce stress corrosion cracking (SCC) susceptibility as specified in fig. 2;

(d) protection of the weld region by a coating.

The most practical solution to this problem without producing a significant reduction in strength was considered to be (d). A coating was required which could still provide protection even though it might crack in service letting in corrosive species to the underlying weld. Thermal spray coating with commercially pure aluminium was considered, but this would only provide "barrier" type protection and would not offer any sacrificial action.

In some earlier work by Birley et al[1] thermally sprayed aluminium-4.5 weight % zinc coatings were applied which could act sacrificially. Zinc loss during arc spraying reduced the effectiveness of the coating. Attention was then turned to the possible use of small additions of mercury, indium or tin to aluminium which can act as "activators" and have been used as electrode material for cathodic protection of structures[2]. Reboul et al[3] has claimed that the presence of mercury or indium gives a similar effect to that produced by scratching and breaking the protective oxide film. Whilst mercury shows[2] the largest single effect, the greatest attention has focused on second placed indium because it is less detrimental from an environmental standpoint. Indium is sparingly soluble in aluminium in the solid state (0.01 wt.% at 20°C) and it has been used in conjunction with tin as it is claimed that ternary alloys are more effective. Holroyd et al[4] has developed a quaternary alloy that can be drawn into a wire-form and from which thermally sprayed coatings can be produced. This could be sprayed using either a flame or arc pistol at high deposition rates on to large surface areas. The aluminium alloy contained 1 wt.% zinc, 0.18 wt.% indium and 0.1 wt.% tin.

The work described in this paper refers to flame and arc spray coatings produced from the quaternary alloy under different conditions of pistol operation. Chemical compositions of the coatings were determined for each condition together with an assessment of their resistance to atmospheric corrosion and to pitting attack on exposure to a chloride solution.

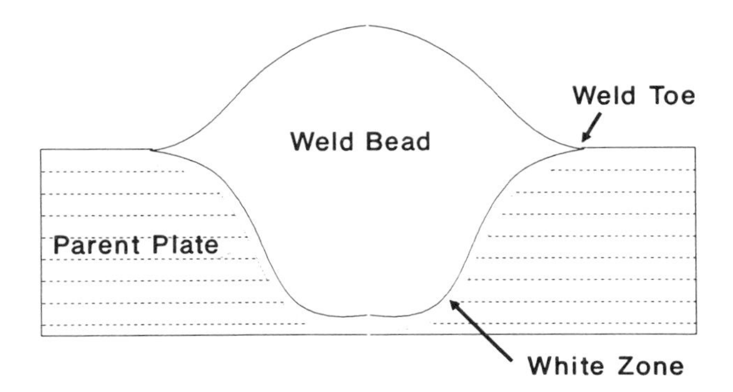

Figure 1 Schematic diagram of a section through a weld bead

2 EXPERIMENTAL

A programme of work was carried out using an aluminium plate alloy (7019) as a substrate material with specimen dimensions 80 mm x 60 mm x 10 mm thick. The alloy composition was Al-4.0 wt.%Zn-2 wt.%Mg-0.35 wt.%Cr-0.15 wt.%Fe-0.30 wt.%Si-0.05 wt.%Zr. Al-1%Zn-0.18%In-0.1%Sn wires of 2 mm diameter were used as the feed stock material for flame and arc spraying. Prior to coating the samples were grit blasted using 40/60 grade alumina.

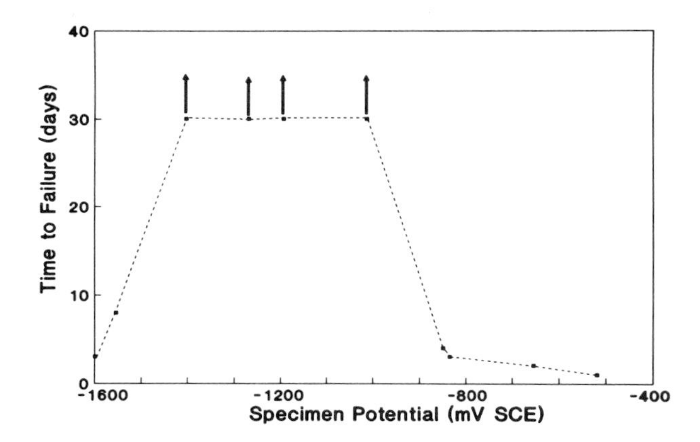

Figure 2 Stress corrosion cracking (SCC) susceptibility (time to failure) versus potential for a 7000 series aluminium alloy in a chloride solution

Spraying Conditions

Flame sprayed coatings were produced using a Metallisation Mk60 spray pistol. The pistol was mounted on a servo-controlled unit which allowed the spray beam to be rastered across the flat plate specimens and so producing coatings of controlled thickness. Coatings were produced under different flame conditions, i.e. oxidising, reducing and neutral and at two different atomising air pressures as shown in Table 1.

Flame parameters		Chemical analysis		
Flame	Atomising air	% Zn	% In	% Sn
Neutral	High	0.97	0.19	0.10
Neutral	Low	0.94	0.19	0.09
Reducing	High	0.97	0.19	0.10
Oxidising	High	0.99	0.19	0.10
Feed Wire		0.98	0.20	0.10

Table 1 Spraying parameters and chemical analysis for flame sprayed coatings

The majority of arc sprayed coatings were produced using a Metallisation 375 pistol powered by a 400 MkII energiser unit. The current (or wire feed rate) could be varied between 80 A (0.6 g/s) and 200 A (1.33 g/s) and arc voltage range was between 24 v to 36 v. Higher voltages, i.e. 45 volts, and increased feed rates up to 2.42 g/s were achieved from a Metallisation arc spray 528E pistol. The arc spray pistol was mounted on the traverse of a lathe so as to allow control over the traverse rate (50 mm/s) over the substrate. To control the air flow in the pistol and the size of the hot metal particles emitted two types of air cap were used for the work, a coarse and a fine air cap with aperture sizes of 9 mm and 6 mm. A complete listing of these conditions is shown in Table 2. Coating thicknesses between 150 and 200 μm were achieved.

Spec. code	Spray parameters			Chemical analysis and element loss				
				Zn		In		Sn
	Volt. (V)	Current (Amps)	Air Cap	wt%	%loss	wt%	%loss	wt%
AA04	26	140	C	0.24	76	0.10	50	-
AA05	26	160	C	0.18	82	0.09	54	-
AA03	31	140	F	0.06	94	0.06	72	0.06
AG04	45	140	F	0.06	94	0.07	67	0.07
AG01	43	400	F	0.02	98	0.04	78	0.11
2411	24	80	F	0.38	62	0.13	35	0.12

C = Coarse, F = Fine

Table 2 Spraying parameters and chemical analysis for arc sprayed coatings

Chemical Analysis

Samples of flame and arc sprayed coatings for chemical analysis were produced by spraying onto a substrate polished to 1 μm allowing the deposit to be easily removed. Chemical analysis was performed on duplicate samples using a plasma emission technique. This had the capability of quantitative analysis for zinc, indium and tin to an accuracy of > 0.01 wt.%.

Atmospheric Corrosion

A few hours after the completion of the flame spraying, evidence of corrosion products was consistently observed on the surface of the specimens. Also certain coatings produced in preliminary trials with the arc spray pistol had demonstrated some evidence of changes in surface condition a few weeks after spraying. These observations indicated the need for a controlled exposure experiment on flame and arc sprayed coatings. Coated specimens have now been left in laboratory air over an eight month period with the temperature (20 ± 2°C) and relative humidity (65 ± 2%) in the room being constantly monitored. A further set of samples were also brushed with an aqueous solution of 3% sodium chloride prior to exposure. The appearance of the coatings was recorded by periodically photographing the exposed surface.

Aqueous Corrosion Behaviour

Samples of the flame and arc sprayed coatings were cut and the edges were masked off using an acrylic based varnish (Fortilac), leaving approximately 100 mm^2 area of coating. The samples were then exposed to an aerated 3% NaCl solution for a period of fifty hours. The free corrosion potentials (FCP) of the coatings and of the uncoated 7019 alloy were monitored using an Oasis data logging system with readings being taken at intervals of ten minutes. All electrode potentials were referenced to the saturated calomel electrode (SCE).

3 RESULTS

Chemical Analysis

The analysis data is presented in Table 1 for the flame sprayed coatings. Virtually no change in the composition of the coatings was found compared to the original feedstock material. However, as seen in Table 2, significant changes in composition were apparent in the arc spray coatings. These varied with changes in the spraying parameters. Increasing the heat content either by changing wire feed rate or voltage reduced both zinc and indium levels in the coating, the zinc losses were considerably higher, i.e. up to 98%. The concentration of tin in the arc spray coatings increased marginally.

Atmospheric Corrosion

Figures 3-5 show a series of photographs of flame and arc sprayed samples exposed to laboratory air for various times up to 16 days (336 h). Samples treated with the sodium chloride solution prior to exposure are also shown in fig. 3-5. It can be seen that certain arc coatings were unaffected by exposure after 336 hours of exposure with and without the presence of chloride, see fig. 5. of exposure. Other arc coatings were attacked locally after 5 h exposure and more generally after 336 h. The presence of chloride has intensified the level of attack, which took the form of non-adherent oxide scales, see fig. 4. All the flame sprayed coatings show signs of attack within five hours of exposure, with and without the chlorine being present, see fig. 3. Oxidation of these specimens continued so that after seven weeks(1176 h) the coating was completely undermined. Continued exposure of the arc sprayed coatings which were initially unaffected has now taken place for 8 months (5400 h) without any significant attack.

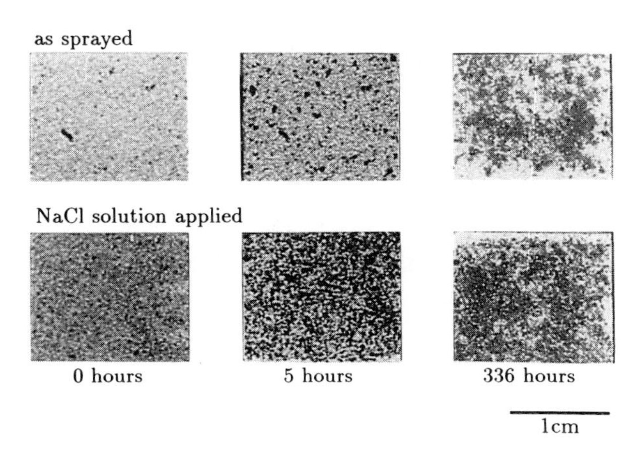

Figure 3 Surface condition of a flame sprayed coating after (i) atmospheric exposure and (ii) pre-treatment with chloride and atmospheric exposure for the times shown

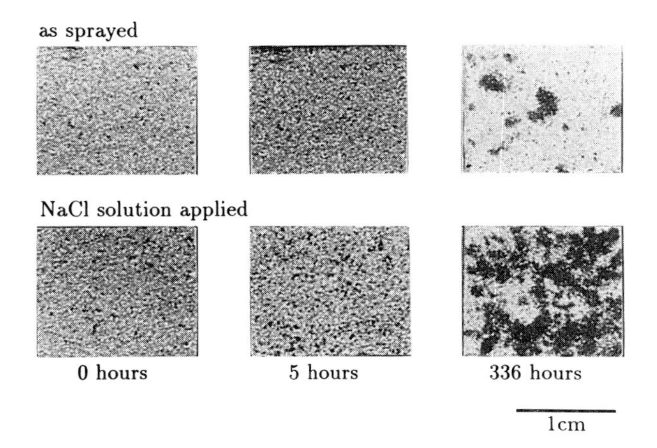

Figure 4 Surface condition for an arc spray coating (AA04) produced under low voltage conditions after (i) atmospheric exposure and (ii) pre-treatment with chloride and atmospheric exposure for the times shown

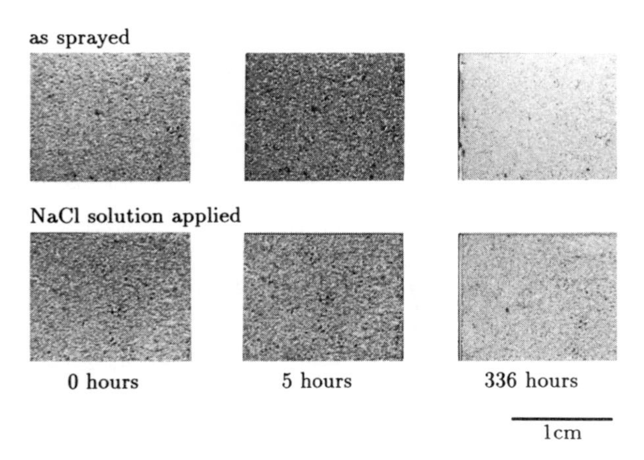

Figure 5 Surface condition for an arc spray coating (AA03) produced under high voltage conditions after (i) atmospheric exposure and (ii) pre-treatment with chloride and atmospheric exposure for the times shown

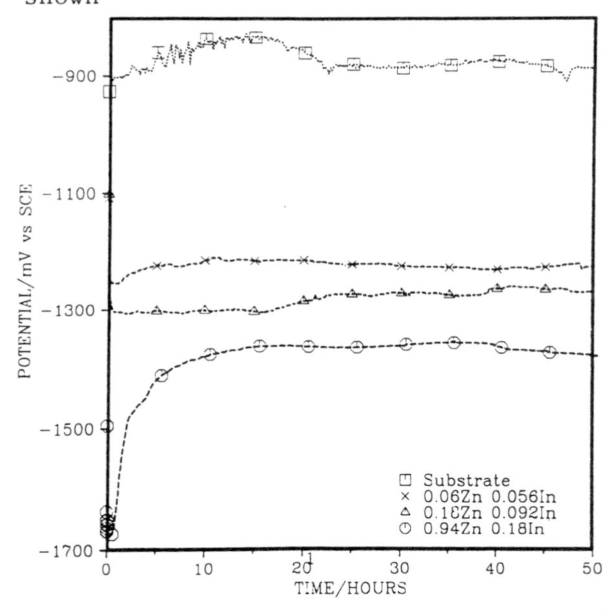

Figure 6 Free corrosion potential versus time for the 7019 substrate, a flame sprayed coating (0.94Zn 0.18In), and two arc sprayed coatings when immersed in aerated 3% sodium chloride solution

Corrosion in Chlorine Solutions

Figure 6 demonstrates the influence of coatings as measured by free corrosion potential (FCP) over a period of 50 h immersion. The 7019 alloy substrate gave a consistent potential of -0.90 V over the major part of the test period. With the coatings a change in potential takes place quite rapidly in the first few minutes of exposure. Thereafter, the arc sprayed coatings maintain roughly consistent potentials for the remainder of the tests. These potentials are approximately 0.30-0.40 V more cathodic than the substrate material. The greater shift of potential from the substrate takes place for the flame sprayed coatings. With these coatings, the FCP is 0.70 V more cathodic than the 7109 alloy in the early stages of test. Such values are maintained for several hours before settling down approximately 0.45 V from the value of the substrate.

Examination of the coatings after exposure revealed that the flame sprayed coating had started to spall and that a gelatinous precipitate was present around the sample. The arc sprayed coatings were all dulled by the exposure but none showed any significant surface corrosion after 50 h exposure.

4 DISCUSSION

Influence of Coating Process Variables on Coating Composition

The chemical analysis results (see Table 1) obtained on the flame sprayed coatings clearly show that only small changes in composition have taken place when they are compared to that of the original wire irrespective of the conditions in which spraying has taken place. Maximum losses are zinc at 4% and indium at 5% with virtually no change in the tin content. Arc spraying produced larger changes in zinc concentration; under all conditions at least 62% of the zinc originally in the feed wire was lost. Significant reductions in the indium content also occur in these coatings, i.e. >35% of original indium content was removed. Tin contents rise in the arc spray coating as the other elements are removed.

Arc spray conditions clearly do influence the amount of elemental loss. The use of a low voltage, i.e. 24 v and a reduced current (or feed rate) of 80 A produced the least removal of both zinc and indium whilst voltages in excess of 40 v and a current of 400 A removed the greatest amount, i.e. 98% of zinc and 78% of indium. The arc pistol can produce finer hot particles through the use of a finer air cap to collimate the air flow stream before the liquid droplets in the arc are stripped away. The finer particles in the spray beam have a greater surface area to volume ratio. Such particles are capable of promoting effective elemental loss by:

(a) preferential oxidation, or

(b) volatilisation of the elements.

The boiling points of the elements concerned are in the ascending order Zn < In < Al < Sn and this would suggest that volatilisation of the elements is a major factor in the arc and subsequently in particle flight in this loss. Alternatively oxidation of certain elements may effectively reduce their concentration either by inclusion in the coating or again by evaporation. The low levels of elemental loss found with the flame spraying operation are no doubt due to the lower temperatures which are involved in this process. Since there is no real difference between the use of the oxidising and reducing flame then oxidation of the elements does not appear to be a serious problem.

<u>Atmospheric Corrosion</u>

The flame sprayed samples which began to show evidence of uneven oxidation after exposure for a few hours (<5h) (relative humidity of 65% at 20°C), all contained the same zinc and indium contents as the feed wire. Previous observations of the unsprayed wire showed up the potential for oxidation but in a longer period (months). The increased activity of the coating may arise from the initial formation of oxides at elevated temperatures whilst cooling after spraying or from the irregular and extended surface which prevents a continuous film from growing. The application of a chloride solution to the flame coated surface may again assist with the early and continuous exposure of the metal to the atmosphere through disrupting continuous oxide formation on what must be a highly 'activated' surface.

As the indium and zinc concentrations fall in the arc sprayed coatings, i.e. with those produced at higher voltages and currents using a fine air cap, the atmospheric oxidation in the presence and absence of chloride becomes less dominant. With samples which have lost 94% and 72% of zinc and indium respectively there was evidence of exfoliating oxide after 8 months exposure to the atmosphere. Thus the reduction in the concentration of the activating element, indium, in the arc coatings has removed to date the highly deleterious attack on the coating which was found with the flame sprayed coatings. These findings are relevant to the potential applications of the coatings to the protection of welds in 7000 series alloys in humid atmospheres. By controlling the arc spraying process it may be possible to maintain a coating for a significant service period.

<u>Corrosion Behaviour in Chloride Solutions</u>

Fig. 6 demonstrates the open circuit potential-time curves in aerated 3% NaCl solution for a sample of the flame sprayed coating and several samples of arc sprayed coatings with different zinc and indium contents. The flame sprayed coatings start initially with a potential of -1.65 v which approaches the potential of unprotected pure aluminium which is -2.3 v. With time this potential becomes less cathodic until after 10 h exposure reaches a stable value of -1.36 v. As the zinc and indium contents reduce in the arc sprayed coatings then the steady state potential becomes less cathodic, i.e. -1.2v in the coating with 94% and 72% loss of zinc and indium. This is still significantly more cathodic than the value (-0.88 v) obtained on the unprotected 7000 series alloy.

5 CONCLUSIONS

1. Thermal spraying can be used to produce dilute aluminium alloy coatings containing zinc, indium and tin, which have the ability to provide cathodic protection.
2. These coatings when flame sprayed retain the concentration of zinc, indium and tin which was associated originally with the feed wire.
3. Arc spraying produced conditions which can promote the loss of zinc and indium from these coatings.
4. Certain arc spraying conditions, e.g. high voltages and faster wire feed rates promote the loss of zinc and indium probably through the evaporation of these elements in the arc and in flight.
5. The atmospheric corrosion performance of the arc sprayed coatings are superior to those produced by flame spray, because the latter have high indium contents.
6. It has been demonstrated that coatings produced by arc spraying containing lower indium concentrations do maintain the potential for sacrificial corrosion protection of high strength aluminium alloys.

ACKNOWLEDGEMENTS

The authors would like to thank the Defence Research Agency for financial support for this work. They are also indebted to Dr. S.S. Birley and Dr. D.B. Bartlett of RARDE, Chertsey, for many helpful discussions.

REFERENCES

1 J.C. Greenbank, P. Andrews and S.S. Birley, "Prevention of exfoliation corrosion in welded Al-Zn-Mg structures by thermally sprayed coatings", 2nd Int. Conf. on Surface Engineering, Stratford upon Avon, Welding Institute, 19.

2 J.T. Reading and J.J. Newport, "Aluminium anodes in sea water", Materials Protection, 1966, 5, 15.

3 M.C. Reboul, P.H. Gimenz and J.J. Rameau, "A proposed activation method for aluminium anodes", Corrosion, 1984, 40, 366.

4 H.J.H. Holroyd, C.R. Wiseman and W. Hepples, "Improved aluminium alloys for spray coatings", Technical Report, FVE 13A/2843, Alcan International Ltd., 1987.

3.1.3
Developments in Hardsurfacing Operations and Materials by the Vacuum Fusion Surfacing Process

N. H. Sutherland and D. Wang

PROVACUUM LIMITED, TELFORD, SHROPSHIRE TF7 4PP, UK

1 INTRODUCTION

Increasingly stringent demands are being made of modern plant and equipment. Enhanced performance is expected, operating in conditions of extremes of temperature and pressure, handling fluids which present abrasion or corrosion challenges formerly regarded as impossible for cost-effective solutions. The financial penalty of interruption of producton becomes a prime consideration when capital expenditure and operating costs are high, so that end-users are constantly seeking improvements in the reliability of plant which they are to operate.

This leads design engineers towards sophisicated systems whose increased costs are judged worthwhile if they achieve reductons in plant downtime. An integral part of these systems is the choice of materials of construction. More advanced alloys have been developed, designed to cope with combined wear problems. The cost of these advanced materials is often considerable, but such financial exposure is accepted when balanced against the commercial implications of loss of production. An inherent risk of this policy is that more established means of wear protection, often significantly less expensive, may be neglected or ignored.

Surface engineering, the selection of a surfacing material and substrate together as a system, providing cost-effective wear performance of which neither is capable on its own, makes an increasingly valuable contribution.

Vacuum fusion of thermally-sprayed surface coatings dates back some thirty years. It cannot therefore be regarded as a new process. However, its fundamental characteristics result in wear protection properties which are unique and in many ways superior to alternative processes. Recent advances in further understanding of its underlying principles, continued developments in operations and control techniques, emerging recognition of the importance of designing the component, substrate and coating as a single integrated system, progress in powder manufacture and the design of post-fusion heat treatment cycles matched in detail with the mechanical requirements of the finished component, have all combined to transform the vacuum fusion option in terms of quality and versatility.

2 THE VACUUM FURNACE FUSION SURFACING PROCESS

The process is based on three distinct operations :

Preparation of the substrate

The component surface is carefully prepared prior to the metal spraying step, so that satisfactory adhesion and entirely uniform and intimate contact between the deposit and substrate can be achieved. Vapour degreaser is used for primary cleaning, and the surface is then roughened. This is normally achieved by blasting with angular steel grit. With certain sensitive substrates, the gritblast medium can be varied, or other means of surface deformation can be used. Surface roughness is closely examined before proceeding to the spraying step.

Thermal Spraying

The self-fluxing alloy powder is applied by spraying. The powder is normally injected into an oxyacetylene flame using air as propellant. Deposit thickness is increased in spray passes until the target is achieved. Sprayed thicknesses can be in excess of 2 mm, but are varied with the component design or choice of powder. At this stage the deposit is in a conditon which could be machined to final size, to provide a certain level of wear protection. The deposit however is porous, and is attached mechanically to the component. Subsequent fusion of the coating to the substrate can be achieved by a torch, but compared with this traditional flame fusion procedure, completing the procedure in a furnace under vacuum conditions is a highly controllable process which can be closely monitored during the critical period of fusion.

Vacuum Fusion

Sprayed components are inspected for dimension and surface roughness, prepared and loaded to the the furnace. During the fusion cycle, furnace temperature is carefully controlled having regard to the physical properties of the substrate and the coating material. Furnace vacuum is maintained at no more than 1/1000 millibar during the cycle. This removes gaseous contaminants and allows more effective "wetting" of the substrate by the sprayed coating. Conditions are monitored at fusion temperature (somewhat higher than solidus temperature of the coating) to soak the furnace charge uniformly and allow sufficient time for the counter-diffusion of elements between deposit and the substrate. Furnace cooling rates are controlled with particular reference to base material properties subsequently required. For those materials which require special mechanical properties in their final form, post-fusion heat treatment cycles are designed for the particular substrate/coating combination.

3 HARDSURFACING ALLOY POWDERS AND THEIR CHARACTERISTICS

The composition of self-fluxing hardsurface powders is based on two matrices, cobalt or nickel. These are combined with a variety of other elements in different proportions, most commonly carbon, chromium, tungsten, silicon and boron, the latter two elements performing the crucial fluxing role in the process. In addition, dispersions of tungsten carbide particles in significant amounts can be introduced. Wear properties of the coating can be modified by altering powder composition, and from an availability of more than twenty standard powders, seven examples with the following compositions are chosen for consideration :

Table 1 Chemical Composition and Nominal Hardnesses of Coatings

Alloy	Co	Cr	W	C	Ni	Fe	Si	B	HRc
1	-	8.5	-	.35	bal	2.4	2.4	1.85	40
2	-	11.5	-	.5	bal	3.6	3.6	2.4	50
3	-	14.5	-	.65	bal	4.4	4.4	3.0	60
4	bal	19	8.0	.75	13	3	3	1.75	45
5	bal	19	13.5	1.35	13	3	3	2.5	56
6	bal	19	14.5	1.5	13	3	3	3.0	62
7	(50% Alloy 2 + 50% WC)								

Microstructure of Coatings

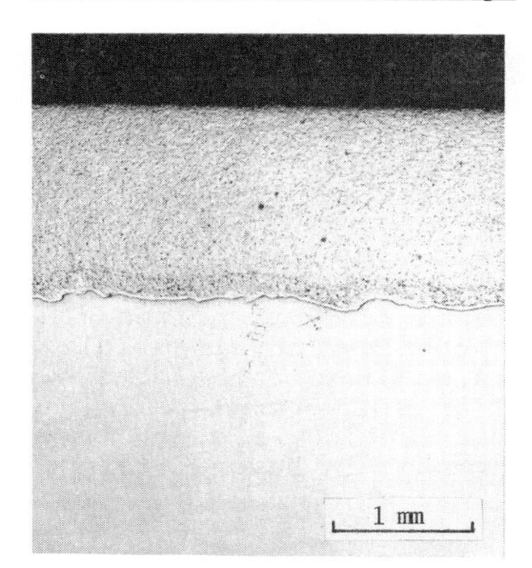

Figure 1 Microstructure of a vacuum fused-coating (as polished)

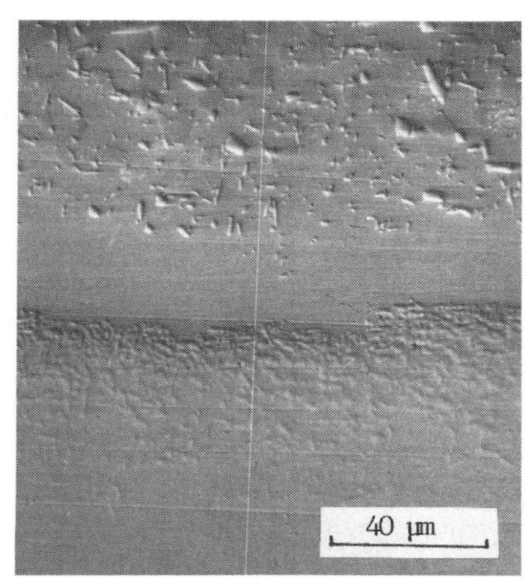

Figure 2 Diffusion zone (Normarski enhanced)

A typical cross-section of a vacuum-fused coating is illustrated in Figure 1. Vacuum fusion is in fact a high temperature diffusion or brazing process. At the critical phase of the process, counter-diffusion takes place and a diffusion zone or alloyed layer is formed between the coating and the substrate, shown clearly in Figure 2. While fusion of the coating can be achieved at any temperature between its solidus and liquidus, experience has shown that the diffusion process will be optimised by control of temperature within a range of 2-4 deg C for a defined period of time at the critical level, to provide maximum coating density and bond strength. These bond strengths are extremely high; a comparison among some common hardfacing processes is presented in Table 2.

Table 2 Comparison of Hardsurfacing Bond Strengths

Hardsurfacing Process	Bond Strength (ksi)
Flame spray	3-5*
Plating	6-8
Plasma spray	7-10
HVOF	7-14*
Vacuum fusion	53-79 **
Weld overlay	69-103**

Source : * Metco Limited
** Deloro Stellite Limited

During fusion, alloy diffusion proceeds within the coating itself, resulting in the precipitation of hard particles like carbides and borides. Because these are formed by diffusion and precipitation, extremely strong cohesion between particles and matrix is achieved, providing the ultimate spalling and erosion resistance. Undoubtedly these hard particles play a crucial part in the protection against abrasion wear.

Cobalt-based alloys

The microstructure of a cobalt-based alloy comprises a fine dispersion of carbides in a cobalt-rich alpha solid solution matrix (Ref 2). Detailed studies of the complex phases reveal that the matrix is an alpha-complete solution with a face-centred cubic lattice (f.c.c.) structure. The carbides can be present in several forms.

(a) Alloy 6 (back-scattered) (b) Alloy 4

(c) Alloy 5 (d) Alloy 6

Figure 3 Microstructure of cobalt-based coatings (b)-(d) Normarski enhanced images

Figure 3(a) is a back-scattered electron image, and clearly shows finely-dispersed tungsten carbides as white particles, and carbides and borides of chromium (dark grey) of a greater length:width ratio. These primary carbides are most likely to be present in the complex stochiometric forms. Some other extremely fine dispersions of carbides and borides are also seen in

Figure 3(a). These phases are likely to be present in a pseudo-ternary eutectic, precipitated in the cobalt-rich solid solution, (Co,Cr,Ni,W)C/ (Co,Cr,Ni,W)B + alpha. The content of these constituents can be varied by adjusting the proportions of tungsten, carbon and boron, Figures 3 (b)(c)(d).

Nickel-based alloys

In the range of nickel-based alloys the matrix is an alpha phase containing nickel, chromium and iron (Ref 2). A eutectic phase is also present, its composition depending on the coating elements, particlarly nickel and chromium. The primary carbides are normally needle or block-like particles and are always the largest in size, shown in Figure 5(a). At the grain boundaries fine acircular carbide and boride complexes are precipitated. The microhardness of the carbide is in the range 2150-3000 Hv and the boride is 3600-4600 Hv while the matrix microhardness range is 420-470 Hv (Ref 1). Again the proportion of these hard particles can be controlled by modifying the content of chromium, carbon or boron, Figures 4(b)(c)(d).

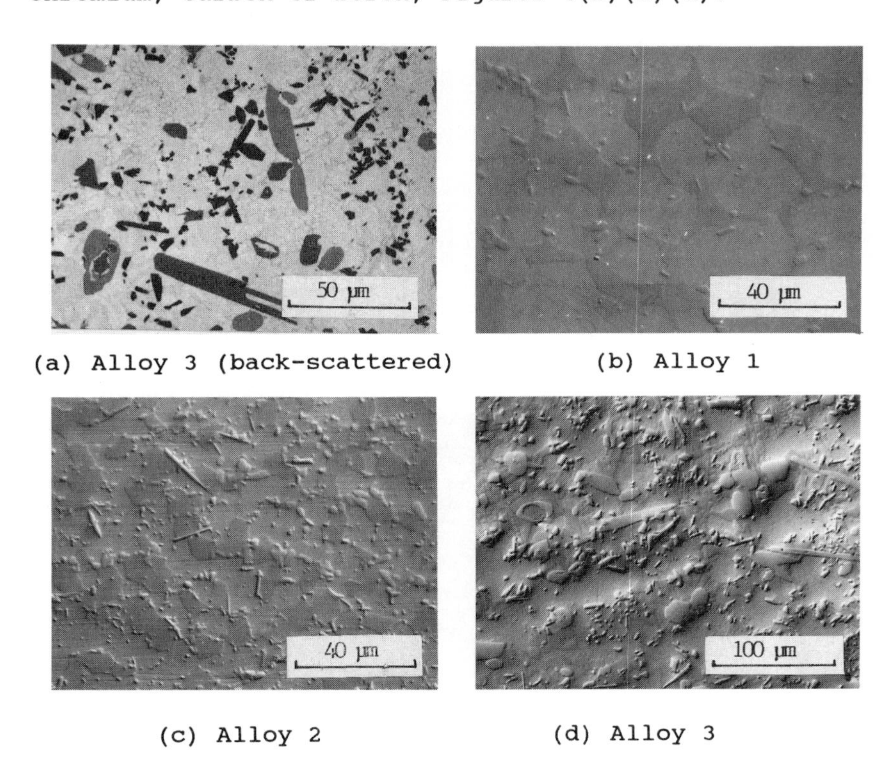

(a) Alloy 3 (back-scattered) (b) Alloy 1

(c) Alloy 2 (d) Alloy 3

Figure 4 Microstructure of nickel-based coatings (b)-(d) Normarski enhanced images

Carbide Dispersions

A further way of introducing hard particles like tungsten carbide (WC form) is by the addition and blending of these particles into a nickel-based alloy. Different types of tungsten carbides can be used, such as crushed, crushed and sintered or agglomerated and sintered particles. During vacuum fusion these are "brazed" with the matrix alloy. The diffusion between the two phases is clearly shown by Figures 5(a)(b)(c).

It has been found that decarburization of WC to form W_2C and other complex phases is detrimental in most wear applications (Ref 3). Evidence has been given which shows both high energy plasma (HEP) and high velocity oxyfuel (HVOF) sprays introduce similar amounts of secondary carbides, 14 to 20%, in the coatings (Ref 4). There are further indications that some of the secondary carbides formed by high temperature reactions are not stable at temperatures below 1250 deg C (Ref 5) but are quenched in by the high cooling rates associated with such spray processes. In contrast, an austemper treatment (austenitising and salt quenching) is successful in recovering WC to a significant level (Ref 6), though the extent has not been quantified.

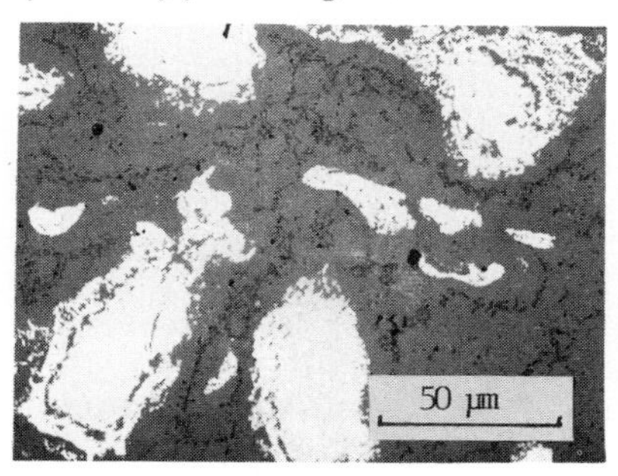

(a) SEM back-scattered

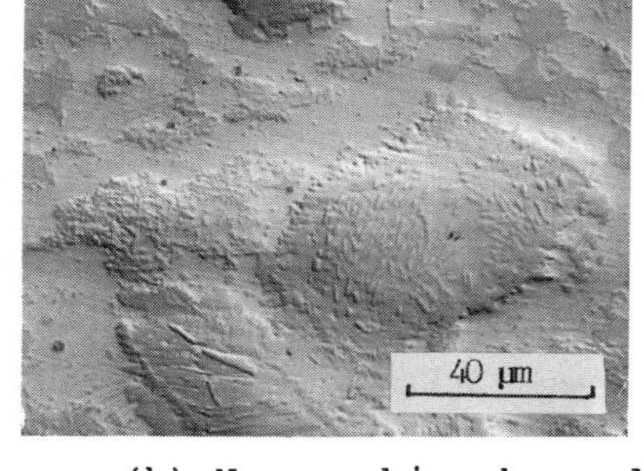

(b) Normarski-enhanced

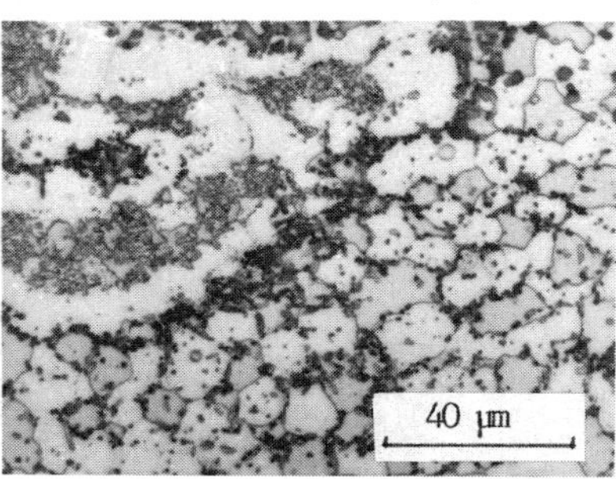

(c) As etched

Figure 5 Tungsten carbide (WC) dispersion in nickel based matrix (Alloy 7)

In comparison with thermal spray processes, vacuum furnace fusion results in the added tungsten carbide particles taking the most stable form, due to the relatively slow post-fusion cooling rates, in the absence of oxygen. In addition to providing superior wear characteristics, it is noteworthy that the carbides precipitated from the solid solution should display the ultimate cohesion strength with the matrix, and provide extremely high erosion and spalling resistance.

4 SUBSTRATE MATERIALS AND MECHANICAL PROPERTIES

Fusion between coating and substrate is completed at a temperature between the solidus and liquidus temperatures of the coating material. This fusion temperature depends on the chemical composition of the alloy, but is normally higher than the austenitisation temperature of most steels. At this level, most substrate materials transform their microstructure. When fusion is complete the substrate can transform to a microstucture different from its original one, depending on the cooling rate achieved. This transformation is a general result, except for austenitic stainless steels such as AISI 316, whose microstructure does not change throughout the complete fusion cycle.

Hardfacing alloys are by their nature generally hard and brittle, and are therefore susceptible to cracking during cooling. The major factors affecting cracking are coating hardness and thermal expansion differences between coating and substrate. Cracking susceptibilty is increased further if the substrate material transforms into martensite, due to the expansion of the steel while the coating is contracting. The degree of martensite expansion is inversely related to the carbon content of the substrate. In general cracking is not a problem with austenitic stainless steels since no phase transformation is involved. For this reason these steels are entirely compatible with the vacuum fusion process, and are commonly used.

In the case of hardenable steels the position is different. Isothermal annealing after fusion is conventionally used, in order to minimise expansion due to phase transformation. Required times and temperatures for this type of treatment of standard steels are generally available (Ref 1,7). While the isothermal annealing treatment is successful in protecting the coating, it does not provide yield and tensile strengths of the substrate material that are normally obtained by a hardening and tempering treatment. This effect is generally not acceptable for the end-user who requires sufficient strength in the component when it encounters extreme conditions.

Recent progress in development of heat treatment techniques (Ref 8) suggests that most of such steels can be hardened without cracking the coating, by direct control of furnace cooling rates, or by a post-fusion heat treatment step through an appropriate hardening route. The substrate is hardened by a programme which involves austempering and marquenching, the exact cycle being selected on the basis of substrate carbon content and transformation characteristics. In this way, steels whose strength requirements are some of the most stringent (AISI 4130, AISI 410 and Inconel 718 are examples) have been successfully heat treated. Figures 6 and 7 illustrate this result.

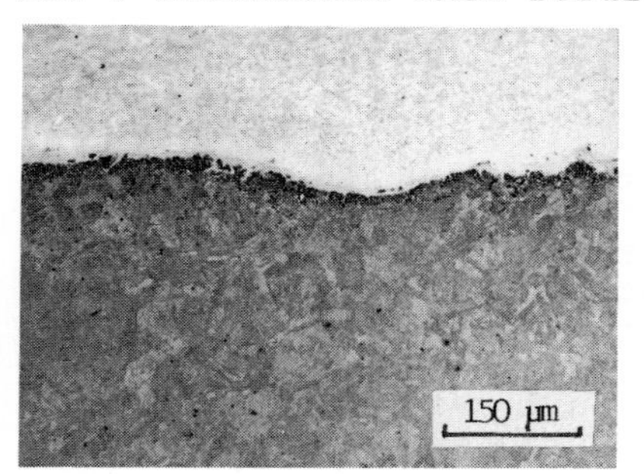

Figure 6 Microstructure of AISI 4130 substrate, austempered (bainite)

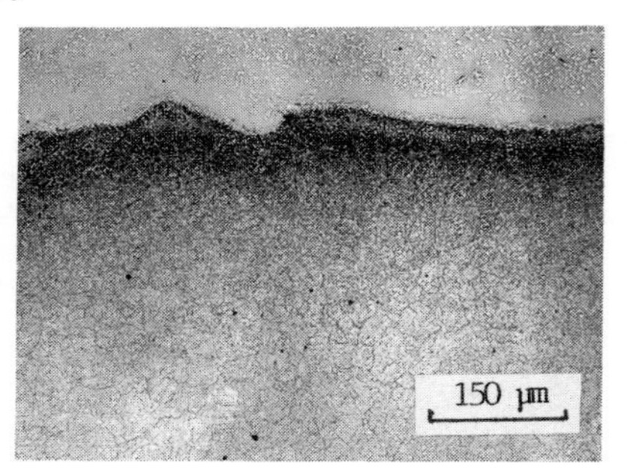

Figure 7 Microstructure of AISI 410, hardened and double tempered (tempered martensite)

As an example, Table 3 presents typical mechanical properties of AISI 410 martensitic stainless steel, achieved by vacuum fusion hardfacing followed by post-fusion heat treatment. In all measured parameters the results exceed the stringent oil industry specification API 6A requirements, by a considerable margin. Consistent achievements of this nature have long been regarded as impossible.

Table 3 Mechanical Properties of AISI 410 Substrate, Hardfaced and Heat Treated

	Actual	Specification (API6A 75K)
Yield (ksi)	84	75
UTS (ksi)	106	95
El (%)	24	18
RA (%)	52	35
Charpy (ft lb at -20 F)	33	15

Programmes are under way to address the varied problems associated with other substrate materials. Of particular interest are those materials which are claiming attention for particular applications due to their enhanced performance in a wide range of operating conditions, extremes of temperature, chloride or sulphide concentrations, or low or high pH environments. Such materials include F6NM, Nitronic 50, 17/4 PH, and the ranges of Inconel and superduplex stainless steels.

5 VACUUM FUSED COATINGS IN OPERATION

Vacuum fused coatings exhibit a number of key features which influence their performance. In contrast to the range of mechanical applications of coatings, they are metallurgically bonded to the substrate. This bond strength virtually eliminates spalling and enhances resistance to galling during operation. Compared with welded deposits, their microstructures are considerably finer, with consequent increased resistance to wear and corrosion. They cause no associated heat-affected zone in the substrate, are not diluted by base material, so that they do not suffer lower hardness as a result. Heating and cooling rates of the fusion cycle are slow, reducing significantly the risk of distortion of the coated components. As-fused surfaces are extremely uniform, providing easier and less costly finishing, often eliminating the need for final grinding.

6 APPLICATIONS OF VACUUM FUSION SURFACING

Vacuum fusion surfacing finds its most appropriate applications where maximum quality is crucial in tackling the most hostile operating conditions. Combinations of corrosion and abrasion, often in extremes of temperature are widely encountered in the oil and gas exploration and in the process industries, and these are among the prime areas for vacuum fusion surfacing. Other areas include chemicals, process, automotive, metal extrusion, food processing and construction industries. When batch quantities can be matched to furnace capacity, coating costs are optimised and are highly competitive in respect to the quality provided.

Continuing advances in the precise understanding of the technical processes involved in vacuum fusion is resulting in further recognition of the appropriate choice of coating, substrate and component as an integrated system to match the exact requirements of the application. Collaboration with powder supplier, equipment manufacturer and end-user is the key to continuing the success of meeting the challenges of industry.

ACKNOWLEDGEMENTS

The authors would like to register appreciation to Mr I R Davies, AEA Industrial Technology, Risley, for his assistance with the preparation of the photographic work of this paper; also to Gerald Bell and Stan Grainger for valued comments on the text.

REFERENCES

1. G R Bell "Sprayed and Fused Coatings-Some Metallurgical Aspects", Metal Fabrication, October 1962.

2. Project Report, Deloro Stellite Limited "An Investigation into the Characteristics of Spray Fused Alloys".

3. S Rangaswamy and H Herman (editors) Proceedings of the Eleventh International Thermal Spraying Conference, 1986 (pp101-110).

4. T P Slavin and J Nerz "Material Characteristics and Performance of WC-Co Wear Resistant Coatings", Proceedings of the Third National Thermal Spraying Conference, 1990 (pp159-164).

5. Edmund K Storms "The Refractory Carbides", The Academic Press, New York, 1967 (p146).

6. W J Lenling, M F Smith and J A Henfling "Beneficial Effects of Austempering Post-Treatment of Tungsten Carbide Based Wear Coatings", Proceedings of the Third National Thermal Spraying Conference, 1990 (pp227-232).

7. "Metco Flame Spray Handbook" Vol II, Powder Process, p42.

8. D Wang "Hardfacing Hardenable Steels by Vacuum Furnace Fusion", Project Report, Provacuum Limited, December 1991.

Section 3.2 Laser Processing

3.2.1
Laser Physical Vapour Deposition of Thin Films

P. Spiliotopoulos,[1] K. Kokinoplitis,[1] Y. Zergioti,[2] E. Hontzopoulos,[2] and D. N. Tsipas[3]

[1] AMT LTD., ATHENS, GREECE

[2] FORTH, CRETE, GREECE

[3] ARISTOTELES UNIVERSITY OF THESSALONIKI, DEPARTMENT OF MECHANICAL ENGINEERING, PHYSICAL METALLURGY LABORATORY, THESSALONIKI, GREECE

1 INTRODUCTION

In the field of materials the beneficial properties and improved behaviour offered by coatings has touched almost every industrial sector. These include dielectric films for microelectronics, optical and magnetic applications, hard nitrides, carbides and borides for cutting and forming tools, sulfides for solid state lubrication and space applications, etc. (1)

A number of techniques exist for synthesizing these high quality films. However the three main types of processes that are widely used are, CVD (Chemical Vapor Deposition), PVD (Physical Vapor Deposition) and thermal spraying.(2)

The thermal spaying technique involves heating a material and then directing it at very high velocities onto a surface. The hot particles condense and adhere on the substrate upon impact. Coating thicknesses may vary from a few microns up to several millimetres. The main disadvantages of this method are the high porosity of the coatings, high intrinsic stress of the films, rough surface due to unmelted globules of the sprayed alloy and limited reproducibility of results.

The other two routes, namely PVD and CVD deposit coating from the gas phase (Vapor Deposition) by an atom by atom transfer process. During these processes elements are ejected from the sources using heat, electrons or ions. Thus, since these are atomistic techniques it follows that the thicknesses of the films attainable usually do no exceed 100μ.

Film growth by any vapor deposition method can be seen in terms of three basic steps:

1) generation of species to be deposited.
2) transport from source to substrate.
3) film growth on the substrates.

Also there are interactive parameters one has to understand in these processes such as plasma characteristics (if plasma is employed) and process parameters (evaporation, sputtering rate, bias, etc.).

Concerning the CVD (3) technique in comparison to the PVD route its main disadvantages are:

1. the elevated operating temperatures (i.e. 900°C) which in several occasions can exceed the limit of the substrate.
2. the corrosiveness and toxicity of the process gases, causing porosity, poor adherence and contamination of the coating.
3. it is an "equilibrium" process, so metastable materials cannot be deposited.

The PVD techniques may mainly be separated into the following three classes:

i) **Sputtering:** The material is bombarded by energetic ions and atoms are ejected and condense on a solid substrate to form a coating.
ii) **Evaporation:** It involves heating and evaporating the constituents (with thermal, radiation, arc discharge or electron beam means) in a vacuum and condensing them on an appropriate substrate.
iii) **Laser Ablation and Vapor Deposition** (LVD). This technique consists of firing a pulsed laser at a target in a vacuum. The particles ejected from the pellet condense on an appropriately located substrate to form the coating. Although the mechanism for the generation of the species to be deposited employing this technique is mainly considered to be thermionic evaporation, several other mechanisms are involved which in relation to its special plasma characteristics make this process unique.

This process is fast, efficient, relatively inexpensive and can produce exceptionally smooth films with the desired crystalline phase, and the correct chemical composition.

Also due to the high energy delivered, alloys can

be deposited without a change in composition as in other techniques. Thus, polycomponent ceramic films may be produced at high depositon rates contrary to other techniques.

The LVD process is considered to be an evaporation process but the deposition characteristics differ from the conventional thermal evaporation process. The reason for this is that while in the conventional method the mechanism for the removal of the material is thermionic emission, on the LVD process different ejection processes of the target material may take place depending on the power of the laser and the ejected material. Also the ablated species in the laser-generated plasma are characterized with kinetic energies in the range of 20-400 eV. which are two to three orders of magnitude greater than the thermal energies (3).

These features of the LVD process are attributed to the complex nature of the interaction of the incident laser beam with the evaporant.

A detailed study (6) showed that only the near surface layers (30-400 nm) of the target are rapidly heated and evaporated. However this thermal shock applied to the evaporant, as well as atomistic level beam-solid interactions, gives rise to various ejection processes of the target material. These so-called mechanisms of laser sputtering have been investigated in detail (7) and are the following:

1) Collisional sputtering, which is due to momentum transfer from the beam to the target.
2) Thermal sputtering, normal thermal evaporation.
3) Electronic sputtering, due to ion explosions(8).
4) Exfoliational sputtering, flakes detach due to repeated thermal shock (Thermal stress).
5) Hydrodynamic sputtering, meterial is removed due to volume changes arising from the thermal expansion of the liquid with melting.

The evaporated particles are further heated by the incoming beam resulting in an enhanced ionized plasma. It has been observed (9, 10) that approximately 30% to 50% of the ejected particles are ionized.

Thus the evaporating beam consists of fast moving atoms and ions (10^6 $cm.s^{-1}$) and slow-moving clusters and droplets (10^3 $cm.s^{-1}$).

The substrate surface exposed to the laser driven plume is bombarded by energetic neutrals, ions and electrons. The nature and energy of these bombarding species initiates a variety of reactions on the substrate that may lead to heating, surface chemistry changes, re-emission of deposited material, modification of film morphology, grain size growth etc.

Thus, substrate bombardment has a pronounced effect on the properties of the film. The bombarding energies of the incident particles depend on the ionization efficiency of the plasma which in turn depends on the power and especially the type of laser used. Also a bias voltage applied on the substrate or an independent ion beam source may be used as alternatives to enhance the bombarding effect.

However there are several other deposition variables that control the properties of the films.
In the case of films deposited by laser albation techniques the main variables are:

- The temperature of the substrate during deposition.
- The arriving vapor flux density.
- The angle of incidence of arriving vapor.
- The thickness of the film.
- The kinetic energy of the arriving vapor.
- The pressure and nature of the ambient gas.

2 EXPERIMENTAL PROCEDURE

The Experimental arrangement employed is depicted schematically in Figure 1. It involves a laser system, a high vacuum chamber with a rotating/moving target base and a substrate heater system.

The chamber was constructed from stainless steel and its dimensions were I.D.=40 cm and height=30cm.
The material used as the evaporant was a TiAlV alloy, having 6% Al, 4% V and the rest Titanium (wt.%). The target was fixed on a rotating target holder.

The excimer laser beam passed through a window into the high vacuum chamber and was focused so that the beam strikes the target at 45° angle.

The substrate material is D2 steel, which is a very common tool steel used for cutting and forming applications. The heat treatment that was applied to the D2 steel was composed of the following steps:

1. Preheating at 600°C for 15 min. in order to avoid distortion during the high temperature austenitizing.
2. Austenitizing at 1020°C for 1 hr.
3. Quenching in oil at 50°C to produce martensite. Some austenite is also formed.
4. Tempering at 550°C to produce the desirable combination of strength and toughness.

The microstructure after tempering consists of tempered martensite and two carbide dispersions, the primary and the secondary carbides. They are both Cr-rich carbides, the primary resulting from the solidification while the secondary carbides precipitate

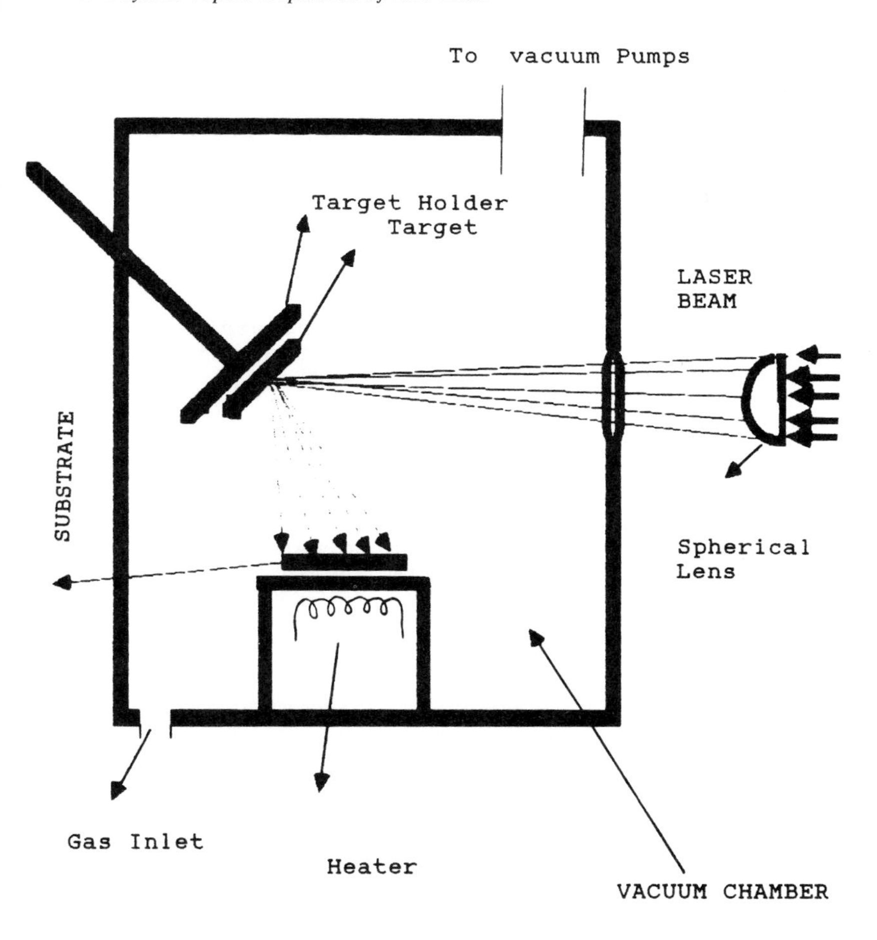

Figure 1 Schematic diagram of the laser vapor deposition system employed

during tempering. The secondary carbides are responsible for the phenomenon of secondary hardening in the steel which occurs at 550°C due to the dissolution of cementite and formation of finer Cr-rich carbides.
The hardness of the steel reaches 62 Rockwell C after tempering at 550°C . Representative micrographs using optical microscopy are shown in figure 2. The white phase is the primary chromium carbides dispersed in the termpered martensite phase.

Small coupons of D2 material carefully ground and/or polished and thoroughly cleaned with Cl_3-ethylene were rinsed with methanol and dried with hot air before being placed in the chamber. They were placed on a substrate fixture which could be heated with an electrical resistance mechanism. The distance between substrate and target was kept constant in this experimental work at 4cm. This distance was selected after experimentation which showed that acceptable deposition rates were obtainable. In future experimental work this distance would also be a variable.

The chamber pressure was lowered to 10^{-4} mbar, which was the actual operating pressure during deposition.

On the basis of the above configuration the most suitable conditions of the deposition of TiAlV were determined to be:

EXCIMER LASER:
- -Wavelength = 248 nm
- -Energy per pulse = 280 mJ
- -Repetition rate = 20 Hz
- -Lens = 50cm

DEPOSITION CHAMBER:
- -Pressure = 10^{-4} mbar (Argon)
- -Distance between target-substrate= 4cm
- -Temperature of substrate= 25 to 350°C
- -Deposition time= 30-40 min.
- -Purging gas= High purity Argon

During deposition a small portion of the area of the samples to be coated had been masked with an aluminium foil in order to create a sharp step on the substrate. This was necessary to measure the thickness of the films with an interferometer attached to a 1000 magnification optical microscope. A green light source of 1092 nm wavelength was employed for these measurements.

Film characterization was done using hardness, electrical conductivity measurements, x-rays and electon microscopy.

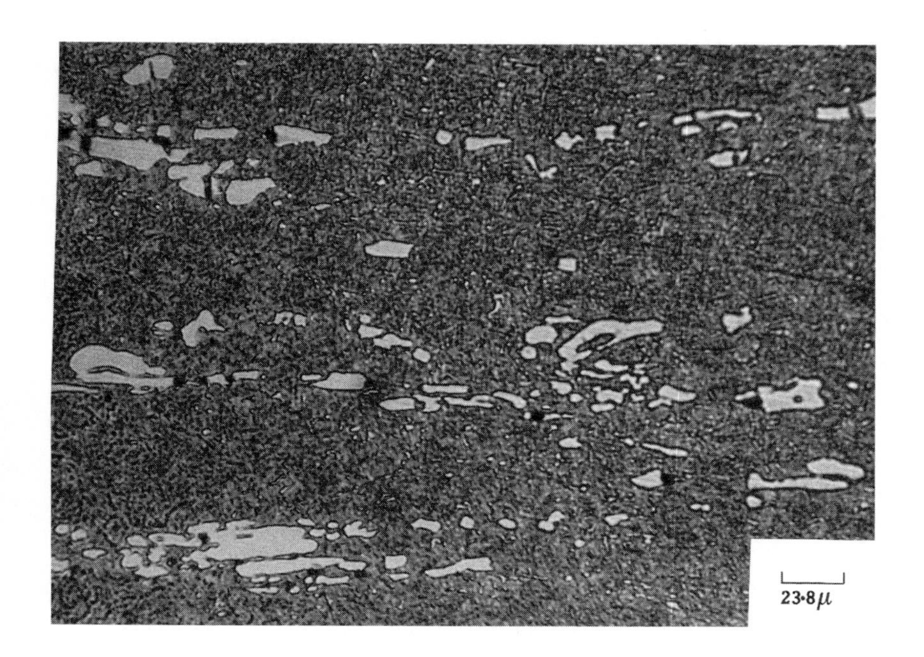

Figure 2 Microstructure of D2 Steel

3 RESULTS AND DISCUSSION

The experiments carried out at the conditions previously mentioned appear to be of greater importance and as such the films deposited at these conditions were characterized.

During LVD the laser beam strikes the target and as a result of this interaction a plasma is generated in the vacuum chamber. This high temperature plasma extends from the target to the substrate surface and it is responsible for a great number of the deposition parameters involved in the process.
The beam parameters, i.e. irradiated spot size, pulse energy, density and number of pulses, influence greatly the layer growth.

Thus, optimization of the laser parameters in relation to other system parameters such as substrate temperature, target to substrate distance, chamber pressure and type of gas, must take place in order to produce high quality films.

The three main types of lasers used for deposition are, excimer [pulse duration (r=15-50ns, λ=193-308 nm)], Nd-YAG [(r=5-15ns, λ=1,064nm)] and CO_2 [(r=25ns to continuous, λ=10.6μm)]. Infrared emitting lasers (CO_2 and Nd:YAG) induce lattice and material vibrations in the irradiated workpiece effecting a thermal process. This is the reason why long pulse duration and long wavelength lasers produce low quality films, including nonstoichiometric composition and a high density of clusters of different sizes.

Compared to CO_1 and Nd.YAG-lasers, excimer lasers enable another kind of energy transfer mechanism caused by extremely high maximum pulse power and emission of radiation in the UV range. The excimer laser beam is absorbed in the near surface layers of the target and it also produces a greater degree of ionization in the plasma (6). Thus the removal process is an interacting process without direct thermal influence and described by the so called photochemical ablation model (4,5). This unique type of ablation process gives rise to stoichiometic evaporation and high quality multicomponent films. Also the increased ionization efficiency is an important parameter in determining the nature of the vapor and the properties of the resulting films. This is the reason why in our experiments the excimer type laser beam has been employed.

The film thickness obtained for the different experimental conditions varied from 0.3 to 0.6 μm, for deposition times ranging between 30-40 minutes, and substrate temperatures between 25 - 350°C.
The microhardness (Vickers) of the thin films has

been attempted to be measured with a 25 gr and a 50 gr load on the indenter. These attempts proved to be mostly unsuccessful.

The through thickness resistance was in the range of hundred ohms indicating conductive films, while the surface resistance of deposited film was in the range of Kohm/cm indicating, in comparison with the through thickness resistance value, again conductive films.

The X-Ray Diffraction patterns indicated the characteristic peaks of D2 substrate and in some samples after long exposure Ti metal was observed.

In a separate experiment where the least adhesive film was removed by extractive replica, i.e. free standing film, and analysed by X-Rays, no crystalline phases were detected indicating that the Ti deposition is largely amorphous in nature with possibly small microcrystalline regions depending on the depositing conditions.

This information obtained by X-Ray diffraction was also confirmed by Transmission Electron Microscopy where the obtained electron diffraction patterns showed diffused rings indicating the existence of amorphous films of approximately 0.5 μm.

Samples were also examined in a scanning electron microscope. Secondary and backscattering electron images were taken in order to observe the surface morphology of the films and identify other film characteristics.

A number of photomicrographs are shown in figure 3. During observation of the samples under the electron beam a spectrum of X-Rays was acquired with an EDS detector attachment. Thus, spot microanalysis was performed at specific areas of the coating and at various substrate locations.

A ZAF correction was applied on the spectrum analysis results for the following elements Fe, Cr, Mo, V, Al, Ti, Mn, Si, S, P. The chemical analysis results for most specimens confirmed the existence not only of Ti but of the other elements Al and V in the coating.

From the above observations it can be said that in most cases a smooth film has been deposited on the majority of the specimens. The adherence of the film depended strongly in the process parameters and the quality and temperature of the substrate. Ti was clearly identified as one of the most predominant elements in most films followed by smaller amounts of Al and V. All these three elements are present in the target which was a Ti-Al-V alloy.

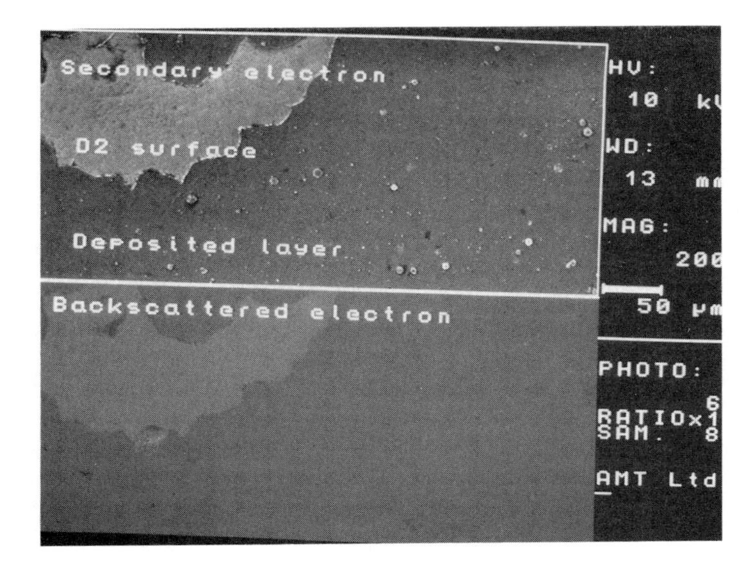

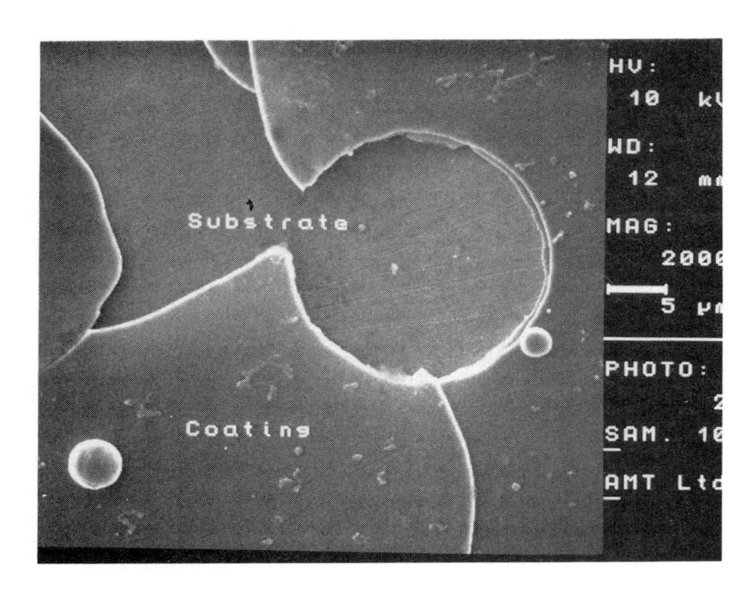

Figure 3 Typical Morphologies of Deposited Coatings

4 CONCLUSIONS

An Excimer Laser, wavelength 248 nm, has been successfully used to deposit TiAlV films in the microns range by the LPVD method. Under the conditions of pressure energy per pulse, temperature of the substrate etc. used, the procedure seems very simple to apply reproducibly and with high rates of deposition. By optimizing the process parameters, smooth, uniform in composition and very adherent films can be obtained. In the present experiments, the Ti films were amorphous with possibly small microcrystalline zones.
It is recommended to further explore and optimize the process for depositing multilayer coatings.

REFERENCES

1 R.F. Bunshan and C. Deshapandey. Vacuum V.41, no.7-9, pg.2190, 1990.
2 R. Barrell and D.S. Rickerby, Surface Engineering, pg.468 Aug.1989.
3 T. Besman et.al. MRS BULLETIN November 1988, pg.45.
4 H. Tonshoff and R. Butje. Proc.5th Int.Conf, Lasers in Manufacturing Sept.1988 pg.35.
5 K.Reicheler and X. Jiang. Thin Solid Films 191 (1990) 91.
6 R.K. Singh and J. Narayan. Journal of Materials pg.13, March 1991.
7 R. Kelly et al. Nucl.Instr. and Methods in Phys. Res. B9 (1985) 329.
8 R.F.Fleischer, P.B. Price and R.M.Walker. J.Appl.Phys. 36, (1965) pg.3645.
9 Eryu, K. Murakami and J.K. Masude, Appl.Phys.Lett ,54 (26) (1989) pg. 2716.
10 F. Gagliano and V. Peak Appl.Opt. 13 (1974) 274.

3.2.2
Effect of Plasma Sprayed Layer Thickness and some Laser Parameters on Laser Sealing of 8.5 wt% Yttria Stabilized Zirconia Alloy

K. Mohammed Jasim,* R. D. Rawlings, and D. R. F. West

DEPARTMENT OF MATERIALS, IMPERIAL COLLEGE OF SCIENCE, TECHNOLOGY AND MEDICINE, LONDON SW7 2BP, UK

* ON LEAVE FROM SCIENTIFIC RESEARCH COUNCIL, BAGHDAD, IRAQ

1 INTRODUCTION

Yttria partially stabilized zirconias (YPSZ's) have been widely investigated for thermal barrier coating systems and contain approximately 6-12 wt% yttria. Recent publications by the authors have discussed some of the processing parameters that need to be controlled to seal layers and thus to reduce porosity[1-8] and hence to increase the resistance to high temperature corrosion and oxidation. However, the effects of power density, interaction time and the plasma sprayed thickness have not been fully investigated. The only study of the effect of power density and thickness was by Zaplatynsky[9] who reported the performance of laser sealed and plasma sprayed samples of 8 wt% YPSZ during cyclic oxidation testing and cyclic corrosion testing. Laser treatment was shown to have no effect on cyclic oxidation performance whereas the corrosion resistance of the sealed samples was at least four times higher than that of the plasma sprayed layers. Increasing the thickness of the layer from 200 to 400μm increased the corrosion resistance by at least five times for both plasma sprayed and sealed layers.

The main processing variables in laser sealing are laser power (P), beam diameter (d) and traverse speed (V). These parameters are commonly considered in terms of power density ($P_A + 4P/\pi d^2$), interaction time ($t=d/V$) and specific energy ($S \approx P/dV$). Operating conditions to obtain high quality laser sealing plasma sprayed zirconia-based layers have recently been reported by the authors in the form of a plot of power density versus interaction time (Figure 1). It should be noted that the regime in Figure 1 which is relevant to this paper, that is the sealing regime, has been specified from data from a range of plasma sprayed ceramics which interact slightly differently with the laser. For example, a power of 0.8 kW, 5 mm beam diameter and 230 mm/s traverse speed produced partial sealing when processing 8.5% wt% YPSZ[10] and 20 wt% YFSZ[8], whereas the same laser parameters gave complete sealing of plasma sprayed layer of 90 wt% 8.5 wt% YPSZ and 10 wt% alumina[10]. Thus the sealing regime in Figure 1 cannot be considered definitive for all ceramics and may only be used as a general processing guide[5]. The extent of the sealing regime given in Figure 1 is more extensive than that previously presented[5]; recent work which is reported in the present paper confirmed the previously quoted specific energy range but found acceptable sealing over a wider range of power density, namely from about 270 W/mm^2 to 10 W/mm^2 rather than to 50 W/mm^2 as previously reported.

In order to specify the sealing regime in Figure 1 it was necessary to assess the quality of the sealing. The assessment of quality is complex as a number of features of the sealed layer have to be considered and quality cannot be fully quantified. The features which have to be taken into account include uniform sealing to a depth typically of ~20-100μm with surface concavity (attributed to vaporisation loss and reduction of porosity) no more than 10-30μm in depth. The crack network, which is characteristic of laser sealed layers should show relatively small crack width (<~5μm for the primary cracks) and small network spacing and the cracks should not extend too far into the sealed region. Surface depressions should be kept to a minimum in terms of number and size. The surface roughness of the sealed layer should be significantly less than that of the plasma sprayed material.

In the work reported here, the effect of plasma sprayed layer thickness was studied in relation to the quality of the sealed layer. Interaction times in the range of 12.5-250 ms were used at a constant power density of 51 W/mm^2; i.e., varying horizontally through and out of the sealing regime(s) shown shaded in Figure 1. Also, an investigation has been made of the influence of power density (51-82 W/mm^2) on the 400μm plasma sprayed layer thickness at a constant interaction time (28.5 ms) for a selected range of specific energies (1.1-1.8 J/mm^2); thus varying vertically through the shaded region. Other data have been obtained for lower power density values and longer interaction times outside the shaded region in Figure 1.

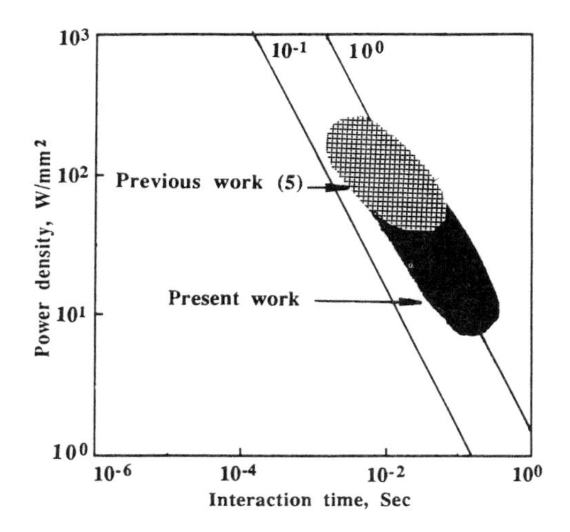

Figure 1 Operating regime for laser sealing process - lines of constant specific energy values (J/mm^2)

Table 1 Laser processing parameters studied

Variable	Effect of plasma sprayed layer thickness	Effect of power density
Power (P), kW	1.0	0.4-1.6
Beam diameter (d), mm	5.0	5.0-7.0
Traverse speed (V), mm/s	22.0-370	7.0-175
Interaction time (t), ms	12.5-227	28.5-1500
Power density ($4P/\pi d^2$), W/mm^2	51.0	10.0-82.0
Specific energy (P/dV), J/mm^2	0.54-9.1	0.5-8.2

2 EXPERIMENTAL PROCEDURES

Plasma sprayed layers of thickness of ~ 200 and 400µm of 8.5 wt% yttria stabilized zirconia were deposited on to a ~100µm intermediate bond layer of Ni-23Cr-6Al-0.4Y (wt%) by Plasma Technik, UK. These layers were applied on to mild steel samples of 6 mm thickness and 25.4 mm diameter.

The effects of layer thickness and power density were studied by varying the process variables as previously described in the Introduction and as also given in Table 1. The main investigation of power density was carried out with a layer thickness of 400μm and an interaction time of ~28.5 ms (d=5mm and V=175 mm/s); power was varied in the range of 1.0-1.6 kW (power density ~51-82 W/mm^2) (Table 1).

All the sealed tracks were examined by scanning electron microscopy (SEM) on the top surface (plan view) and transverse sections. The latter were cut carefully and prepared using standard grinding and polishing procedures and measurements were made of the depth sealing and depth concavity. Roughness measurements were taken in the traverse direction on the laser processed surfaces. The roughness of the plasma sprayed coatings was also measured for comparison purposes. The roughness was measured using a Talysurf 10 and was specified in terms of a centre-line average (CLA). The length of surface monitored was limited by specimen size and was chosen to be either 0.4mm or 1.25 mm. The CLA values presented are the mean of at least five readings.

3 RESULTS AND DISCUSSION

Effect of Plasma Sprayed Layer Thickness

Figures 2(a) and (b) show for both 200 and 400μm thick plasma sprayed material a decrease in the width and depth of the sealed layers with decreasing interaction time (increasing traverse speed); similar trends have been previously reported[8]. For all the traverse speeds used the sealed width is less than the beam diameter (5 mm); this is demonstrated by the graph of Figure 2(a) as well as the plan views of single tracks presented in Figure 3. The data in Figure 2(a) also show that there is a trend to a small decrease in width with increase in plasma sprayed thickness from 200 to 400μm. However, this is not reflected in the sealed depth data, Figure 2(b), which do not show a significant effect of layer thickness. It should be noted that the depth of sealing is measured from the base of the concavity to the plasma sprayed region; if the values of depth of sealing plus concavity depth are compared then the values for the 200μm layer are slightly greater than those for the 400μm layer (Figure 4). Figure

4 also shows that the depth of concavity decreases with traverse speed and reaches almost zero at the highest speed studied.

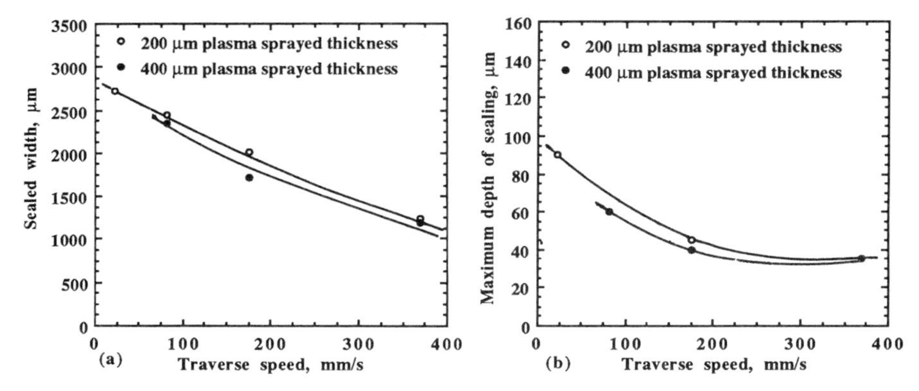

Figure 2 Sealing width (a) and depth of sealing (b) versus traverse speed for different plasma sprayed layer thicknesses (1.0 kW laser power and 5 mm beam diameter)

Figure 3 Low magnification SEM micrographs of as-sealed layers of 8.5 wt% YPSZ processed at different traverse speeds (a) 82 mm/s, (b) 175 mm/s and (c) 370 mm/s respectively (400μm plasma sprayed layer thickness)

The microstructure of the sealed layers as observed on the upper surface was either a dendritic type at relatively high specific energy (ranging from ~1.8-1.5 J/mm^2) or a cellular type at lower values of ~1.5 J/mm^2 specific energy (Figure 5). The cell size increased slightly with increased plasma sprayed layer thickness, in accordance with a decrease in cooling rate, Figure 6(a).

The sealed layers as observed on plan views (upper surface) show crack networks and depressions (Figure 3). The crack width (maximum values are reported), Figure 6(b), and average network spacing, Figure 7(a), decrease with decreasing interaction time (increasing traverse speed) for both thicknesses. The width and spacing values for the 400μm thick sprayed layer were smaller than those for the 200μm layer and transverse sections also showed that the cracks were shallower. However the dependence of the microcrack characteristics on thickness changes for thicknesses greater than about 600μm, e.g. increasing the plasma sprayed thickness from 350 to 700μm, produced an increase in the width and spacing for 20 wt% YFSZ[10]. It is suggested that this is due to increasing residual stress with increasing plasma sprayed thickness above a critical thickness of ~600μm.

The average size of depressions observed on the upper surface decreased with decreasing interaction time with the exception of the 200μm thickness layer processed at the lowest traverse speed, Figure 7(b). Although, there were relatively small differences in the size between the 200 and 400μm thick layers, the percentage area of depressions is less in the 400μm thick layer at traverse speeds greater than 82 mm/s, Figure 8(a). The reduction

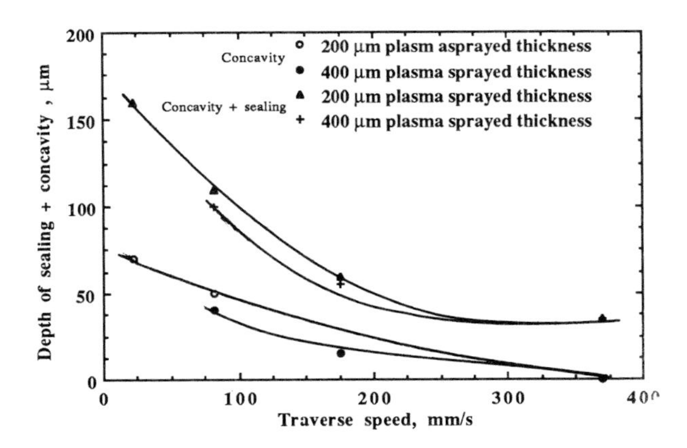

Figure 4 Depth of concavity and depth of concavity + sealing against traverse speed for different plasma sprayed layer thicknesses (1.0 kW laser power and 5 mm beam diameter)

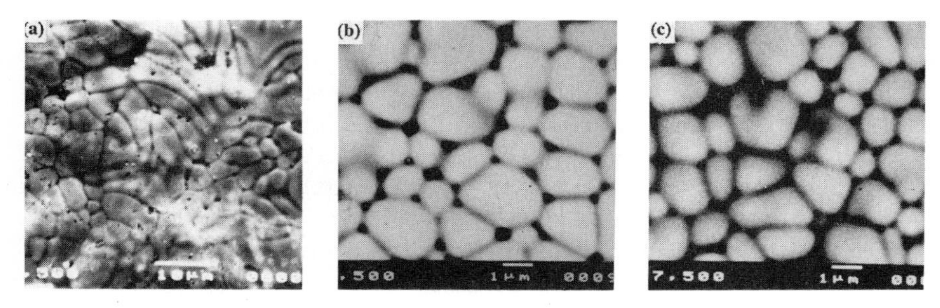

Figure 5 High magnification SEM micrographs of as-sealed layers showing dendritic structure at low speeds and cells at high traverse speeds (400μm plasma sprayed layer thickness), (a) 82 mm/s, (b) 175 mm/s and (c) 370 mm/s respectively

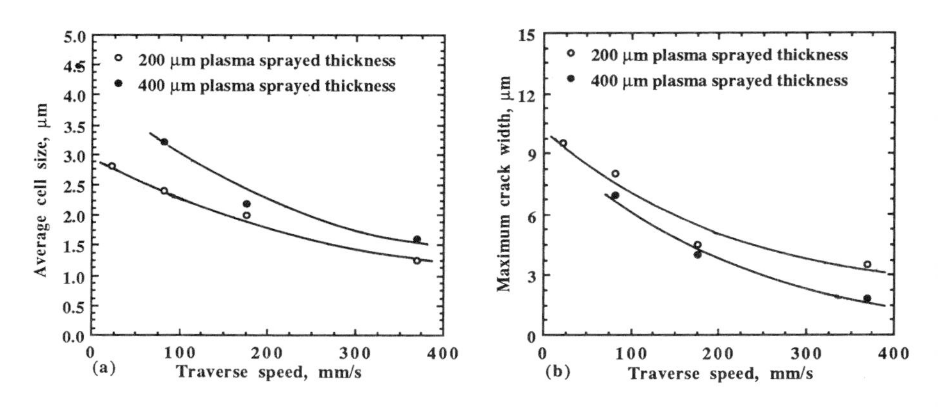

Figure 6 Average size of cells (a) and maximum crack width (b) against traverse speed for different plasma sprayed layer thicknesses (1.0 kW laser power and 5 mm beam diameter)

in proportion of depressions in the 400μm layer leads to a lower roughness than in the 200μm layer, Figure 8(b). The roughness of the sealed layers is 1-3μm which is a considerable improvement on the roughness of 5μm for the plasma sprayed material.

Effect of Power Density and Interaction Time

Figure 9 shows the data for depth and width of the sealed layer for values of power density in the range ~51-81 W/mm^2

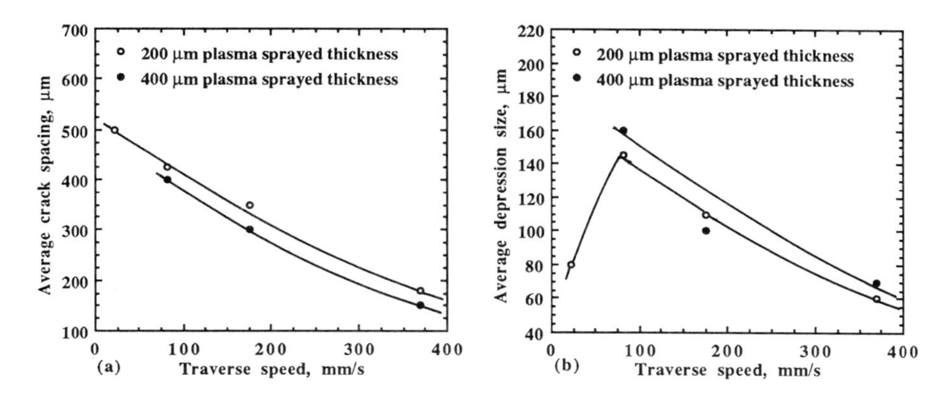

Figure 7 Average spacing of cracks (a) and average size of depressions (b) against traverse speed for different plasma sprayed layer thicknesses (1.0 kW laser power and 5 mm beam diameter)

at a constant interaction time of 28.5 ms. Increasing the power density leads to an approximately linear increase in depth and width.

The increase in depth of sealing with power density (varying vertically through the sealed region in Figure 1) is disadvantageous for the quality of the sealed layer due to the concomitant increase in crack width (Figures 10 and 11). SEM examination of transverse sections also shows that cracks penetrate vertically through the sealed layer at power densities > ~60 W/mm^2 and specific energies of > ~1.4 J/mm^2 (Figure 12). Furthermore, the crack spacing increases with power density in an approximately linear relationship, Figure 9(b).

The depth of concavity values were in the range of ~10-30µm, Figure 9(a), and the percentage depressions/ total sealed area decreased from a maximum of only ~4% with increasing power density (Figure 11). The average size of depressions also decreased with increasing power density, Figure 9(a). Very low roughness in the range of 1-1.3µm was observed which is mainly due to the small proportion of depressions (Figure 11). The cell size as measured on the upper surface increased with power density from 2 to 3µm.

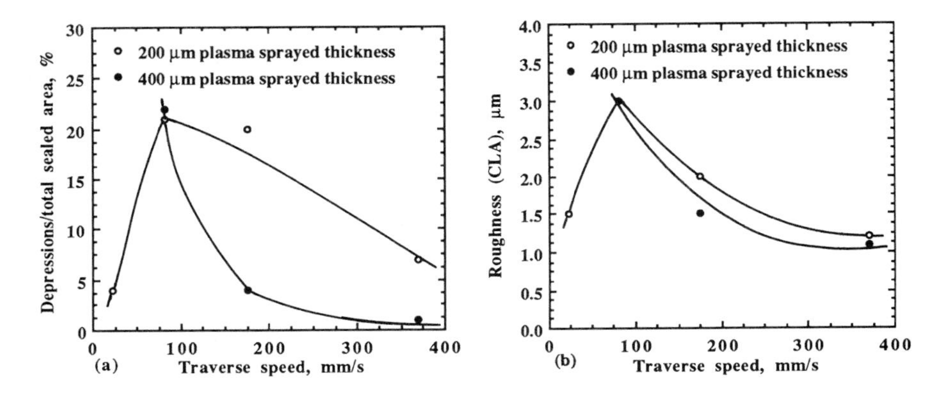

Figure 8 Depressions/total sealed area % (a) and roughness (b) against traverse speed for different plasma sprayed layer thicknesses (1.0 kW laser power and 5 mm beam diamter)

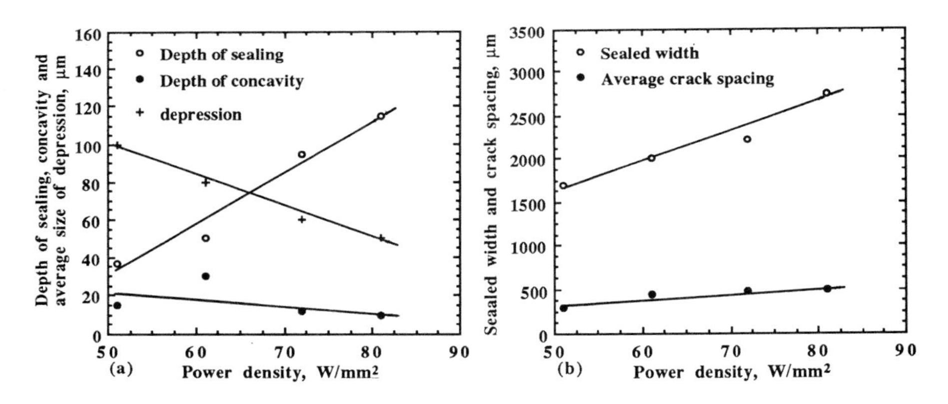

Figure 9 Depth of sealing, depth of concavity and average size of depressions (a) and sealed width and average spacing of cracks (b) against power density (175 mm/s traverse speed and 5 mm beam diameter)

Quality of Sealed Layers -

The effects of plasma sprayed thickness and power density on the quality of the sealed layers are summarized in Figures 13 and 14. Generally, increasing plasma sprayed layer thickness from 200 to 400µm improves the quality (Figure 13). High power density has a significant negative effect on quality; lower quality sealed layers are produced above 61 W/mm^2 as shown by the schematic diagram of Figure 14. To determine the effect of lower power density (<40 W/mm^2), a few experiments were carried out at a power density of ~20 and 10 W/mm^2 and these indicated that there was an improvement in quality with respect to the plasma sprayed layer (Figure 15). In the present work optimum sealing was observed at intermediate traverse speeds.

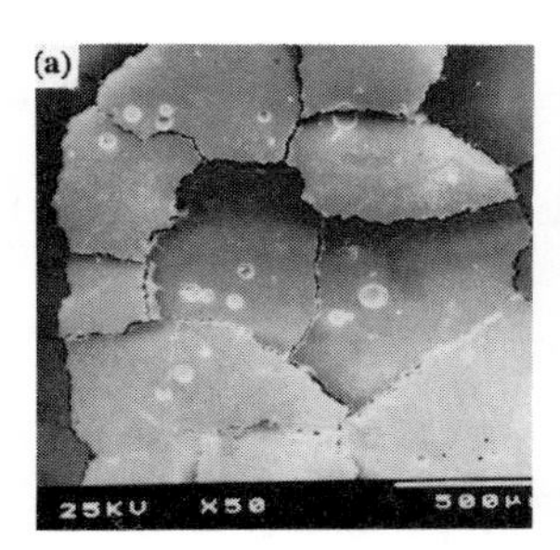

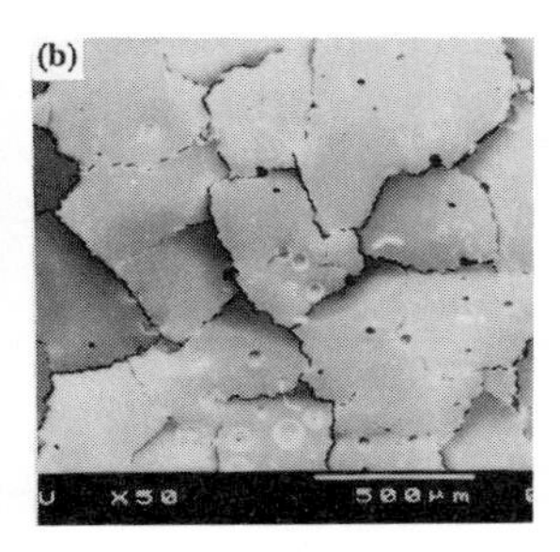

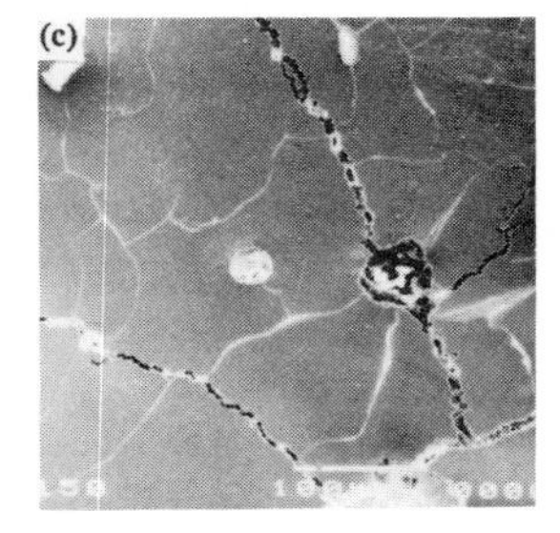

Figure 10 Plan views of as-sealed layers processed at different power densities and specific energies, (a) 61 W/mm^2, 1.37 J/mm^2, (b) 71 W/mm^2, 1.6 J/mm^2 and (c) 82 W/mm^2 and 1.83 J/mm^2 respectively

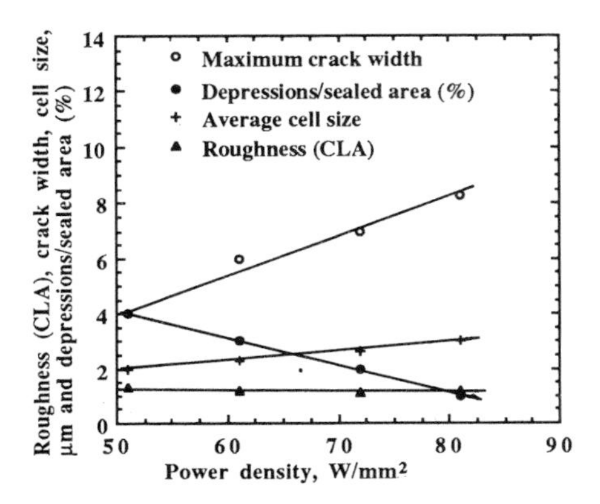

Figure 11 Roughness, maximum crack width, average size of cells and depressions/sealed area (%) against power density (175 mm/s traverse speed and 5 mm beam diameter)

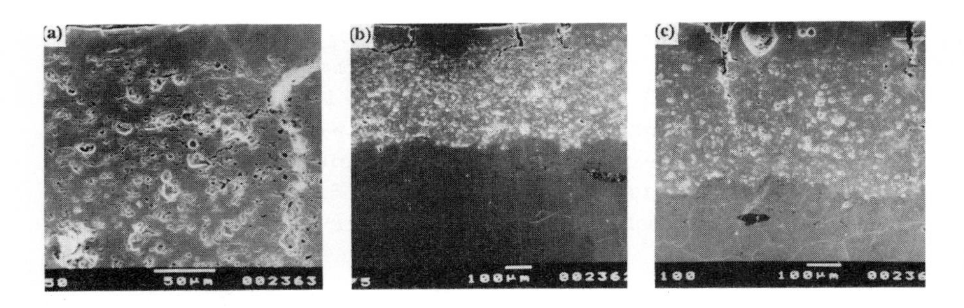

Figure 12 Transverse sections of as-sealed layers showing deep vertical cracks increased with increasing power density (conditions similar to Figure 10)

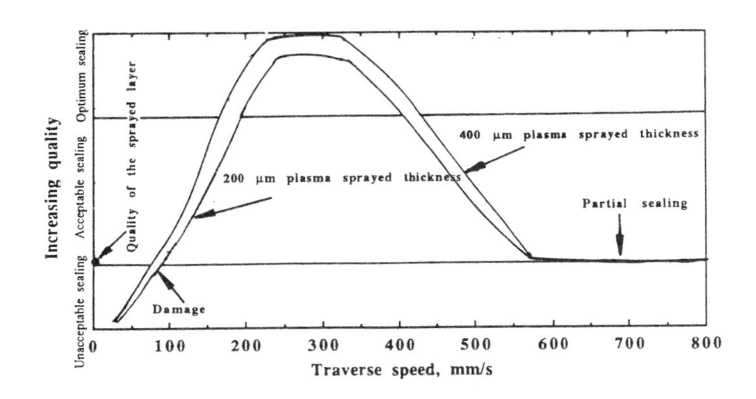

Figure 13 Quality-traverse speed relationship

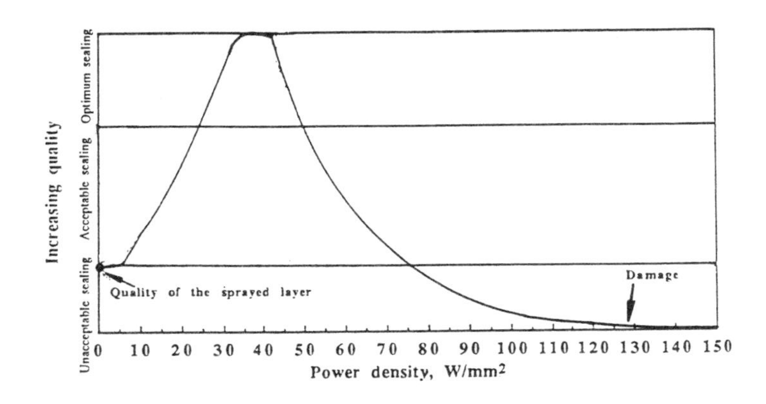

Figure 14 Quality-power density relationship

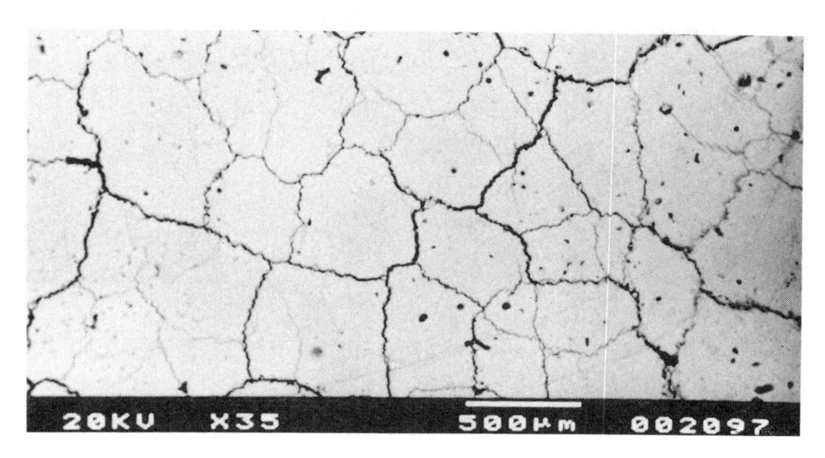

Figure 15 General plan view of as-sealed layer processed at low power density (10 W/mm^2) showing the absence of depressions

Overlapped Tracks

In practical situations laser sealing would require coverage of extensive areas by using overlapped tracks rather than single tracks. Therefore, a few 25 mm diameter samples (400μm plasma sprayed thickness) were covered using a 5 mm diameter beam and ~50% overlapping. No spalling was observed on sealing plasma sprayed layers deposited on a bond layer, Figure 16(a), and the quality of the surface was acceptable, Figure 16(b). The major problem, which was assessed from transverse sections, was the presence of relatively deep vertical cracks formed in the overlapped interfaces; it is suggested that these are mainly due to the high overlap ratio used which is expected to produce high residual stresses, Figure 16(c). Thus, the weak regions are the overlap interfaces and it is desirable either to reduce the overlapping ratio of the tracks or to reduce the depth of sealing by using lower specific energy; further work is needed to determine the optimum conditions.

4 CONCLUSIONS

1. Increasing the plasma sprayed thickness from 200 to 400μm and decreasing the interaction time to ~10-25ms give an important benefit in that sealed layers

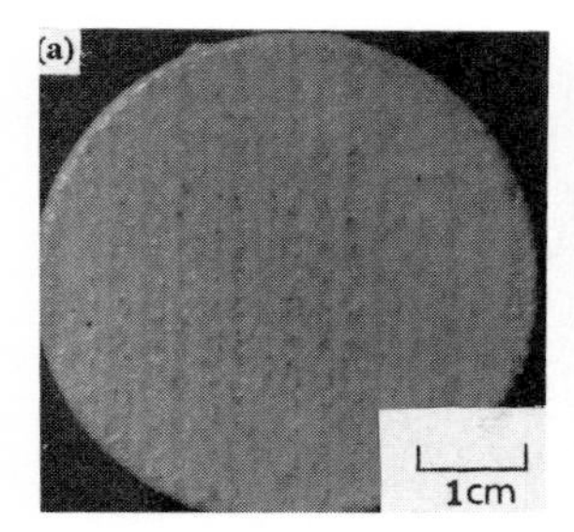

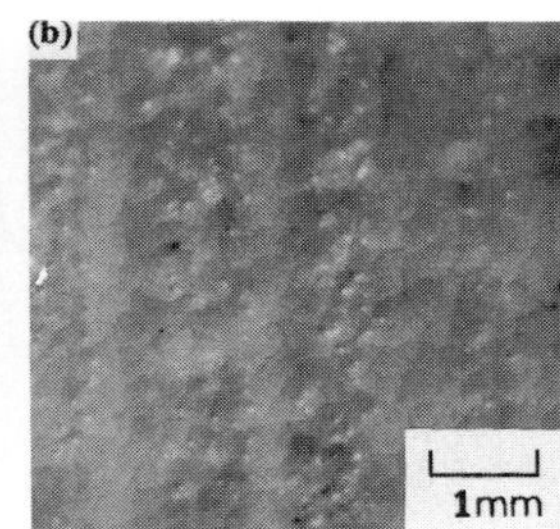

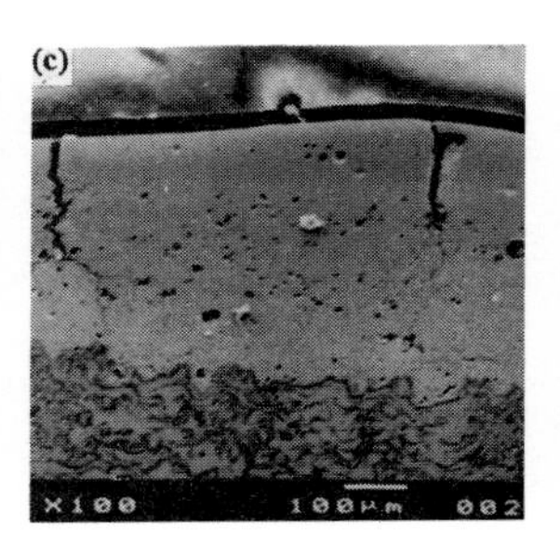

Figure 16 Plan views and transverse section of partially overlapped tracks sealed under shrouding gas, 1kW laser power, 5 mm beam diameter and 230 mm/s traverse speed, (a) low magnification, (b) higher magnification showing the absence of depressions and (c) vertical cracks at the overlapped interface

of the desired thickness can be produced with only a few depressions.

2. A considerable reduction in roughness was obtained after laser sealing.
3. It is possible to seal plasma sprayed layers at low power density (less than 40 W/mm^2) and relatively high interaction time (30-200 ms) to obtain a good quality sealed layer with small amounts of depressions.
4. For single track sealing, it is possible to produce the concavity depth to a very low value e.g., 10μm or less.
5. Exploratory tests on overlapped tracks show the importance of overlap ratio in controlling cracking.

ACKNOWLEDGEMENTS

The authors thank Professor M. McLean, Head of Department and Professor D. W. Pashley for providing the facilities of the laser laboratory. One of the authors, K. Mohammed Jasim, thanks the Government of the Republic of Iraq, and Scientific Research Council, Baghdad for awarding a scholarship.

REFERENCES

1. K. Mohammed Jasim, D.R.F. West and W.M. Steen, J. Mat. Sci. Letts., 1988, 7, 1307.

2. K. Mohammed Jasim, D.R.F. West, W.M. Steen and R.D. Rawlings, 'ICALEO 88' USA, 1989, p.17.
3. K. Mohammed Jasim, D.R.F. West and R.D. Rawlings, 'Laser-5', iitt, France, 1989, p.90.
4. K. Mohammed Jasim, R.D. Rawlings and D.R.F. West, J. Mat. Sci., 1991, 26, 909.
5. K. Mohammed Jasim, R.D. Rawlings and D.R.F. West, ibid, 1992, 27, 1937.
6. K. Mohammed Jasim, R.D. Rawlings and D.R.F. West, ibid, 1992, 27, 3903.
7. K. Mohammed Jasim, R.D. Rawlings and D.R.F. West, Mat. Sci. and Tech., 1992, 8, 83.
8. K. Mohammed Jasim, R.D. Rawlings and D.R.F. West, Surface and Coatings Technology, To be published, 1992.
9. I. Zaplatynsky, Thin Solid Films, 1982, 95, 275.
10. K. Mohammed Jasim, R.D. Rawlings and D.R.F. West, Unpublished work, 1992.

3.2.3

Prevention and Repair of "End Grain Corrosion" of Type 347 Stainless Steel in Hot Nitric Acid Using Laser Surface Remelting

J.-Y. Jeng,[1] B. E. Quayle,[2] P. J. Modern,[3] and W. M. Steen[1]

[1] LASER LABORATORY, DEPARTMENT OF MECHANICAL ENGINEERING, UNIVERSITY OF LIVERPOOL, PO BOX 147, LIVERPOOL L69 3BX, UK

[2] BNFL SELLAFIELD, SEASCALE, CUMBRIA CA20 1PG, UK

[3] BNFL, CORPORATE RESEARCH AND DEVELOPMENT DEPARTMENT, RISLEY, WARRINGTON WA3 6AS, UK

1 INTRODUCTION

High power lasers provide a concentrated energy source, which can be used to modify the surface properties of materials. A range of laser surface treatments is available. These include laser heat treatment, laser surface remelting, alloying, and cladding; furthermore, each technique can be used to improve the corrosion resistance of the material surface. For example, laser surface heat treatment has been used to prevent intergranular corrosion of stainless steel 304[1], the intergranular corrosion of 304L stainless steel can be prevented using LSM techniques[2] and laser surface alloying and cladding processes have been used to prevent or repair the intergranular corrosion of 304L stainless steel[3]. This paper reports on investigations into the laser surface remelting of 347 stainless steel and the effects of this on the "end grain" corrosion resistance in nitric acid.

Stabilised grades of stainless steel were selected for the original nuclear fuel reprocessing plant constructed in the 1950's at BNFL Sellafield. Steel making technology at that time resulted in carbon levels which required the presence of preferential carbide formers in the steel, eg niobium or titanium. This was because at the relatively high carbon levels present in the steels, chromium-rich carbides could be precipitated at the grain boundaries during welding and this could lead to intergranular corrosion problems. Stabilising elements such as Nb and Ti were used to prevent intergranular corrosion sensitisation by preferentially forming niobium carbide or titanium carbide instead of chromium carbide. However plant operational experience indicated that, in very corrosive situations, 347 stainless steel could suffer from preferential attack on tubes, plates, and forgings[4] in a direction parallel to the hot working direction (rolling direction). This feature became known as "end grain" corrosion.

Blom[5] and Stickler[6] reported that "end grain" corrosion appeared to be associated with stringers of stabilising element carbides. NbC stringers were originally assumed to be contributing to "end grain" corrosion in stabilised grades of austenitic stainless steel. However, more recently it has been found that NbC is not soluble in nitric acid. Furthermore some examples of "end grain" attack were also found to occur in non-stabilised grades. For example, Kain[7] reported that the presence of MnS inclusions in AISI 304L stainless steel (observed as "end grain" pitting) makes the material susceptible to attack in nitric acid environments. MnS stringers in non-stabilised grades have been blamed for "end grain" corrosion. It is true that MnS is very soluble in nitric acid, but unless the material adjacent to the stringers is susceptible to corrosion, then all that occurs is the dissolution of the stringers not severe intergranular corrosion. Current theories on the mechanism of "end grain" corrosion centre on the rôle of manganese sulphide inclusions and impurity segregation in the steel during manufacturing. The manganese sulphide inclusions have a low melting point and, as the last liquid to solidify in the steel, collect as films at the grain boundaries[8]. The presence of inclusions indicates the regions where high impurity segregation can be expected in 'dirty' steels. Processing of the cast material to produce plate and pipe elongates the stringers and the associated region of impurities eg S, P, etc, in the rolling direction. The temperature and time used during rolling are not sufficient for homogenisation of the impurities to occur and they remained associated with the stringers. In the final annealed structure the impurities segregate to grain boundaries close to the stringers. These final elongated stringers and associated impurities can have a marked effect on the directional properties[8]. It would be beneficial to prevent and/or repair these "end grain" attacks of 347 stainless steel using a laser surface treatment technique to prolong the service life of older plant.

2 EXPERIMENTAL

Material and Experimental Arrangement

The chemical composition of the 347 grade niobium-stabilised stainless steel used in this work is shown in Table 1. The experimental arrangement for LSM of the 347 stainless steel samples is shown in Figure 1. A long cylindrical nozzle was designed for the prevention of oxidation during the LSM process.

Laser Surface Remelting of 347 Stainless Steel

Table 2 outlines the range of laser processing parameters. Samples 5042 to 5044 were LSM on both the plate surface and "end grain" surface, and samples 5045 to 5047 were only LSM on the "end grain" surface.

Laser Surface Remelting of Corroded 347 Stainless Steel

Four specimens were corrosion tested before LSM and their corrosion rates are also tabulated in Table 2. These four specimens were corrosion tested again after LSM in order to simulate the repair of "end grain" attack of 347 stainless steel. In order to compare the effect of the LSM repair, 10% of the "end grain" surface was not treated so that it was further exposed in the second corrosion test.

Corrosion Assessment and Metallurgical Examination of LSM Samples

The BNFL standard practice, which is a modified version of ASTM 262 practice C[9], was employed to assess the corrosion performance of LSM samples. In particular, the LSM specimens were corrosion tested directly after the LSM process, i.e. without any prior specimen preparation in order to simulate the situation for "on line" LSM

Table 1 Chemical composition of 347 stainless steel

Element	Cr	Ni	C	Mo	Nb	Si	S	P	B	Ti
Weight Percentage	**18.01**	**13.19**	.0503	.127	**.79**	.441	.0138	.022	.0004	.000
Element	Cu	Co	Sn	W	V	Al	As	Mn	Fe1	Fe2
Weight Percentage	.091	.226	.000	.083	.144	.000	.0094	1.17	**65.42**	65.42

Table 2 Experimental data and corrosion rate of LSM 347 stainless steel

Sample No.	Power KW	Traverse Speed(mm/s)	Corrosion Rate(mm/yr) 48	96	144	192	240
5021(As Recieved)			0.265	1.179	2.292	3.182	3.229
5042	1.5	6	0.255	0.478	0.696	0.954	0.164
5043	1.5	18	0.298	0.705	1.288		
5044	1.5	36	0.416	1.899	2.844		
5045*	1.59	6	0.197	0.324	0.508	0.67	0.84
5046*	1.55	12	0.234	0.414	0.674	0.82	0.905
5047*	1.5	24	0.291	0.619	0.855	1.063	
5022^			0.269	0.507	0.7	0.817	1.324
5053	1.56	12	0.367	0.754	1.072		
5023^	0.235	0.417	0.613	0.75	1.204		
5054	1.59	24	0.434	1.026	1.404		
5024^	0.24	0.433	0.642	0.806	1.393		
5035	1.53	12	0.397	0.639	0.886	1.184	1.381
5025^	0.249	0.428	0.641	0.819	1.562		
5054	1.53	24	0.486	1.338	2.423		
5083**			0.251	1	2.097	3.07	3.774
			7.266(288)	4.969(366)	11.379(384)	4.926(432)	4.667(480)
5084**			0.255	1.135	2.464	3.799	3.934
			5.415(288)	7.216(366)	4.719(384)	5.352(432)	6.512(480)

*: End grain treatment only; ^:Corroded Samples before laser surface treatment.

**: 10 period of Huey test of as recieved ML8 stainless steel 347.

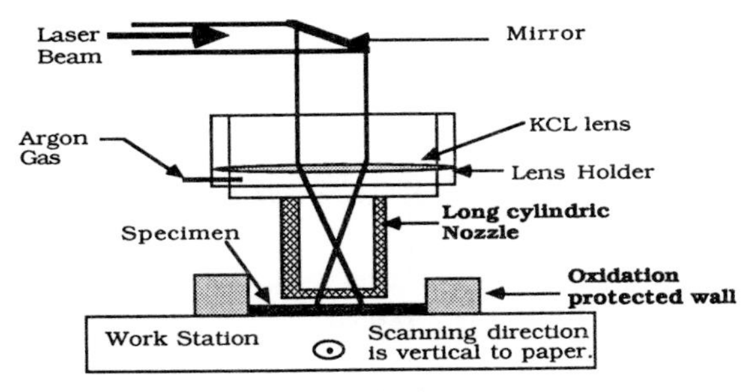

Figure 1 Laser surface remelting arrangement with long cylindrical nozzle for oxidation protection

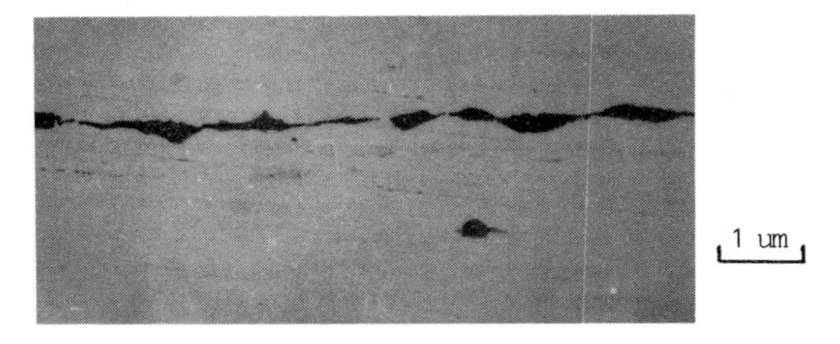

Figure 2 A typical example of stringer in 347 stainless steel

treatment. Optical microscopy and scanning electron microscopy (SEM) were also used to examine the solidification structure and corrosion characteristics of LSM samples and also to characterise the relevant features of "end grain" corrosion of 347 stainless steel samples.

3 RESULTS

Corrosion Performance of 347 Stainless Steel in Nitric Acid

A typical example of a MnS stringer inclusion in 347 stainless steel is shown in Figure 2. Examples of the corrosion performance of 347 stainless steel in hot nitric acid are shown in Figure 3. "End grain" attack of 347 stainless steel is shown in Figure 3(a), and the corrosion performance of the plate surface is shown in Figure 3(b). A higher magnification micrograph of a sample containing "end grain" attack is shown in Figure 4(a). An indication of "end grain" depth is shown in Figure 4(b). It can be seen that many sites of localised corrosion ("end grain" attack) were present at the "end grain" face and on the face parallel to the rolling direction. However, no such localised attack occurs on the "as-rolled" plate surfaces.

Corrosion Performance of LSM 347 Stainless Steel in Nitric Acid

The corrosion rates of corroded; LSM; and LSM corroded 347 stainless steel are tabulated in the right column of Table 2. In Figure 5 it can be seen that the corrosion rate of LSM 347 stainless steel is reduced significantly from 3.229mm/yr for the as-received sample, to 0.84mm/yr after LSM after the 5th period of Huey corrosion testing. SEM examination of the LSM "end grain" face and an untreated area are shown in Figure 6. It can be seen that corrosion has not occurred in the laser surface remelted region.

Repair of "End Grain" Corrosion

The corrosion response of LSM corroded 347 stainless steel is shown in Figure 7. It can be seen that the corrosion rate was reduced after LSM, even though the LSM corroded samples still had 10% of untreated area present at the "end grain" surface. This demonstrates the

(a) 50 um (b) 50 um

Figure 3 Corrosion performance of 347 stainless steel in hot nitric acid,
(a) corrosion performance of "end grain" surface (SEM)
(b) corrosion performance of plate surface (SEM)

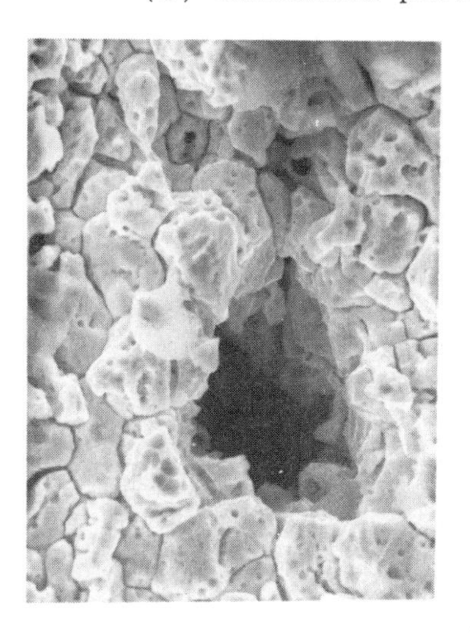

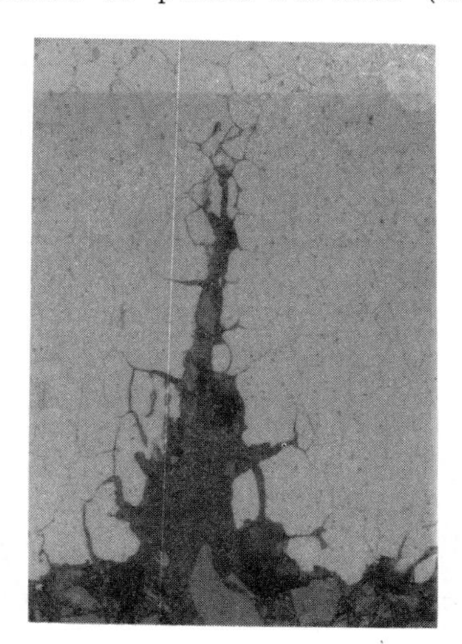

(a) 1 um (b) 1 um

Figure 4 "End grain" attack of 347 stainless steel in hot nitric acid,
(a) higher magnification of "end grain" attack (SEM)
(b) "end grain" attack penetration (optical)

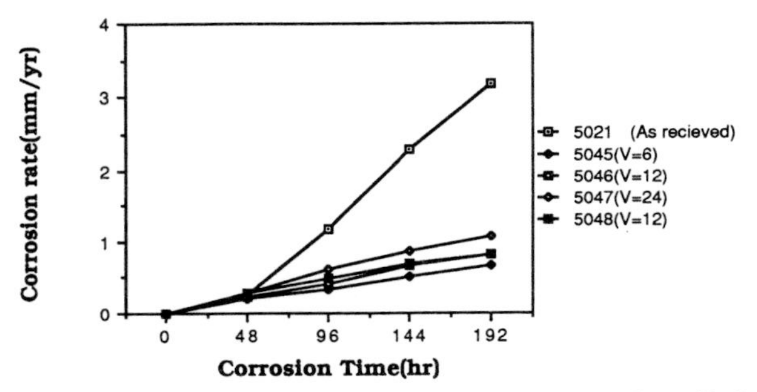

Figure 5 Prevention of "end grain" attack of 347 stainless steel

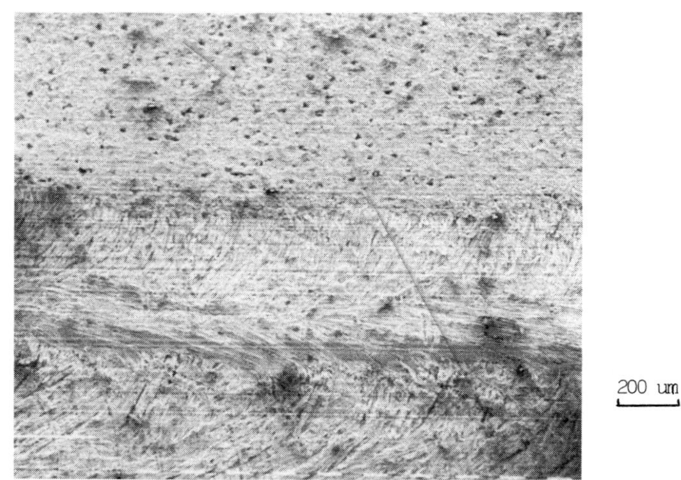

Figure 6 Micrograph showing the prevention of "end grain" attack

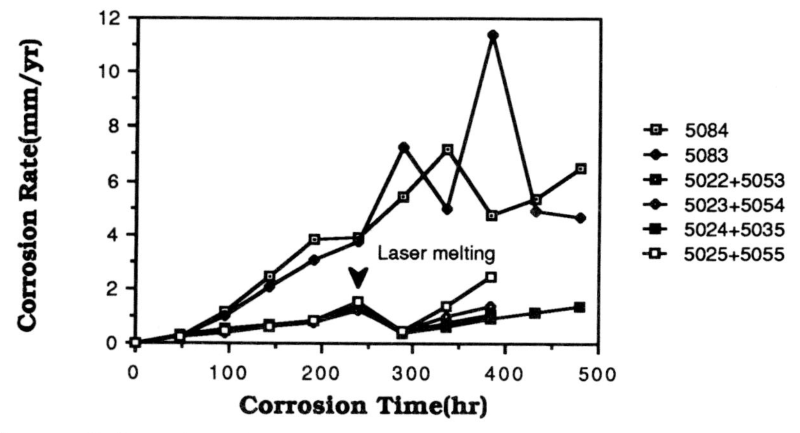

Figure 7 Repair of "end grain" corrosion using laser surface remelting

effectiveness of the process for repair of corroded surfaces. Metallurgical examination, presented in Figure 8, provides further evidence that repair of "end grain" attack on 347 stainless steel is possible using LSM. Comparisons of the corroded sample and LSM corroded 347 stainless steel are shown at the top and bottom of Figure 8(a). A higher magnification view of the interface of LSM and untreated area of the corroded sample is presented in Figure 8(b). The corrosion characteristics of the LSM corroded sample are presented in Figure 8(c). From Figure 8(a), Figure 8(b) and Figure 8(c), it can be seen that the originally existing "end grain" attack in corroded 347 stainless steel has been "sealed" and hence repaired using LSM. Similarly, optical examination of the corrosion penetration in the LSM corroded sample (Figure 9) shows a comparison of "end grain" attack at the untreated area and with "end grain" attack sealed by LSM. It can be seen that an area of the originally existing "end grain" attack has been sealed by the LSM pool, and the penetration of the "end grain" corrosion in the untreated area was much deeper than that of the originally existing LSM "end grain" corroded area.

The Effect of LSM of the Plate Surface on the Corrosion Performance

Samples 5042 and 5045 were treated with similar laser processing parameters except that only "end grain" surfaces of sample 5045 were remelted, and both of "end grain" surface and plate surface of sample 5042 were remelted. The corrosion rates of these two samples are shown in Figure 10. It can be seen that the corrosion rate of LSM "end grain" surface only is lower than that of LSM both "end grain" surface and plate surface. The implication from this is that the LSM plate surface had a lower corrosion resistance than the original plate surface.

4 DISCUSSION

"End Grain" Corrosion Mechanism of 347 Stainless Steel

A schematic outline of the "end grain" corrosion mechanism is shown in Figure 11. During the casting process the impurity elements such as sulphur, manganese and the range of trace elements segregate and, being of

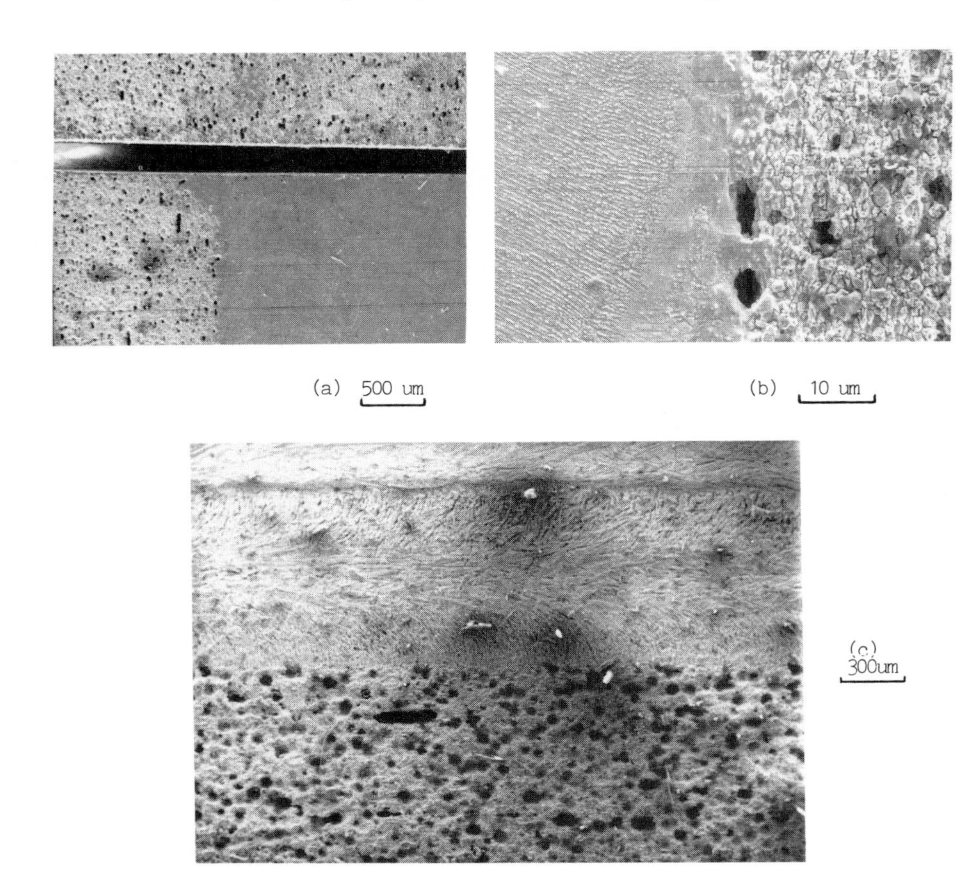

Figure 8 The micrograph showing the repair of end grain attack using LSM.
(a) Comparison of Corroded sample and LSM corroded sample.(SEM)
(b) The sealing of end grain attack after LSM.(SEM)
(c) The corrosion performance of LSM corroded sample and untreated area.(SEM)

Figure 9 Corrosion penetration of LSM corroded sample

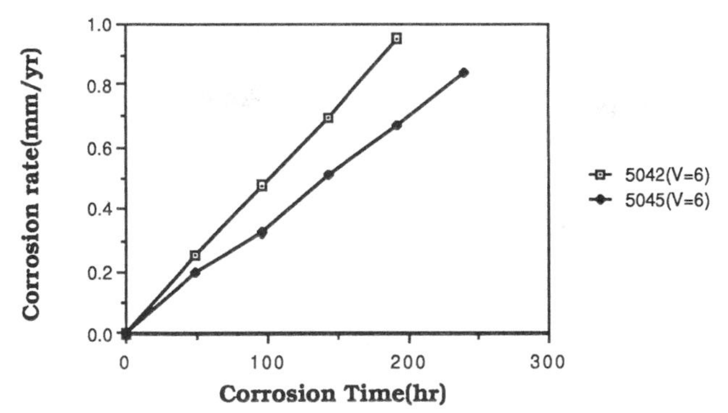

Figure 10 The effect of LSM plate surface on the corrosion performance

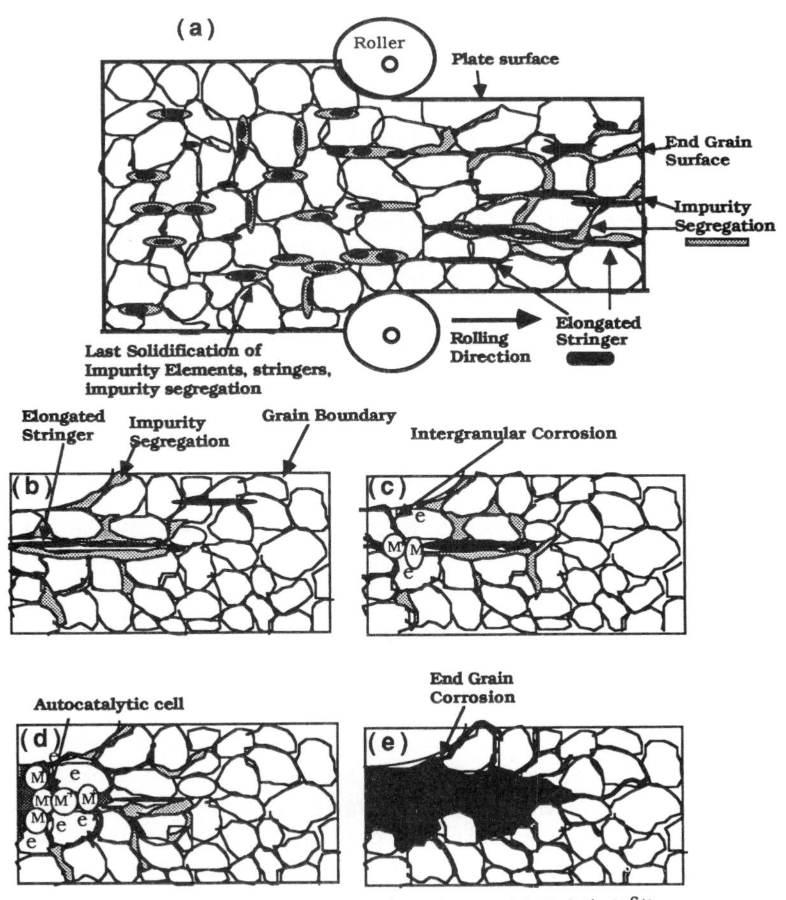

Figure 11 Mechanism of end grain corrosion of stainless steel 347. ($M^+=Cr^{6+}$)
(a) The formation of elongated stringers and segregation of impurities.
(b) Formation of stringers and impurity segration parallel to the rolling direction.
(c) Dissolution of stringer and intergranular corrosion.
(d) Drop out of grain near end grain surface.
(e) End grain attack.

relatively low melting point, are the last materials to solidify at the grain boundary, Figure 11(a). During the subsequent rolling operations, these impurity elements are rolled and plastically deformed into an elongated film in a direction parallel to the rolling direction producing stringer inclusions as shown in Figures 11(a) and 11(b). The outer regions of the casting are the first to solidify and hence impurity segregation and stringer formation do not occur in these regions. Thus no localised attack occurs at the as-formed plate surfaces.

In hot nitric acid environments (modified Huey test[9]), these MnS stringers dissolve quickly because they are anodic with respect to the passive stainless steel surface[10]. The associated segregation at the grain boundaries close to these stringers is also susceptible to attack. Dissolution results in Cr^{6+} ions collecting in pits formed as shown in Figure 11(c). These pits are then enlarged by rapid intergranular attack induced by the Cr^{6+}ions, because Cr^{6+} ions in the Huey solution move the corrosion potential into the transpassive range where rapid grain boundary corrosion can occur[10,11]. Therefore, as the concentration of Cr^{6+} ions increases so does the corrosion producing more Cr^{6+} ions producing accelerated corrosion, etc. In this confined space, a local corrosion cell is set up, which is autocatalytic as shown in Figure 11(d). Thus "end grain" corrosion develops along the direction parallel to the rolling direction as shown in Figure 11(e).

Principle of "End Grain" Attack Prevention and Repair Using LSM

The principle of "end grain" attack prevention and repair using laser surface remelting is outlined schematically in Figure 12 and Figure 13. A theory of the mechanism of "end grain" corrosion has been outlined in Figure 11 and has been presented above. The key reason "end grain" attack occurs is believed to be due to the regions of the impurity segregation and the stringers elongated in the rolling direction during the rolling operation. During LSM, refinement and homogenisation of the microstructure occur. The segregated regions and stringers (either NbC or MnS) are remelted within the melt pool as shown in Figure 12(a) and once the melt pool has solidified the impurities are uniformly redistributed

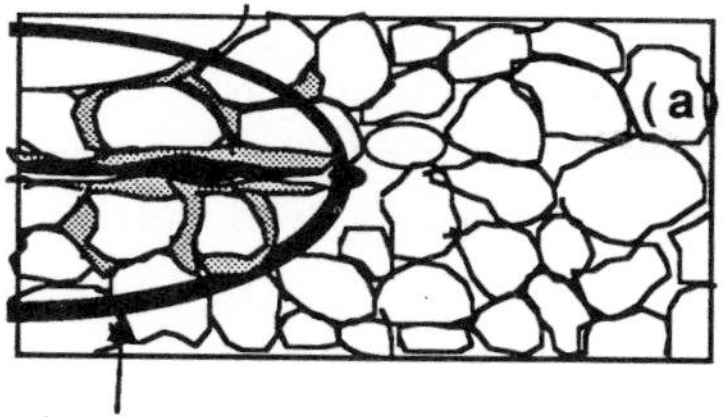

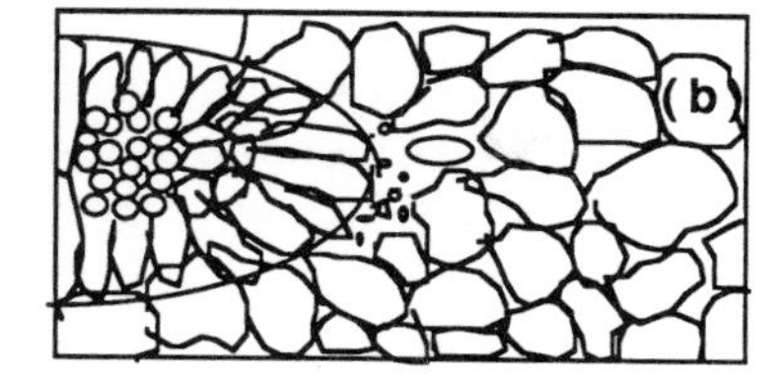

Figure 12 Mechanism of "end grain" corrosion prevention
(a) stringers and impurity segregation were remelted within molten pool
(b) redistribution of stringers, impurities and all the materials within melt pool

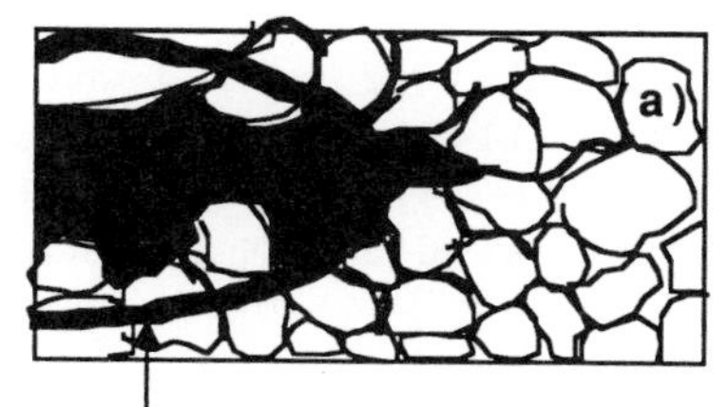

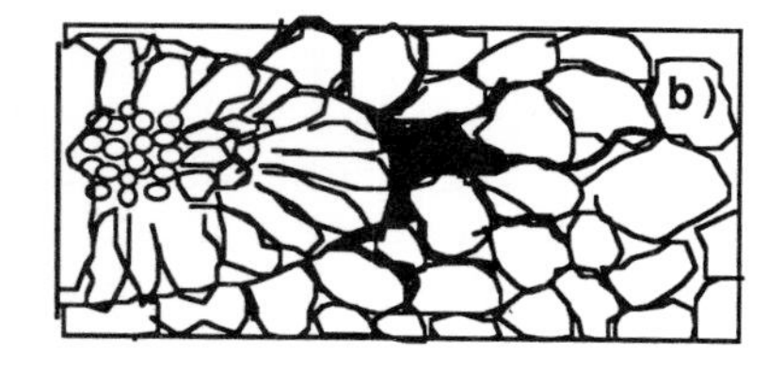

Figure 13 Mechanism of "end grain" corrosion repair
(a) "end grain" attack was remelted and filled with surrounding material within molten pool
(b) "end grain" attack was sealed within molten pool

within the melt pool owing to the high convection and mixing within the melt pool region as shown in Figure 12(b) [12]. As a consequence of this the prevention of 347 stainless steel "end grain" attack is possible using the LSM process.

The mechanism of the repair of "end grain" attack of 347 stainless steel is shown in Figure 13(a) and Figure 13(b). During the solidification process of LSM the material with "end grain" attack would be resolidified and redistributed uniformly within the molten pool as shown in Figure 13(b). Hence, the original "end grain" attack will be filled with the surrounding material or sealed within the molten pool. The autocatalytically localised cell would no longer exist. Therefore the 347 stainless steel "end grain" attack can be repaired using a LSM technique.

The Effects of LSM Plate Surface on the Corrosion Performance

The corrosion results of LSM plate surface show that the LSM plate surface made the corrosion resistance of 347 stainless steel worse. The plate surface contained few impurities and in the 'as-rolled' condition showed good corrosion resistance. When they are LSM the cast structure produced will contain chromium segregation and hence make the corrosion resistance worse.

5 CONCLUSIONS

1. The use of laser surface remelting techniques for the repair or prevention of "end grain" attack in 347 stainless steel "end grain" faces has been shown to be feasible.

2. Corrosion rates have been reduced from the original 3.229 mm/yr to less than 1 mm/yr after laser surface remelting of the "end grain" corroded surfaces. Further "end grain" attack was not observed on the laser surface remelted "end grain" surfaces.

3. Laser surface remelting of the as-formed 347 stainless steel plate surfaces leads to a reduced corrosion resistance level for these areas and is not recommended.

REFERENCES

1. Y. Nakao and K. Nishimoto, Trans. of the Japan Welding Society, 1986, 17(1).
2. T.R. Anthony and H.E. Cline, J. Applied Physics, 1976, 15, 2181.
3. J.Y. Jeng, Q. Brian, P. Modern and W.M. Steen, 'Reconstitution of Corroded Stainless Steel 304L by Laser Surface Remelting, Alloying and Cladding,' to be published in the International Conference, "Advances in Corrosion and Protection," UMIST, Manchester, UK, 28th June - 3rd July, 1992.
4. R.D. Shaw, British Corrosion J., 1990, 25(2), 97-107.
5. U. Blom and B. Kvarnback, Material Performance, July 1975, pp. 43-46.
6. R. Stickler and A. Vinckier, Trans. Am. Soc. Metals, 1961, 54, 362.
7. V. Kain, S.S. Chouthai and H.S. Gadiyar, British Corrosion J., 1992, 27(1), 59-65.
8. H.E. Boyer and T.L. Gall, 'Secondary Refining Processes,' Metals Handbook Desk Edition, ASM, 1985, p. 22.7.
9. NF Standard, 'Test Procedure for Nitric Acid Corrosion Test for Nitric Acid Grade 18/10L Stainless Steel and 25/20L Stainless Steel,' NF 0107/1, British Nuclear Fuels plc, Risley, Warrington, Dec. 1986.
10. A.J. Sedriks, International Metals Review, 1983, 28(5), 295-307.
11. F.G. Wilson, British Corrosion J., 1971, 6, 100-108.
12. C. Chan, J. Mazumder and H.H. Chen, J. of Appl. Phys., 1988, 64(11), 6166-6174.

3.2.4
Laser Claddings for Fretting Resistance

Y. X. Wu, Y. M. Zhu, Z. X. Zhu, K. L. Wang, and L. Z. Ding

DEPARTMENT OF MECHANICAL ENGINEERING, TSINGHUA UNIVERSITY, BEIJING 100084, PEOPLE'S REPUBLIC OF CHINA

1 INTRODUCTION

The austenitic 1Cr18Ni9Ti stainless steel is one of the most important metals that has been widely used in the environment of vibration and impact. Effective fretting resistance is essential for stainless steel parts working in such conditions. So, improvement in fretting resistance is very important for the exploitation of this steel. Generally, surface modification, such as chemical conversion coatings and solid (or fluid) lubricants[1,2] is used to protect the surface against fretting wear, but still, it is not always satisfactory in the case of severe working conditions. As fretting is a type of composite wear involving abrasion, adhesion, oxidation and fatigue[3], the fretting resistant coatings have to possess combined properties of high strength and toughness. Laser claddings of nickel-based self-fluxing alloys are expected to meet the demand.

Laser claddings of METCO 16C and METCO 31C on 1Cr18Ni9Ti stainless steel substrates were investigated and carried out by laser melting of flame sprayed coatings. Their microstructure and properties were analysed and tested. A 50 hours intensive vibration test proved that the fretting resistance of laser clad coatings is far better than that of a carbonitrided layer.

2 EXPERIMENTAL MATERIALS AND PROCEDURE

The chemical composition and melting points of METCO 16C and METCO 31C powders are shown in Table 1. The cylindrical specimens of 1Cr18Ni9Ti stainless steel (Φ7mm x 26mm) were cleaned and roughened by lightly grit

Table 1 Chemical composition and melting points of alloy powders

Powder	Chemical composition (wt.%)									Melting point (^{0}C)
	Cr	B	Si	C	Fe	Cu	Mo	Co-WC	Ni	
METCO16C	16	4	4	0.5	2.5	3	3	-	bal.	1010
METCO31C	11	2.5	2.5	0.5	2.5	-	-	35	bal.	1040

blasting before flame spraying. The thickness of the sprayed coatings was from 0.5 to 0.6mm.

Under the protection of nitrogen, a CO_2 laser system with maximum output power of 2kW, was used to remelt the as-sprayed specimens. The optimized processing parameters are as follows:

Laser power: 1.2-1.6kW Beam diameter: 4 mm
Scanning speed: 300-400 mm/min Overlap: 1 mm

The surfaces of laser clad cylinder and sleeve specimens were ground to have roughness of Ra~ 0.32-0.63 µm. The laser clad cylinder specimens were matched with nitrided and carbonitrided sleeves.

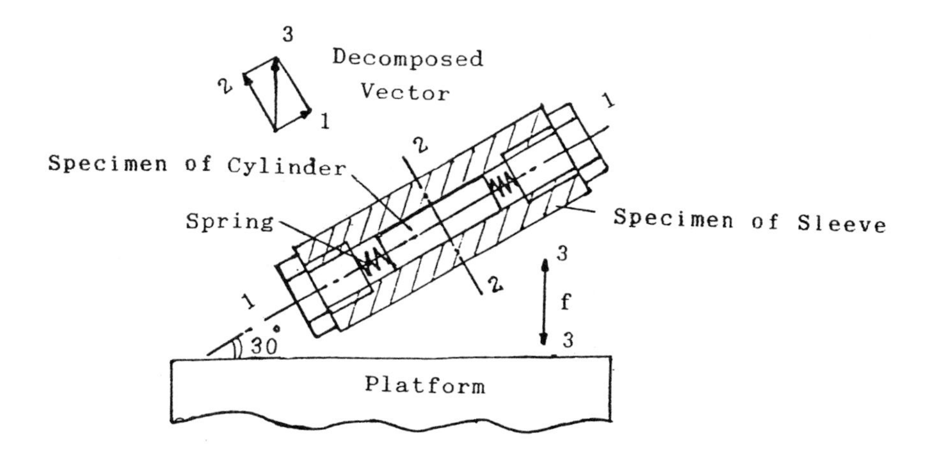

Figure 1 Schematic illustration of additional apparatus on Instron vibrator

The intensified vibration test was carried out on an Instron 1506 Electric Vibrator. Figure 1 shows the experimental apparatus for vibrative specimens. The clearance between the interacting pair of cylinder and sleeve was 0.25-0.30mm. The springs at both ends allow the cylinder to slide axially. The sleeves were clamped on the vibrator's platform with an angle of 30 degrees. Therefore, the movement of the cylinder inside the sleeve was of two dimensions: axial sliding and radial vibration or impact. The vertical vibration vector comprises an axial and a radial component.

The vibration of the platform had a frequency of 120 Hz, and an amplitude of 0.5mm. Both the materials and surface treatments of the cylinders and sleeves are given in Table 2.

Table 2 Frictional couples used in vibration tests

Couple	Cylinder (1Cr18Ni9Ti)	Sleeve (1Cr11Ni2W2MoV)
A-A'	La.cla. coating of METCO16C	Carbonitrided layer
B-B'	La.cla. coating of METCO31C	Carbonitrided layer
C-C'	La.cla. coating of METCO16C	Nitrided layer
D-D'	Carbonitrided layer	Nitrided layer

3 RESULTS AND DISCUSSION

Microstructural Characteristics of Laser Clad Coatings

Laser Clad Coating of METCO 16C. The homogeneous, pore- and crack-free fine microstructure of laser clad coating of METCO 16C, includes three constituents: white dendritical crystals, black eutectic, and flake or star-like precipitates (Figure 2). Their compositions measured by EDAX and WDAX are listed in Table 3 and Table 4. Combined with the results of X-Ray Diffraction Analysis (XRDA) in Figure 3, these constituents may be inferred as follows. The white dendritical crystal is a nickel-based solid solution (γ) with f.c.c. structure strengthened by Fe, Cr etc. As the matrix in the microstructure of laser clad coating of METCO 16C, γ(Ni,Fe) endows the coating

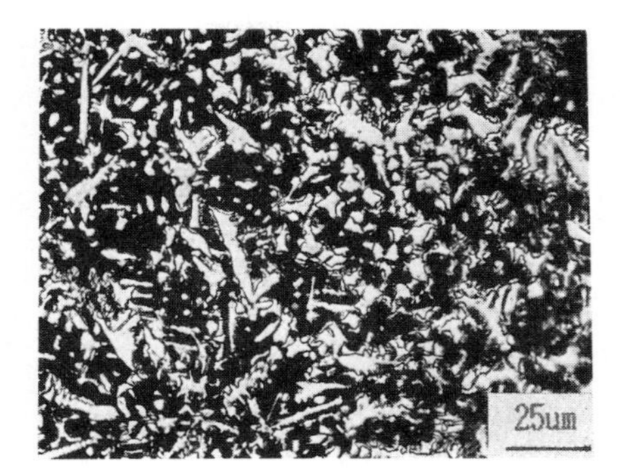

Figure 2 Microstructure of METCO 16C coating

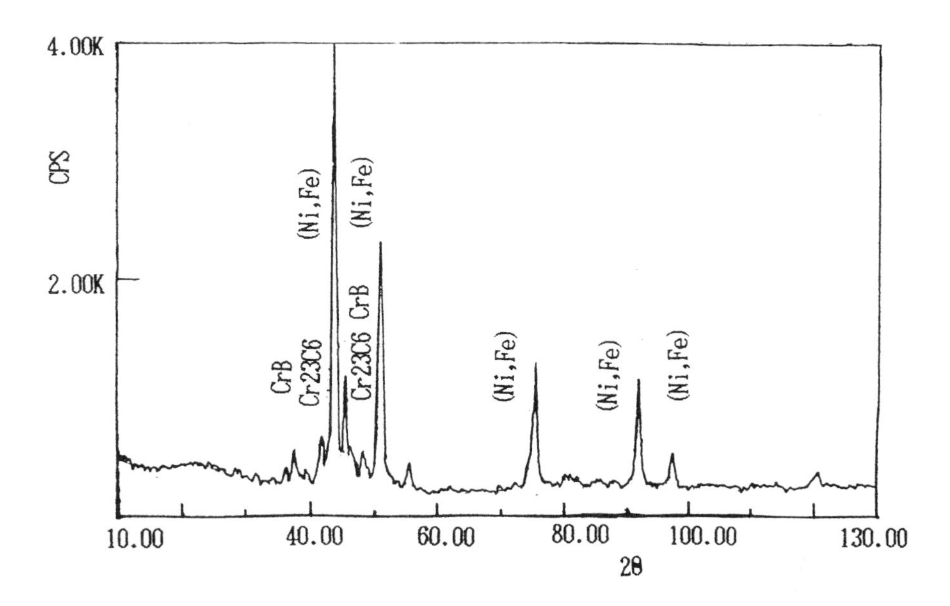

Figure 3 X-ray diffraction spectrum of METCO 16C coating

with better strength and toughness. As indicated from the data, the main elements in the flake or star-like precipitation are chromium and boron. Therefore, the flake and star-like precipitates are intermetallic compounds of CrB and $Cr_{23}C_6$ which act as hardening phases in the coatings. The black eutectic is composed of solid solution γ(Ni,Fe) and intermetallic compounds of carbides and silicides. The phases in the laser clad coating are determined not only by alloying elements, but also by the

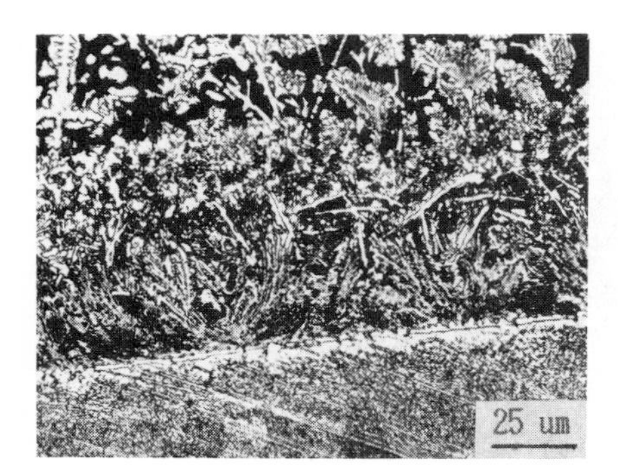

Figure 4 Bonding area of METCO 16C coating with substrate

Figure 5 Micrograph of METCO 31C coating

parameters of the laser process which greatly affect the cooling rate of the coating. The lower the cooling rate, the more prevalent is the equilibrium crystalline phase. Because the specimens used in our experiment were small, the cooling rate was low, so a larger amount of eutectic crystal was observed in their microstructure.

Figure 4 shows the metallurgically bonded area of laser clad coating with the substrate. The white band of 4-5μm, formed by mutual diffusion of elements between laser clad coating and substrate, consists of Ni, Fe and other elements, the proportion of which is related to the melting depth to substrate. Metallurgical bonding is one of the advantages of laser clad coating: it enhances the bonding strength of the laser treated coating with the substrate.

Laser Clad Coating of METCO 31C. Figure 5 shows the homogeneous and pore-free microstructure of laser clad coating of METCO 31C. According to the XRDA results (Figure 6), the phases in the microstructure are γ(Ni, Fe, W), WC, Ni_4B_3 etc. The γ-phase is a nickel-based solid solution of f.c.c. structure, including strengthening elements of Fe, Cr, W etc. The regularly shaped white flakes are unmelted tungsten carbide (WC), acting as a hardening phase in the coating; the matrix is an eutectic composed of γ(Ni,Fe), carbides and borides. In addition, there are some blocks of WC enveloped by Co.

The bright solid solution band in Figure 7 indicates the firm metallurgical bonding between coating and substrate.

Table 3 Elemental analysis of METCO 16C coating by EDAX

Phase	Composition (wt.%)					
	Ni	Cr	Fe	Mo	Si	Cu
Star-like	9.97	70.81	8.36	10.86	-	-
Flake	8.41	71.88	8.85	10.86	-	-
Dendrite	70.75	7.38	11.03	-	7.68	3.16
Eutectic	70.38	4.33	4.18	1.00	18.44	1.67

Table 4 Pulse number of WDAX for different phases

Phase	Pulse number of WDAX			
	Cr	Ni	Fe	B
Flake	3600	1000	260	23
Star-like	2500	1000	660	12
Matrix	1400	3000	1150	3

Microhardness of Laser Clad Coating. As shown in Figure 8, the hardness of the laser clad coating against the depth from the surface is nearly constant and there is a steady transition towards the coating/substrate interface. It indicates that both the microstructure and the composition of the coatings are homogeneous, and that a firm metallurgical bond was obtained. The highest microhardness of METCO 16C and METCO 31C reaches Hv 580 and Hv 900 respectively. The higher hardness of the laser clad coating of METCO 31C than that of METCO 16C is mainly due to the strengthening effect of the hard phase WC in the coating.

Results of Vibration Test

Macrographic Examination of Worn Surfaces. During the 50 hours vibration test, a macrographic examination of worn surfaces was carried out regularly. During the first

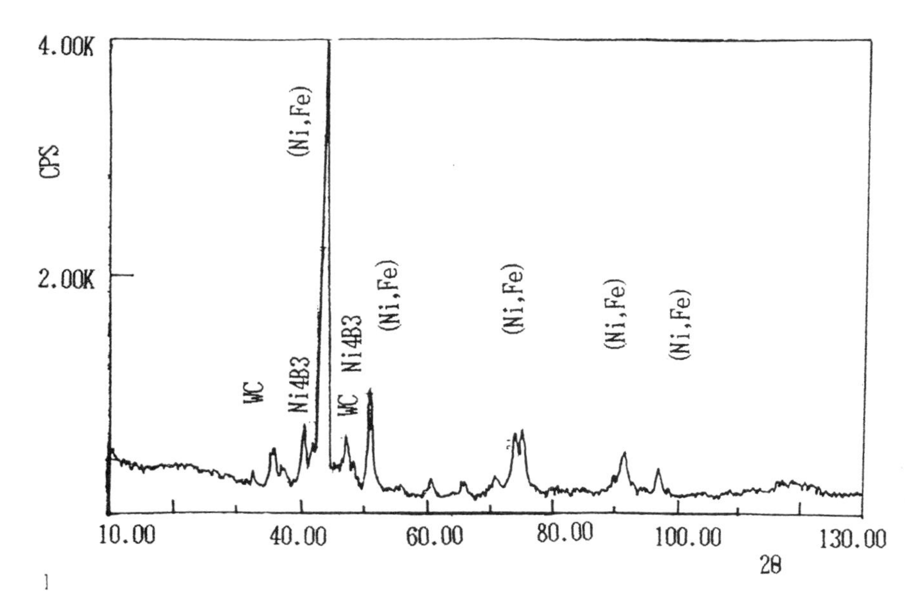

Figure 6 X-ray diffraction spectrum of METCO 31C coating

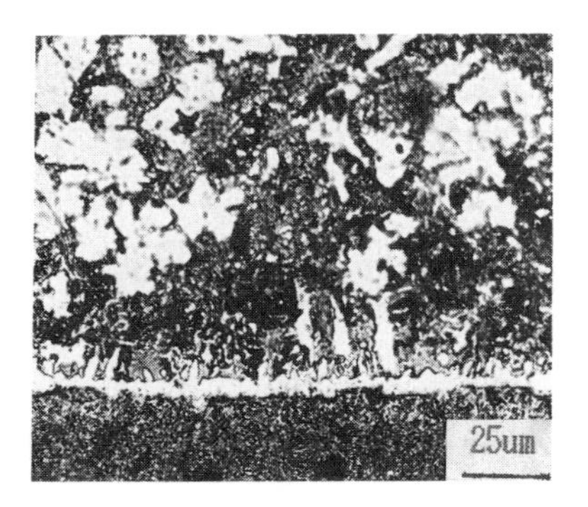

Figure 7 Bonding area of METCO 31C coating with substrate

15 hours, the worn area of the cylinder increased; a black and red-brown wear debris, identified as Fe_3O_4 and Fe_2O_3, was obtained. After 35 hours however, the worn area had not increased considerably, and furthermore, the worn area after 50 hours vibration was almost the same as that after 35 hours. It shows that after a period of running-in, the fretting wear between interacting surfaces enters into a steady stage, where the wear rate is maintained at a lower level[4].

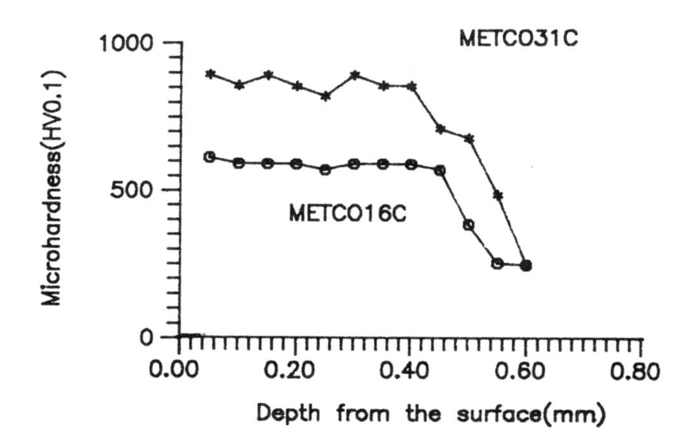

Figure 8 Microhardness distributions of METCO 16C and METCO 31C coating

Width of Wear Track. The width of the most severe wear track on the cylinder surface was measured by morphometer. The data and graph are given in Figure 9. Wear of laser clad cylinders (A, B and C) of 1Cr18Ni9Ti stainless steel is less than that of carbonitrided sleeve (D). When matched with the nitrided sleeve of 1Cr11Ni2W2MoV steel, the laser clad cylinder of METCO 16C is the best for wear resistance, nearly four times better than that of the carbonitrided layer (D).

In addition, the wear sleeves matched with laser clad cylinders decreased simultaneously. This may be owing to the compatibility of the interacting pair.

Morphology of Wear Track. Two kinds of movement, i.e.vibration (or impact) and sliding between the cylinder and sleeve enabled the worn surfaces to be characterized as fretting. The red-brown oxide powder of Fe_2O_3 and Fe_3O_4 as well as the thin flakelets of wear debris were generated during the vibration test.

The worn surfaces of cylinders were analysed under SEM (Scanning Electron Microscope). It can be seen from Figure 10 that flaking and micro-ploughing occurred on the cylinder coating in varying degrees. Flaking caused by the fatigue component of fretting, led to coating damage;

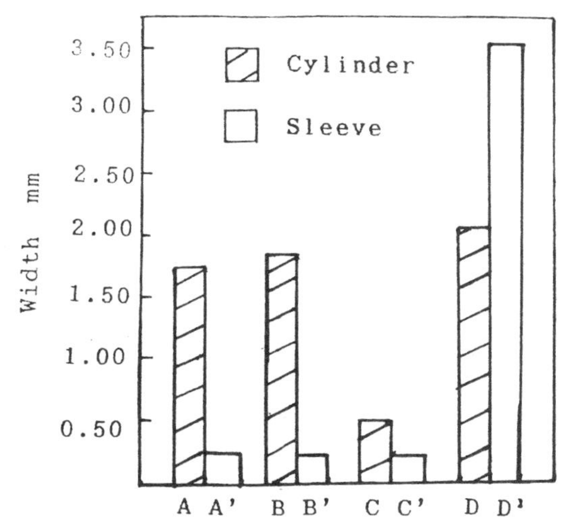

Figure 9 Width of worn surfaces

(a) Cylinder A (b) Cylinder B

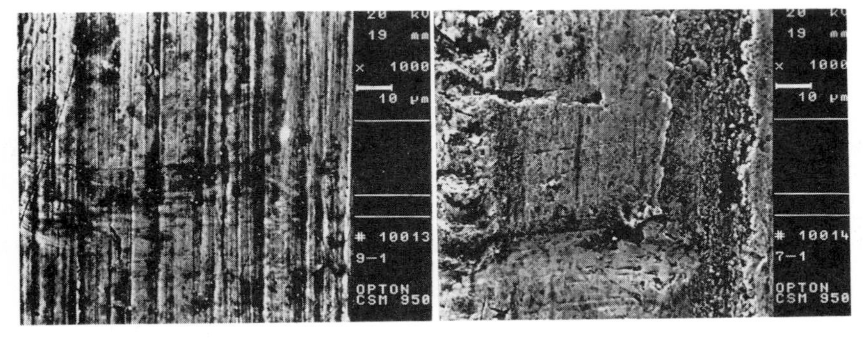

(c) Cylinder C (d) Cylinder D

Figure 10 Morphology of wear track

in turn, the flaked particles gave rise to abrasive wear. But the laser clad coatings of cylinders (A, B and C) showed less flaking than that of the carbonitrided layer of 1Cr18Ni9Ti stainless steel (i.e. cylinder D) which suffered severe flaking and ploughing. This is in accordance with the width of worn track in Figure 9.

Thus, it can be inferred that the laser clad coating of nickel-based self-fluxing alloys on the 1Cr18Ni9Ti stainless steel substrate has a higher fretting resistance than that of a carbonitrided layer.

4 CONCLUSIONS

1. Laser claddings of nickel-based self-fluxing alloys (METCO 16C and METCO 31C) on 1Cr18Ni9Ti stainless steel substrates are homogeneous and pore-free. The nickel-based solid solution is the main micro-structural feature of the coatings. The others are the intermetallic compounds of carbides and borides.

2. Laser clad coatings of METCO 16C and METCO 31C have better anti-fretting properties, which are markedly higher than that of a carbonitrided layer of 1Cr18Ni9Ti stainless steel.

3. The combination of strength with toughness and the metallurgical bonding with the substrate play a key rôle in improving fretting resistance of the laser clad coatings.

ACKNOWLEDGEMENT

To Professor Fangze Li and Shiliang Dai, Central Laboratory of Strength and Vibration, Tsinghua University, for supplying the vibration device and helpful proposals.

REFERENCES

1. S. Li and X. Dong, 'Erosion and Fretting of Materials,' The Publishing House of Engineering Industry, Beijing, 1987, p.461.
2. S.F. Murray, 'Wear Resistant Coatings and Surface Treatments,' CRC Handbook of Lubrication, 1984, Vol.2, p.623.

3. J. Sato, M. Sato and S. Yamamoto, Wear, 1981, 69, 167.
4. J. Liu, 'An Introduction to Wear, Corrosion Resistance of Materials and Surface Treatments,' The Publishing House of Engineering Industry, Beijing, 1988, p.15.

Section 3.3 Electroplating

3.3.1
Electroplated Metal Matrix Composites

P. R. Ebdon

PERA INTERNATIONAL TECHNOLOGY CENTRE, NOTTINGHAM ROAD, MELTON MOWBRAY, LEICESTERSHIRE LE13 OPB, UK

1 INTRODUCTION

There is a general awareness in the Surface Engineering industry of the techniques, if not the art, of electroplating. What may not be such common knowledge are details of the relatively new materials that are known as metal matrix composites (MMC's). These are a range of materials that have a conventional metallic matrix which is reinforced with particulates, short whiskers or continuous fibres. This second phase is normally ceramic in nature, such as silicon carbide or alumina, although it could be metallic or graphite fibres.

Up to 20-30% by volume of the reinforcing phase may be included, depending on the properties required which can be tailored for particular applications. Metal matrix composites perform very well in terms of specific strength and stiffness when compared against other common structural materials, Figure 1, as well as possessing improved elevated temperature characteristics.

The majority of MMC's in production today have aluminium as the matrix material, with either SiC or alumina particulate reinforcement. Other MMC's with alternative matrix metals are under development, but these have yet to reach the stage of significant volume production.

The advantages that are gained from the incorporation of reinforcement accrue from the improved physical properties such as stiffness and strength, with only minor penalties to pay in terms of density when looking at the aluminium based MMC's.

With the commercial availability of aluminium composites from companies that are now producing the material at prices that are comparable with conventional aluminium alloys, interest has been focusing on the potential end uses for these advanced materials.

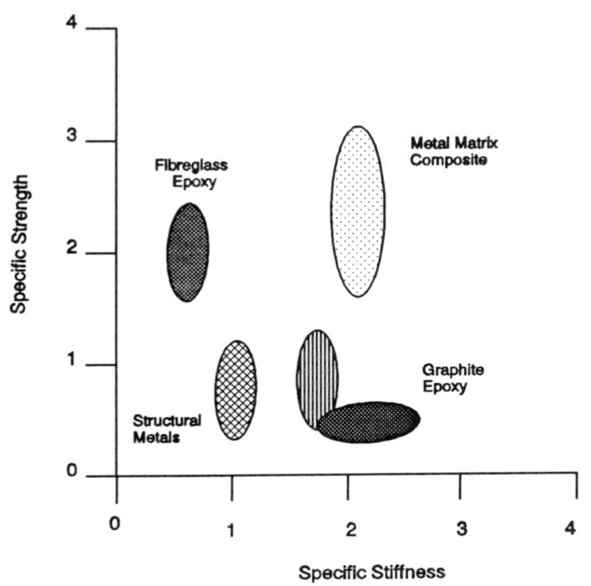

Figure 1

Obviously the enhanced mechanical performance holds promise for both the aerospace and automotive industries, where for instance weight advantages may be made without sacrificing stiffness or strength, or resistance to thermal shock can be improved, and here a lot of effort is being devoted to identifying the precise physical properties of the various types of composite that are available.

As these progress and the utilisation of this class of materials becomes more widespread, greater attention is being directed to the methods by which the composites may be fabricated and converted into finished components.

As part of a European research initiative under the COST programme, Pera International is reaching the latter stages of a project that has been studying the machinability of aluminium MMC. This has revealed, not surprisingly, that the material exhibits both considerable abrasion resistance and abrasivity.

This study and others has lead to the conclusion that situations are likely to occur with the materials where wear can be a factor, either to the MMC or to the counterface.

In addition to the tribological aspects of the application of these materials, there can be coupled the effect that the second phase additions will have on the corrosion resistance of the composite. Here the reinforcing material can act as a noble phase, increasing susceptibility to corrosion.

Both the wear and corrosion resistance of these materials may be modified by engineering the surface of the MMC, with most of the conventional and newer metal finishing technologies having a potential role.

This paper will concentrate on the potential possibilities of aqueous technologies in addressing these twin problems of corrosion and wear.

2 CORROSION

Aqueous Pretreatments

With conventional wet treatments, whether an electroplated coating or a conversion coating, problems during pretreatment are often caused by non-metallic inclusions, areas of inhomogeneity or porosity in the substrate material. The utilisation of a coating that is substandard, either through porosity or suspect adhesion, can also cause enhanced problems through localising corrosion or contributing to wear debris problems.

Electroplated Coatings. Of the conventional electrodeposited coatings, typical selections that may be made could include chromium, nickel (both electrolytic and electroless) and composite coatings containing either hard or lubricating particles.

Although, as outlined earlier, the MMC material will cause difficulties when processing, the performance of the coating on composites could well be enhanced through the higher modulus of the substrate. This would give greater support to relatively thin coatings of low ductility that can give problems under impact loading when the substrate plastically deforms.

By far the most popular pretreatment for aluminium prior to electroplating is the zinc immersion process, which is then either directly plated with the final coating, or given an intermediate strike layer of copper.

It is the quality of the immersion coating that is going to substantially determine the performance of the subsequent layers, and this is itself determined to some extent by the pretreatment applied to the raw substrate. A pretreatment sequence that prepares the MMC without exposing or undermining excessive amounts of the ceramic is required, whilst the subsequent immersion treatment must be able to accommodate the small areas of exposed second phase that will be inevitably be present.

Additionally, the strike and/or the top coating should have the ability to cover any porosity remaining from the previous treatments.

Hard chrome is well known as a bearing material and as such, once the coating is properly applied, should perform in a fairly predictable manner. Nickel is unlikely to be used in its common role of resizing worn components in this application, due to the production of near net shapes and the specialised nature of the substrate material, but could have a useful effect in combating fretting corrosion. Likewise, the composite coatings of electroless nickel/PTFE can also assist in this area[1] or in the reduction of the friction coefficient.

Anodising. Anodising is an attractive way of imparting corrosion resistance to aluminium, and this method of treatment has received most attention as a way of protecting the MMC. However, although the workers have examined the corrosion performance of the anodised layers through the use of electrochemical impedance spectroscopy,[2,3] few details are given of the difficulties encountered in producing the oxide layers, although in the case of hard anodising (Type III) it has proved possible to produce an anodised layer with corrosion properties that are an improvement over the bare MMC. Conventional (Type II) coatings have been found by some workers to be detrimental to the corrosion resistance of the MMC.[4] Others have noted a distinct difference in the properties of the anodised layers on composites as opposed to the corresponding wrought alloys. Some researchers have found that alloying elements rather than the presence of silicon carbide is more important in determining corrosion resistance after anodising.[5,6]

Further work is required to produce the anodic coatings under production conditions and examine the effect of pretreatment on their structure, as well as identifying any defects in the coatings that may be associated with the second phase particles. Some results have shown that the anodic coating can tolerate the presence of the silicon carbide and that the particles may be accommodated within the film itself, although at a much smaller proportion than present in the base material. Where the silicon carbide goes to is still the subject of conjecture; it is thought to be either oxidised and incorporated into the coating or lost into the anodising electrolyte through undermining.[6] Conventional corrosion testing to provide standard data for development engineers to enable them to predict service life in comparison with conventional wrought alloys is also required.

Conversion Coatings. Conversion coating or chemical passivation has also been studied as a route to corrosion control in these composites. Here, much of the work has been performed with a relatively new passivating solution, cerium trichloride, with a view to avoiding the environmental and health risks associated with the use of the conventional chromate containing solutions.[7] Whilst the work reported so far is encouraging, a lot of research is required to define the mechanisms by which these coatings are formed and operate. One drawback to their early introduction may be the long immersion times required to form the protective layers; up to a week is not conducive to the rapid turnround of components normally required. However, the processing time for these solutions will doubtless be reduced as more is learnt about their operation.

Chromate conversion coatings have been looked at on these materials, but data on their performance is scant.

Similarly, no references have been found relating to the production of phosphate coatings on MMC's, which is an alternative pretreatment prior to painting.

Non-Aqueous Techniques. Thermal spraying as a coating application method should be less sensitive to the presence of the second phase particles in the substrate. Such coatings can be applied both for tribological and corrosion resistance purposes, and may be followed by organic finishing if required.

The majority of the emerging vacuum techniques for depositing high performance coatings can be utilised in this application, the important limitation here being the melting point of the matrix material. In this area, not only can discrete coatings be applied to the substrate, but the surface can be converted through processes such as ion implantation that could have a direct impact on the MMC's performance. No significant work appears to have been performed in any of these areas.

3 WEAR

Several investigations have been made into the tribological characteristics of MMC's. One of the most recent is outlined in a report that has been issued[8] after work performed under the COST 506 initiative. The work looked at both lubricated and unlubricated pin-on-disc tests as well as abrasive testing.

The unlubricated pin-on-disc results showed considerable material flow under heavy loading conditions both for conventional alloys as well as a series of composite materials. Coefficients of friction of around 0.6 were noted when the composite pins were tested against a conventional bearing steel (AISI 52100) with a hardness of 670-840HV and a surface roughness of R_a=0.1 microns. Analysis of the wear debris revealed the presence of SiC particulates.

Lubrication with a mineral oil considerably reduced the wear, with the reinforcing particles acting as load carriers. Friction coefficients of about 0.12 were noted in this case.

Abrasive testing was carried out using a rubber wheel and quartz sand abrasion tester which was suitable for use for testing in accordance with ASTM G65, although a lower load was used. The results here indicated an improvement of 44% for a SiC particulate reinforced aluminium MMC compared to a conventional alloy, although the results were 7% worse than cast iron.

No references have been seen that examine the tribological effect of coatings produced by any of the methods outlined in the first section of this paper.

4 CONCLUSIONS

MMC's are a range of new materials that require their surfaces to be engineered, not only to protect them from corrosive environments, but also to reduce the problems of wear, both to the MMC and to the counterface.

The challenge of optimising processing conditions to produce a superior product is one that the surface engineering community needs to address to provide the operating data that design and production engineers require when specifying these composite materials.

REFERENCES

1. P R Ebdon, Plating and Surface Finishing, 1988, 75, 65
2. F Mansfield, S L Jeanjaquet, Corrosion Science, 1986, 26, 727
3. F Mansfield, Corrosion, 1988, 44, 856
4. P P Trzaskoma, E McCafferty, C R Crowe, J. Electrochem. Soc. 1983, 130, 180
5. P P Trzaskoma, Corrosion, 1990, 46, 402
6. P P Trzaskoma, E McCafferty, in Aluminium Surface Treatment Technology, ed R S Alwitt and G E Thompson (Pennington, NJ: The Electrochemical Society, 1986), 171
7. F Mansfield, S Lin, S Kim, H Shih, Corrosion, 1989, 45, 615
8. A Jokinen, V Rauta, Final Report COST 506, Sub-Group 6, Project SF3, VTT-MRG B - 9203

Figure 1: Courtesy of Alcan International

Section 3.4 PVD, Sputtering, Plasma Nitriding, and Ion Implantation

3.4.1
Developments in Plasma-assisted PVD Processing

A. Matthews, A. Leyland, K. S. Fancey, A. S. James, P. A. Robinson, P. Holiday, A. Dehbi-Alaoui, and P. R. Stevenson

THE RESEARCH CENTRE IN SURFACE ENGINEERING, THE UNIVERSITY OF HULL, COTTINGHAM ROAD, HULL HU6 7RX, UK

1 INTRODUCTION

The combined benefits of substrate pre-heating and active species provision which plasma processing of PVD coatings allows, were first recognised by Berghaus (1) in the early 1930's. The first practical use of PAPVD coatings was by Mattox (2) in the 1960's - who coined the phrase 'Ion Plating'. Widespread industrial interest in these coating processes was however initiated by the development of plasma-enhancement techniques (3,4) in the late 1970's (nearly half a century after the initial concept was devised), which allowed improved control of the deposition characteristics and reactive deposition of refractory ceramic compounds.

Over the past decade there has therefore been a rapid upsurge of interest in plasma-assisted coating technologies, which has to a large extent formed the basis for the newly emerging discipline of Surface Engineering. Since 1982, research at the University of Hull has been geared towards both a more detailed understanding of PAPVD process phenomenology and (based on this widening knowledge base) the development of an increasingly diverse range of ion and plasma-based surface treatments.

Here we discuss a variety of surface treatment studies which are currently in progress within the Research Centre in Surface Engineering at Hull - and in partner laboratories at Sheffield Hallam University and the University of Northumbria at Newcastle. These studies include :

i.)	Process Fundamentals	- eg. plasma theoretical modelling, process diagnostics and monitoring, coating thickness distribution and uniformity, the development of new hybrid coating techniques.
ii.)	Plasma Diffusion Treatments	- eg. plasma nitriding and carburising, plus combined 'duplex' treatments with PVD hard coatings.
iii.)	Ceramic Coatings for Wear and Corrosion Resistance	- eg. Ti-based coatings, (eg. Ti(B,N), Ti/Ti(B,N) multilayers) and multi-treatment coatings for corrosion resistance (linked with ii.).
iv.)	Thermal Barrier Coatings	- eg. Yttria-Partially-Stabilised Zirconia (PYSZ) - structural modifications for improved mechanical / thermal properties.
v.)	Diamond-Like Carbon (DLC) films for optical and mechanical applications	- eg. plasma-assisted PVD and CVD, Fast-Atom-Beam (FAB) source techniques, coating of polymers, metals, ceramics and glasses.

In addition to the above, extensive work has been carried out on coating evaluation and test methods, and on means to assist in the coating design/specification process.

2 PROCESS FUNDAMENTAL STUDIES

Although PAPVD coating techniques are already quite widely exploited in industry, there remain a number of areas in which the scientific understanding of process phenomena is somewhat lacking, and whose solution would extend the industrial applications-areas for PAPVD. Through applied research, initially using idealised process models, we have developed a theoretical framework from which basic predictions about the deposition process can be made. Feedback from these applied studies provides extended knowledge of process variable interactions, with which to modify and improve existing deposition systems, and also to develop new processing techniques to overcome current technical limitations of certain coating treatments.

2.1 Modelling of Plasma Phenomena

Early PAPVD ('ion-plating') techniques (2) employed a simple D.C. diode plasma configuration, where the vacuum treatment chamber (which is at earth potential) acts as the anode in the system, whilst the workpiece (ie. the surface to be coated) is electrically insulated from the chamber and negatively biased, to form the cathode. Under the action of this bias voltage, the carrier gas in the system (usually argon) becomes ionised - the main volume of the chamber comprising an extended 'negative glow' region under the pressure/voltage conditions commonly employed (figure 1.). The main voltage drop however occurs near to the cathode, forming a cathode sheath across which ions at the edge of the negative glow accelerate to bombard the cathode surface. The ion bombardment induces electron emission from the cathode - these electrons accelerate across the sheath, away from the cathode and into the negative glow, producing impact ionisation of neutral gas species and thus sustaining the discharge, (figure 1.).

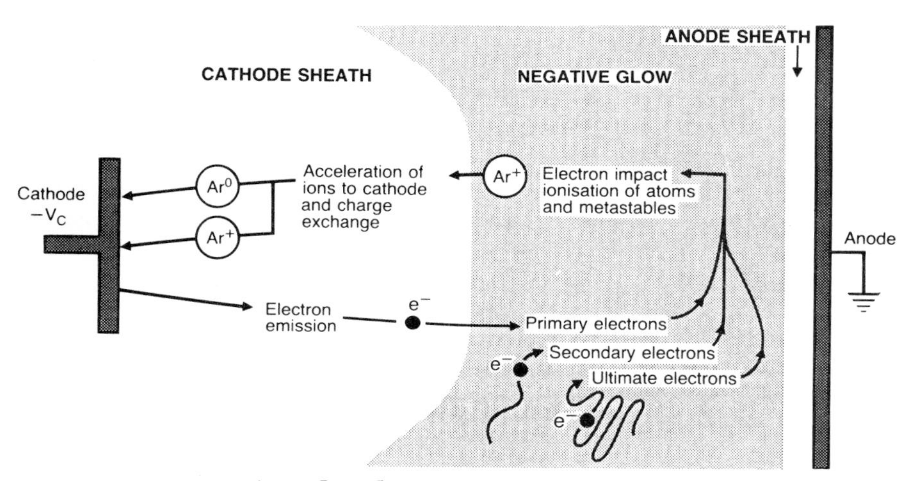

Figure 1 Schematic of a low-pressure argon D.C. diode glow discharge

In present commercial coating systems, some form of ionisation enhancement is commonly employed to improve coating properties and facilitate process control, by increasing the 'ionisation efficiency' (ie. the proportion of the total species, incident at the cathode, which are ionised). We have demonstrated that this and other factors, such as increasing the cathode current density, reducing the chamber pressure and decreasing the cathode sheath thickness (L), all improve the bombarding ion energy spectrum and hence the efficiency of the ion-plating process (5,6). A low L/λ value (where λ is the mean-free-path for the dominant ion charge-exchange collision process taking place in the cathode sheath) has been shown to improve the deposition process by maximising incident ion energies at the cathode, (figure 2.). Ionisation enhancement is usually achieved by increasing the availability of ionising electrons in the plasma (4,8). This is true both for evaporative systems and now also in sputtering, where the 'unbalanced' magnetron system is increasingly employed (9,10). The latter, in effect, allows the release of part of the intensified plasma associated with the magnetron target into the deposition volume - thereby increasing the ion current density to the substrate and considerably improving coating adhesion and structure. It has recently been demonstrated that the beneficial properties of the unbalanced magnetron can be combined with those of the arc source in a hybrid system (9) which incorporates the best features of each process (ie. good bonding, structure and coating uniformity, with high refractory material deposition rates), whilst minimising their disadvantages (eg. evaporant macrodroplets, which give poor surface finish and/or mechanical properties).

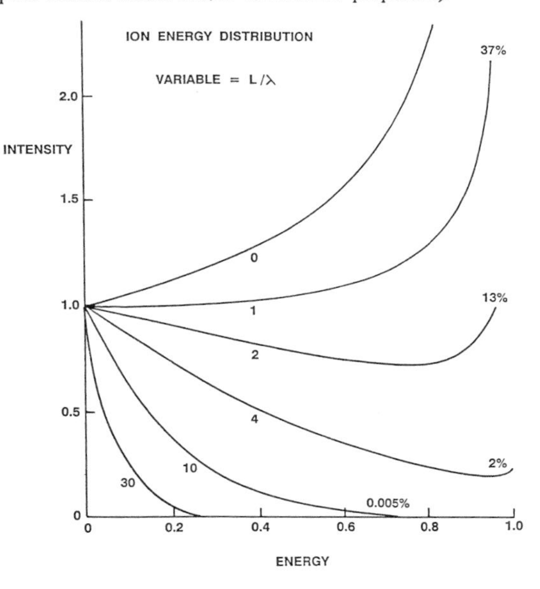

Figure 2

Normalised ion energy spectra for a range of L/λ values - after Davis and Vanderslice (7).

(The ion energy values are relative to the workpiece voltage fraction.)

The percentage values against each L/λ curve (between L/λ = 1 and L/λ = 10) indicate the number of ions arriving at the workpiece with the maximum energy (which corresponds to the workpiece voltage applied) in each case.

2.2 Coating Thickness Distribution and Vapour Transportation Models

The coating thickness distribution in PAPVD systems can be influenced by a number of factors, including bombardment anisotropy (due to differential ionisation enhancement effects (6)), evaporant material distribution and component orientation. We have shown that predictions can be made in terms of a coating front-to-back thickness ratio (R), which takes account of the 'thermalised' and 'non-thermalised' vapour fluxes within the coating chamber (11,12).

The thickness ratio can be expressed as :

$$R = \frac{1 + \exp(-s/l)}{1 - \exp(-s/l)} = \mathrm{Coth}\,(s/2l) \qquad \ldots [1]$$

where s is the source-to-substrate distance and l is an associated vapour mean free path, (figures 3a,b).

Experimental validation of this model using electron beam evaporation (with a range of process conditions and evaporant materials) suggests that the vapour distribution is influenced by the production of a 'virtual' source above the actual source. This virtual source effect is however not

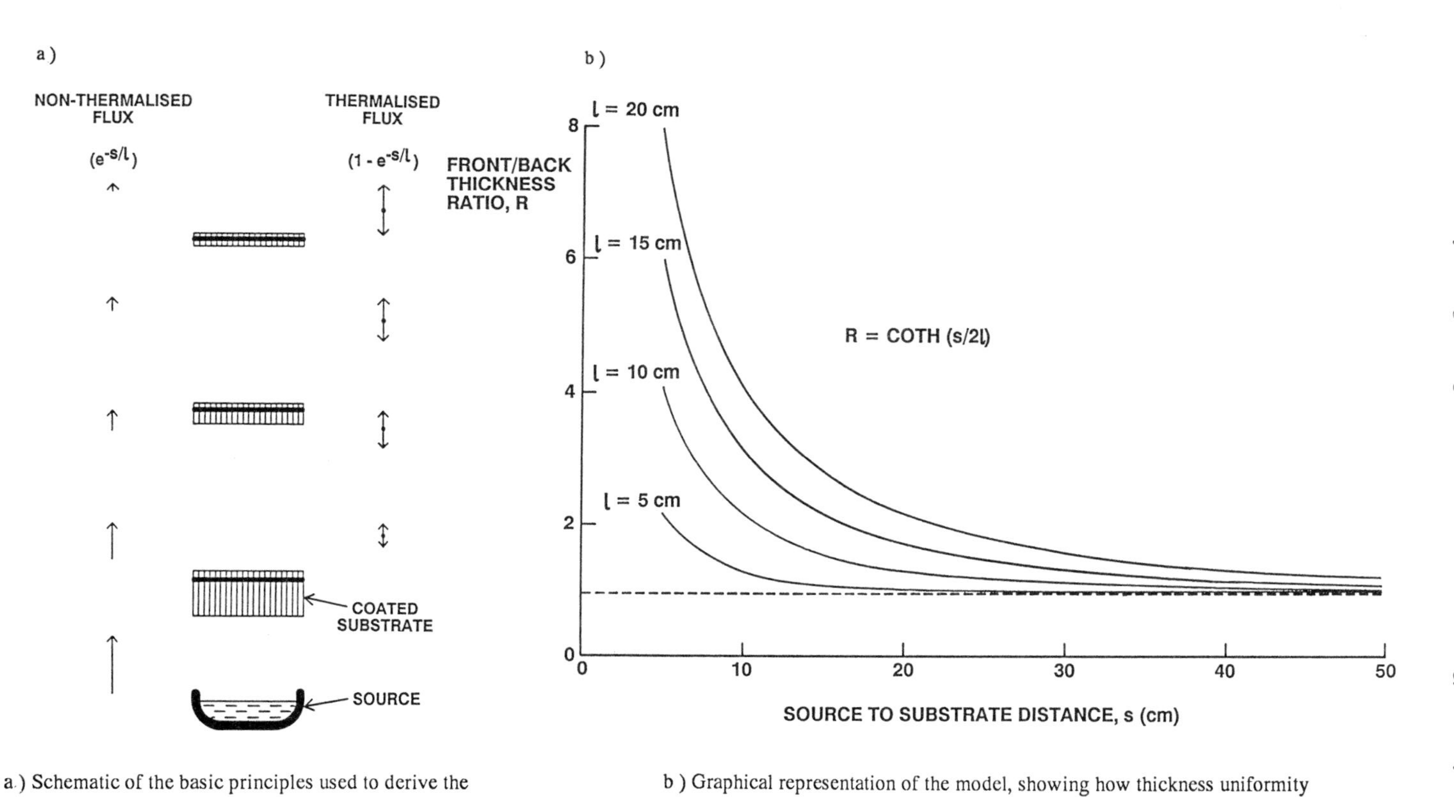

a) Schematic of the basic principles used to derive the coating thickness uniformity model R = Coth (s/2l)

b) Graphical representation of the model, showing how thickness uniformity changes with source-to-substrate distance and gas-scattering parameter, l

Figure 3

detected when magnetron sputtering or cathodic arc evaporation are employed (13). It is thought that the virtual source is a dense vapour cloud from which the evaporant flux appears to emanate. From cathode sheath thickness measurements, under metal evaporation conditions, there is evidence to suggest that this virtual source effect could influence atom cluster formation (14,15). These clusters, which are believed to consist of a minimum of tens of atoms per unit charge, may be formed by homogeneous nucleation. The potential significance of this phenomenon in the modelling of thin film nucleation and growth mechanisms is enormous, and research to further investigate this effect is now proceeding at Hull.

With regard to the lateral thickness uniformity above a PVD vapour source, most theoretical models are based on the Cosine Law of emission - which was originally derived to describe the effusion of gases from an ideal Knudsen cell (16). The more sophisticated of these models take account of factors such as the virtual source effect (17), described above, and the influence of evaporation rate on the lateral spread of the vapour cloud (which can be described by a cosine power law, 18) when predicting thickness distributions. Recent work at Hull has shown that the lateral thickness distribution models can be modified and combined with the thickness-fall-off component of the model described by Equation [1] (figure 3a.), to allow predictions of coating thickness throughout the deposition chamber - with only limited practical data required (19).

3 PLASMA DIFFUSION TECHNIQUES

Based on our interest in applied process fundamentals in PAPVD technology, and with the knowledge gained from the theoretical concepts outlined in Section 2, it was felt that the benefits in ion-energy control available in enhanced-plasma low-pressure systems could be employed to provide significant improvements in the quality and applicability of plasma-based thermochemical diffusion treatments. Comparative studies of various low-pressure plasma nitriding techniques which are on-going at Hull (20,21) have recently been supplemented by a new low-temperature carbon-diffusion process, which has been shown to give improvements in the abrasive wear resistance of stainless steel substrate materials (22).

3.1 Plasma Nitriding

We have used Optical Emission Spectroscopy (OES) studies of nitrogen plasmas (15,20) to develop a new, low-pressure triode-discharge nitriding system. Spatial resolution of nitriding species across the cathode sheath (using a fibre-optic OES probe) in various diode and triode nitrogen discharge configurations has allowed the determination of changes in the dominant nitrogen ionic species incident at the cathode surface (ie. N^+ or N_2^+); this data has been used to calculate L/λ values for each discharge (20,21).

By minimising L/λ for a given workpiece negative voltage, using an additional electron-emitting electrode (in our system a negatively-biased hot tungsten filament) to control the plasma ionisation efficiency, optimum nitriding performance can be achieved for a particular application. A high workpiece voltage (providing high ion energies) gives maximum nitriding efficiency (particularly for 'difficult' materials with adherent surface oxides - such as chromium-alloy and stainless steels), whilst a low workpiece voltage reduces the cathode sheath thickness to provide maximum hole and recess penetration. Low-frequency R.F. plasmas (which, due to the nature of the capacitive cathode sheath below 1MHz, can provide very high ion energies) are also particularly effective for nitriding difficult materials - even at low temperatures (ie. less than 500°C (20), figure 4.). A major benefit of all low pressure nitriding systems is the potential to increase the ion-bombardment to total-flux-impingement ratio at the workpiece, such that 'white layer' formation can be controlled.

In a situation where a 'duplex' nitriding/coating treatment is envisaged, it has been demonstrated that the white layer is frequently detrimental to coating adhesion (23). Our low pressure system enables complete removal of the white layer during processing, without reducing the diffusion layer thickness, (figure 5a.), and eliminating the need for post-treatment finishing prior to coating. The treatment has been shown to considerably improve TiN coating adhesion (as measured by the scratch test method) on tool steel substrates (21, figure 5b.) This is believed to be due to the improved coating load support provided by the diffusion layer - which prevents severe substrate plastic deformation. Even after the coating has spalled, the continuing resistance

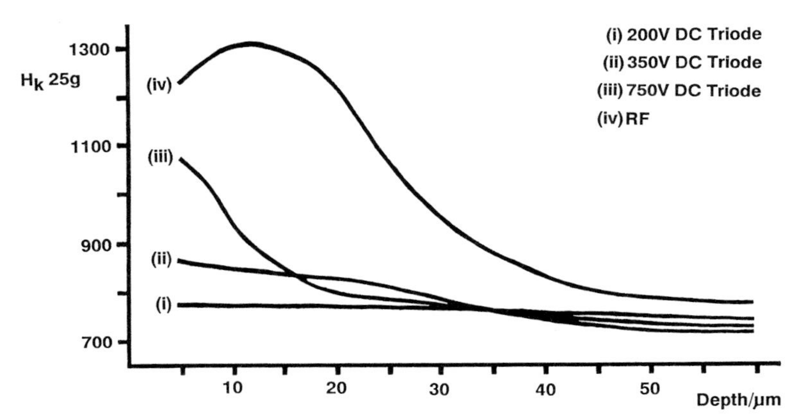

Figure 4 Nitriding hardness-depth profiles for AISI H13 steel treated for 3 hours at 450°C [550°C, (iv)] in various 7.5mTorr [15mTorr, (iv)] D.C. and R.F. nitrogen plasmas

to plastic deformation which the nitrided layer gives is likely to provide substantial practical benefits in terms of delaying the onset of catastrophic component failure in service. These effects can be illustrated by the lower indenter tangential force (F_T), both above and below the upper critical load (L_{C2}) for coating failure, (figure 5c.). Since low-pressure nitriding treatments can be carried out in virtually unmodified industrial TiN ion-plating equipment, there is the potential to combine the white-layer-free nitriding process with PVD hard coatings in a single, cost-effective treatment cycle.

3.2 Plasma Carburising

Further to our nitriding process developments, we recognised the need for a low temperature plasma diffusion treatment which could provide substantial improvements in low-load adhesive and abrasive wear resistance on non-tool-steel substrate materials. In particular, a treatment for stainless steels, which would not totally undermine the substrates' corrosion resistance in aqueous environments, was a primary target for our research. We have therefore investigated a low-temperature carburising treatment (ie. below 400°C) using an Argon/Hydrogen/Methane plasma and a range of stainless steel substrate materials (22). It was found that commonly-used stainless steels such as AISI 316/321 (Austenitic) and 431 (Martensitic) could be effectively treated at temperatures as low as 320°C, to produce diffused carbon layers up to 100μm thick (after 20 hours treatment). The layers produced were typically two-phased, with a thin (3-10μm) compound layer at the surface and a thicker (30-100μm) underlying diffusion layer. Abrasive wear tests using the rubber wheel technique (24) have demonstrated dramatically improved surface properties (22). X-ray diffraction (XRD) analyses indicate that the surface compound layer consists primarily of Chromium-rich $M_{23}C_6$-type carbides, which we believe may be produced by both secondary-phase precipitation and shear transformation mechanisms at the low treatment temperatures and pressures employed (25). It is this layer that provides the main improvement in abrasive wear resistance.

4 CERAMIC COATINGS FOR WEAR AND CORROSION RESISTANCE

In conjunction with investigations into diffusion and duplex treatments, and continuing work on the industrial applicability of 'second-generation' hard coating treatments (incorporating alloys, or combinations of nitrides, carbides, borides and oxides of titanium, other transition metals and aluminium), we are also actively researching new 'third-generation' processes, incorporating multiple layers and/or multiple treatments (26,27). One of the primary objectives of this work is to extend the versatility of

a)

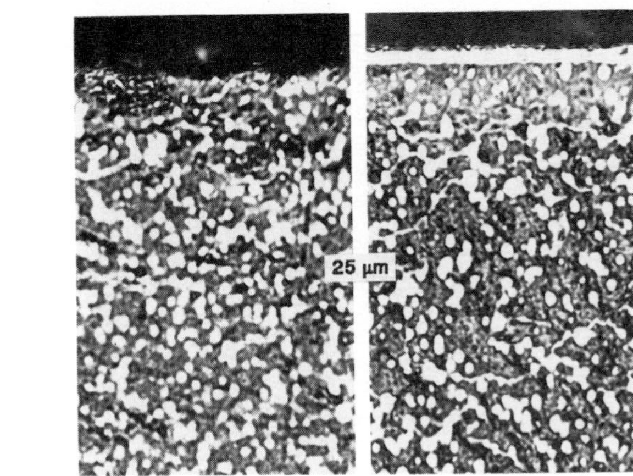
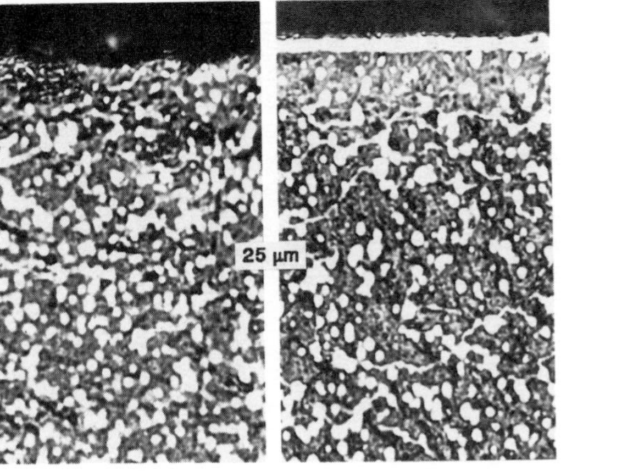

b)

SAMPLE	UPPER CRITICAL LOAD (L_{C2}) /N	ERROR (σ)
A	31.1	±0.58
B	30.2	±2.52
C	33.3	±0.40
D	29.2	±0.56
E	30.8	±0.97
F	32.3	±0.30
G	31.8	±0.83
U	24.7	±1.16

c)

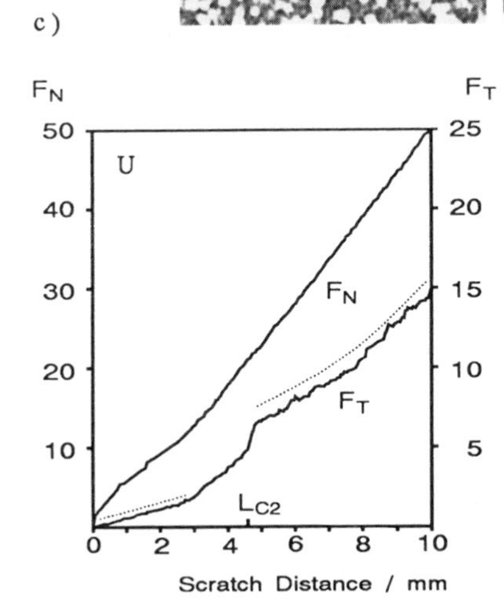

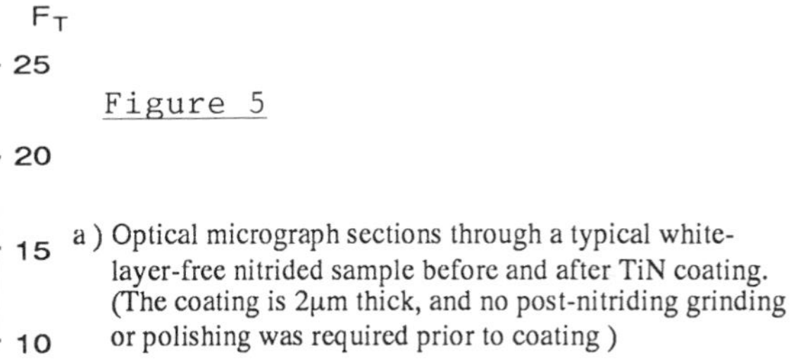
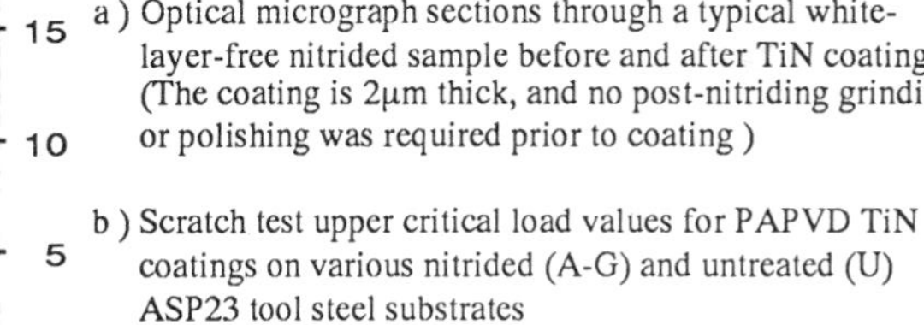

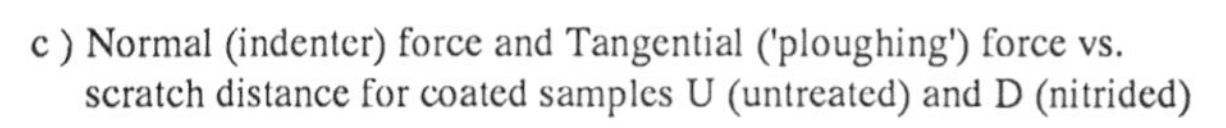

Figure 5

a) Optical micrograph sections through a typical white-layer-free nitrided sample before and after TiN coating. (The coating is 2µm thick, and no post-nitriding grinding or polishing was required prior to coating)

b) Scratch test upper critical load values for PAPVD TiN coatings on various nitrided (A-G) and untreated (U) ASP23 tool steel substrates

c) Normal (indenter) force and Tangential ('ploughing') force vs. scratch distance for coated samples U (untreated) and D (nitrided)

hard coating treatments from traditional cutting/forming/extrusion tooling, to the wider field of machine components (particularly in materials extraction and processing). A combined need for both wear and corrosion performance improvements has, to date, limited the applicability of many PVD wear coatings in 'wet' environments, where the intrinsic residual through-porosity often induces accelerated pitting corrosion through unsuitable galvanic coupling of coating and substrate materials.

Another continuing requirement is for thick PVD wear coatings with improved toughness and resistance to abrasive/erosive particulate wear - both in materials extraction (where - as in other areas - the environmental implications of chrome electroplating are of increasing concern) and in automotive combustion engines and aerospace turbines (where the manufacturing costs and poor thermal shock resistance of bulk ceramics can be prohibitive). Here multilayered treatments with mixtures of refractory ceramics and tougher metallic or intermetallic phases are likely to be of ever increasing importance over the next decade.

With regard to multitreatment coatings for combined wear and corrosion resistance, we have demonstrated that both surface passivation pre-treatments and appropriate metal interlayering have the potential to provide improved resistance to corrosion in, for instance, acidic aqueous environments (26) - without seriously compromising wear performance. In particular, plasma substrate pre-oxidation treatments and PVD nickel interlayering, combined with TiN coatings, has shown improvements on stainless steels under potentiodynamic corrosion testing in 0.1N H_2SO_4 (26). Pre-nitriding and PVD chromium interlayering treatments were less successful in this particular system - due to factors such as redistribution of austenite-stabilising elements in the substrate during the nitriding treatment, and accelerated pitting corrosion of chromium in H_2SO_4 with low dissolved O_2 concentrations (as would be the case at the root of through-coating defects). Their performance with other coating/substrate combinations can be expected to differ however, and there is an increasingly strong requirement to optimise particular PAPVD treatments for specific environmental applications. The use of low temperature plasma carburising treatments (which should have a less detrimental effect on substrate electrochemical properties than, for instance, nitriding) and electroless nickel interlayers (which can be age-hardened during the TiN coating stage) for these applications, are new areas of development for our laboratory, (25,28).

In the case of coatings for abrasion and erosion resistance, there is an increasing requirement for thick PVD hard coatings to replace electroplated coatings and monolithic ceramic components. A common problem with thick PVD coating production is that of internal stresses within the coating, causing brittle fracture in service, or spontaneous spallation/delamination during treatment. This problem is now being overcome by use of thin metallic interlayers to accommodate internal compressive stresses and to deflect or stop through-cracking, before catastrophic failure occurs. For example, TiN coatings have a tendency to spall at a thickness of 10μm or greater, however we have shown that Ti interlayers can be employed to increase thickness up to 60μm, with improved toughness (27). Other studies indicate that these types of coatings provide advantages over electroplated chrome in automotive engine component applications (29).

Recent work at Hull in developing new Ti-based hard coatings for machining and casting operations has shown the promise of alloy coatings such as Ti(B,N) and (Ti,Al)N, (27,30). Ti(B,N) in particular, has shown remarkable levels of hardness (up to 12,000 H_K25g) but is susceptible to brittle fracture. The use of Ti interlayers has given encouraging low-load wear results in pin-on-disc testing of these coatings, however machining trials have been less conclusive when compared to monolayer alloy coatings (27,30, figure 6.). The potential for Ti(B,N) coatings in aluminium automotive component casting has also been investigated, with evidence of improved performance over traditional processes (27).

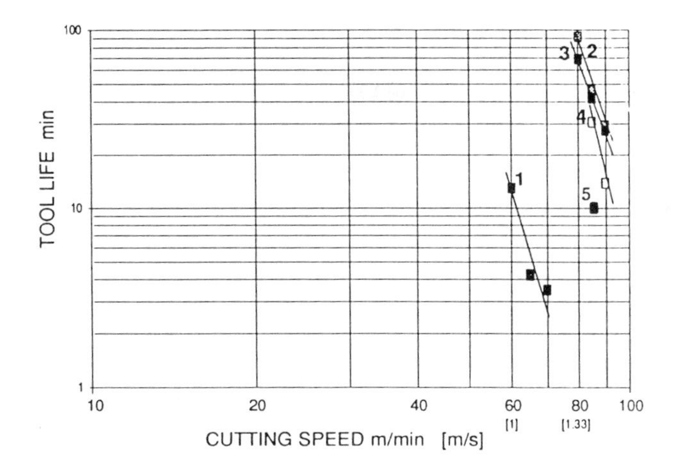

Figure 6

'Taylor' log./log. plots of tool life vs. cutting speed for various combinations of tool and PVD coatings.

1. Uncoated tool insert
2. TiN coated insert
3. (Ti,Al)N coated insert
4. Ti(B,N) coated insert
5. Ti(B,N)/Ti multilayer

(Tests carried out against 21NiCrMo2 cementation steel work material to ISO 3685-1977 standard)

5 THERMAL BARRIER COATINGS

In many thermal combustion engine applications - particularly in the aerospace industry - future developments in engine efficiency are constrained by the inadequate surface properties of existing component materials (which cannot perform satisfactorily at the elevated temperatures required for improved fuel economy and power output). In approaching potential solutions to this problem (using Thermal Barrier Coating (TBC) systems, for instance), it has to be recognised that good thermal insulation is in itself not enough. Factors such as : i.) thermal shock, ii.) particulate erosion, iii.) high temperature corrosion and oxidation, and iv.) aerodynamic efficiency, all play a major role in the overall system performance.

TBCs based on Yttria-Partially-Stabilised Zirconia (PYSZ) have been used on static combustor and after-burner components in many commercial aero-engines since the mid-1970s, however the severe stresses encountered by components which rotate (such as turbine blades) have meant that coating adhesion is inadequate for many of the items which would most benefit from this kind of treatment. Moreover, commercial TBCs have traditionally been produced by plasma spraying techniques, which require macroscopic surface roughening prior to coating to maximise 'mechanical' adhesion. This invariably leads to a rough finished product with poor aerodynamic efficiency, which must then be polished before use. Accurate thickness control on complex blade geometries is also difficult, and the interfacial 'chemical' adhesion and oxidation resistance are compromised.

Electron Beam (EB) PVD has emerged with some success as an alternative process to Plasma Spraying (31), with large increases in thermal shock durability - due in part to the segmented structure of the vacuum-evaporated deposit. Despite the additional benefits in improved erosion and oxidation resistance EBPVD, as practised, is however still largely a line-of-sight process, requiring extensive component manipulation to achieve uniform coatings In addition, suitable coating micro-structures can only be achieved by deposition at high substrate temperatures. At Hull we have (since 1982) developed an R.F. plasma-assisted EBPVD process for PYSZ (ZrO_2, 6-8%Y_2O_3), which has provided substantial improvements in adhesion (32,33), and in structure control (19,34,35). We have recently demonstrated that a mixed-structure coating, with a thin, dense plasma-assisted layer at the coating-substrate interface (for improved adhesion and oxidation/corrosion resistance), followed by a thick, columnar gas-evaporated layer (for improved strain tolerance and thermal properties), dramatically enhances thermo-mechanical performance in thermal cycling durability trials (figure 7.), when compared to individual coatings produced by each of these techniques (19). Only PAPVD-based methods allow such controlled structure changes in a single treatment process, whilst maintaining a low substrate temperature.

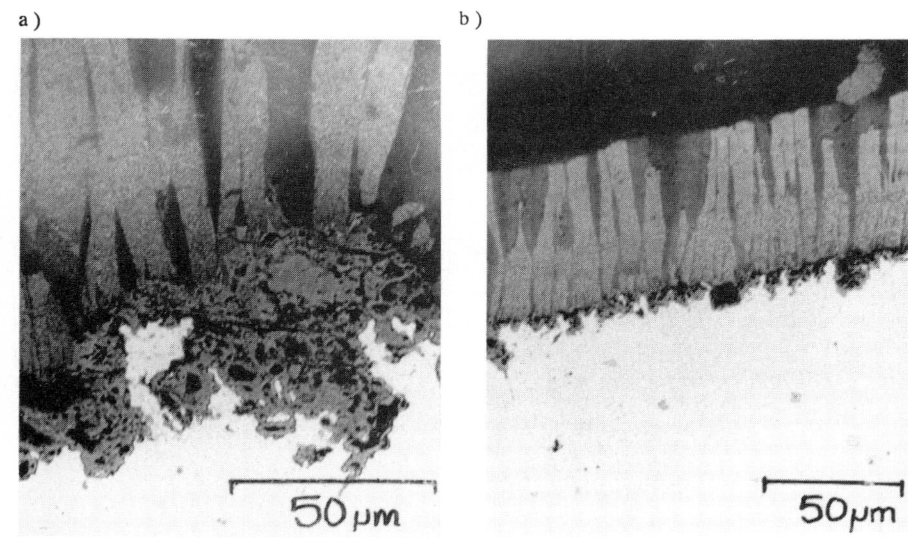

Figure 7 Corrosion of Nimonic substrate material after 'Burner-Rig' thermal cycling trials: a) Single layer EBPVD coating, b) Twin layer (EBPVD + PAPVD interlayer) coating

6 DIAMOND-LIKE CARBON FILMS FOR OPTICAL AND MECHANICAL APPLICATIONS

Diamond and Diamond-Like Carbon (DLC) coatings are receiving increasing research and industrial interest - due to their unique range of physical, chemical, electrical, optical and mechanical properties. A variety of innovative ion and plasma-based DLC deposition systems using R.F. and D.C. PAPVD, PACVD, Fast-Atom-Beam (FAB) sources and other hybrid processes, have been studied at Hull, (36-41). The possibility of depositing DLC on a wide range of substrates, including ferrous materials (36,37), silica glass (37,38) and plastics (39), has been demonstrated. A model for DLC coating deposition, which relates the optical properties and relative concentration of bonded hydrogen in the films (and thus, indirectly, the mechanical properties) to the plasma parameters (ie. C : H ratio, discharge voltage and degree of ionisation) has been developed (37,40). An improved discharge layout which can provide increased substrate current densities (by several orders of magnitude), across a wider pressure range, has also recently been investigated (41). Preliminary results suggest that this technique can produce ultrahard DLC films at a deposition rate of up to 3μm/hour on stainless and mild-steel substrates.

The influence of hydrogen additions to the plasma, in modifying the internal bonding and structure of the films, has also been researched (37). Techniques such as FAB source deposition have (due to the directional nature of the active species provision) shown strong dependence of optical properties on substrate orientation (38); thus further work is now in progress at Hull in scaling-up the DLC process to encompass complex-shaped components.

7 CONCLUSIONS

We have reported various process developments which are relevant to improving the range and versatility of PVD-based plasma treatments - the ultimate objective being to increase the industrial applicability of these types of Surface Engineering techniques. With this in mind, we have also been particularly involved in the standardisation of materials testing (30,42), such that recognised test procedures can be established, allowing the collation of quantitative, rather than comparative data, to predict coating and treatment performance in particular prospective applications.

Additionally, we are developing computer expert systems (43-45) to implement newly available (and increasingly reliable) test information, with the aim of assisting the Design Engineer to select and specify Surface Engineering treatments at the component design stage - rather than as a solution for pre-existing wear and corrosion problems. We are working in conjunction with major European companies to facilitate the wider objective of Surface Engineering, ie. - the design of substrate and treatment together to complement each other, and thus provide a cost-effective performance solution of which neither material is capable individually.

ACKNOWLEDGEMENTS

The diverse range of work reported here has been financed by a number of industrial sponsors, the SERC and the EC Brite/Euram scheme; that support is gratefully acknowledged. We also thank colleagues and co-workers in the Research Centre in Surface Engineering at Hull, the Surface Engineering Research Group at the University of Northumbria at Newcastle and the Materials Research Institute at Sheffield Hallam University, for their invaluable assistance.

REFERENCES

1. **B. Berghaus** German Patent DRP 668,639 (1932).
2. **D.M. Mattox** *Electrochem. Technol.* 95 (1964) p.2.
3. **R.F. Bunshah, R. Nimmagadda, W. Dunford, B.A. Movchan, A.V. Demchishin & N.A. Chursov** *Thin Solid Films* 54 (1978) p.85.
4. **A. Matthews** *Ph.D. Thesis* University of Salford (1980).
5. **K.S. Fancey & A. Matthews** *Surf. Coat. Technol.* 33 (1987) p.17.
6. **K.S. Fancey & A. Matthews** *IEEE Trans. Plasma Sci.* 18 (1990) p.869.
7. **W.D. Davis & T.A. Vanderslice** *Phys. Rev.* 131 (1963) p.219.
8. **A. Matthews** *J. Vac. Sci. Technol.* A3 (1985) p.2354.
9. **P.A. Robinson & A. Matthews** *Surf. Coat. Technol.* 43/44 (1990) p.288.
10. **W.D. Sproul** *Surf. Coat. Technol.* 49 (1991) p.284.
11. **K.S. Fancey & A. Matthews** *Surf. Coat. Technol.* 36 (1988) p.233.
12. **K.S. Fancey & A. Matthews** *Proc. 1st Int. Conf. on Plasma Surface Engineering,* Garmisch-Partenkirchen, Germany, Sept. 1988. pub. DGM, Oberursel, Germany (1989) p.61.
13. **K.S. Fancey, P.A. Robinson, A. Leyland, A.S. James & A. Matthews** *Mat. Sci. Eng. A* 140 (1991) p.576.
14. **K.S. Fancey & A. Matthews** *Appl. Phys. Lett.* 55 (1989) p.834.
15. **K.S. Fancey & A. Matthews** *Vacuum* 41 (1990) p.2196.
16. **A. Leyland & A.S. James** in D.S. Rickerby & A. Matthews (eds.) *Advanced Surface Coatings* pub. Blackie, Glasgow (1991) p.66.
17. **T.C. Reiley** *Ph.D. Thesis* Stanford University (1974).
18. **E.B. Graper** *J. Vac. Sci. Technol.* 10 (1973) p.100.
19. **A.S. James** *Ph.D. Thesis* University of Hull (1991).
20. **A. Leyland, K.S. Fancey, A.S. James & A. Matthews** *Surf. Coat. Technol.* 41 (1990) p.295.
21. **A. Leyland, K.S. Fancey & A. Matthews** *Surf. Eng.* 7 (1991) p.207.
22. **P.R. Stevenson, A. Leyland, M.A. Parkin & A. Matthews** Presented at *ICMCTF '92,* San Diego, CA. USA, April 1992. ....submitted to *Surf. Coat. Technol.*
23. **Y. Sun & T. Bell** *Mat. Sci. Eng. A* 140 (1991) p.419.
24. **G.A. Saltzman** in R.G. Bayer (ed.) *Selection and Use of Wear Tests for Coatings ASTM STP679.* pub. ASTM, (1982) p.71.
25. **J.S. Cawley, B. Lewis, P.R. Stevenson, A. Leyland & A. Matthews** - paper to be presented at PSE '92, Garmisch-Partenkirchen, Germany, October 1992.

26. **M.J. Park, A. Leyland & A. Matthews** *Surf. Coat. Technol.* 43/44 (1990) p.481.
27. **A. Leyland** *Ph.D. Thesis* University of Hull (1991).
28. **M. Bin-Sudin, M. Kalantary, P.B. Wells, A.S James, A. Leyland, A. Matthews & B.L. Garside** - paper to be presented at PSE '92, Garmisch-Partenkirchen, Germany, October 1992.
29. **V.V. Lyubimov, A.A. Voevodin, A.L. Yerokhin, Y.S. Timofeev & I.K. Arkhipov** *Surf. Coat. Technol.* 52 (1992) p.145.
30. **A. Matthews, A. Leyland, B. Matthes, E. Broszeit, H. Ronkainen & K. Holmberg** EC EURAM Project, MAIE-0028-UK - extended Final Report (1991).
31. **R.F. Demaray, R.W. Fairbanks & D.H. Boone** ASME Report 82-GT-264 (1982).
32. **K.S. Fancey, & A. Matthews** *J. Vac. Sci. Technol.* A4 (1986) p.2656.
33. **A.S. James, K.S. Fancey, & A. Matthews** *Surf. Coat. Technol.* 32 (1987) p.377.
34. **A.S. James & A. Matthews** *Surf. Coat. Technol.* 41 (1990) p.305.
35. **A.S. James & A. Matthews** *Surf. Coat. Technol.* 43/44 (1990) p.436.
36. **A. Dehbi-Alaoui, A.S. James & A. Matthews** *Surf. Coat. Technol.* 43/44 (1990) p.88.
37. **A. Dehbi-Alaoui, P. Holiday & A. Matthews** *Surf. Coat. Technol.* 47 (1991) p.327.
38. **A. Dehbi-Alaoui, A. Matthews & J. Franks** *Surf. Coat. Technol.* 47 (1991) p.722.
39. **A. Dehbi-Alaoui, B Ollivier & A. Matthews** *Proc. 11th Int. Conf. on Vacuum Metallurgy, ICVM' 92* Antibes, France, May 1992. pub. SFV, (1992).
40. **A. Dehbi-Alaoui & A. Matthews** *Diamond & Related Materials* 1 (1992) p.445.
41. **A. Dehbi-Alaoui & A. Matthews** Presented at *ICMCTF '92,* San Diego, CA. USA, April 1992. submitted to *Surf. Coat. Technol.*
42. **H. Ronkainen, S. Varjus, K. Holmberg, K.S. Fancey, A.R. Pace, A. Matthews, B. Matthes & E.Broszeit** in D. Dowson, C.M. Taylor & M. Godet (eds.) *Proc. 16th Leeds-Lyon Symp. on Tribology*, Lyon, France, Sept. 1989. pub. Elsevier, Amsterdam, Netherlands (1990) p.453.
43. **A. Matthews** *Proc. Institute of Metals* Autumn '89 Meeting, Sheffield 'A Cutting Edge for the 1990s' pub. IoM. London, (1989).
44. **C.S Syan, A. Matthews & K.G. Swift** *Surf. Eng.* 2 (1986) p.249.
45. **C.S Syan, A. Matthews & K.G. Swift** *Surf. Coat. Technol.* 33 (1987) p.105.

3.4.2
Supersaturated Metastable Alloys by Unbalanced Magnetron Sputtering

D. P. Monaghan, R. D. Arnell, and R. I. Bates

CENTRE FOR THIN FILM AND SURFACE RESEARCH, DEPARTMENT OF AERONAUTICAL AND MECHANICAL ENGINEERING, SALFORD UNIVERSITY, THE CRESCENT, SALFORD M5 4WT, UK

1 INTRODUCTION

In the past, increased solid solubility of alloy mixtures has been investigated by using several techniques, some of which date back to the last century[1]. Such departures from the predicted equilibrium phase diagrams have relied in the main on rapid quenching from the melt. Modern *Rapid Solidification Processing* (RSP) involves cooling rates in the range 10^4-10^7 K/s. Most of the present day developments were initiated by Duwez[2]. Duwez concluded that if sufficiently high cooling rates were achieved in the Cu-Ag system, the melt could be frozen fast enough to prevent the nucleation of two distinct FCC phases. A process was developed known as the *Duwez gun,* which projected small particles of liquid metal with sufficient velocity onto a good heat sink, to form splats or thin foils. Duwez was able to create a continuous, metastable series of Cu-Ag alloys without any two phase region. At around the same time Duwez created the first metallic glasses by rapid quenching in the Au-Si and Au-Ge systems[2]. The main RSP techniques that are employed today are dependent upon the following: single roller melt-spinning[3]; twin-roller melt-spinning[4,5,6]; melt-extraction[7]; drop-smashing (piston and anvil quencher)[8].

All the above methods, which rely on a source melt, have been used extensively to study metastable and glass structures. Another technique which has also has received attention is deposition from the vapour phase onto substrates which are, in some cases, cooled to liquid nitrogen/helium temperatures. The source vapour can be thermally evaporated[9], or more usually cathodically sputtered[10]. However from a metallurgical point of view, despite the wider ranges of solution and glasses that are possible, the thin film deposition of metastable solid solutions has received much less attention.

Alloys formed by rapid solidification may exhibit one or more of four forms of metastability[11].

1. An extension of solid solubility beyond the equilibrium value, which may be partial or complete.

2. The formation of one or more metastable crystalline phases.

3. Lowering of the M_s temperature when a martensitic transformation takes place, sometimes to such an extent that the martensitic is replaced by a different phase.

4. Formation of a metallic glass.

The main advantages that are perceived from RSP are based on the following[11]: more homogeneity of composition; finer grain, cell, and dendrite sizes; the ability to incorporate more solute than by conventional methods (leading to increased age hardening) and improved mechanical properties at all temperatures. If these benefits can be incorporated into a surface by using a technique that can process items at speed and does not require the sometimes complicated substrate cooling requirement, then practical applications can be expected.

2 PHYSICAL VAPOUR DEPOSITION

The physical vapour deposition process is a term applied to a variety of techniques that create a vapour of material under vacuum, which is then deposited upon a substrate to form a coating. The physics of the process will determine the microstructure of the film and this can vary from open and columnar to fully dense and non-columnar. This surface coating can then be used for a variety of applications depending upon the nature of the deposited film. Excluding the films deposited for microelectronics, the main present day application is for wear protection and other tribological properties. The most widely used films rely on a reactive deposition to form compound coatings, such as titanium nitride, that have very high hardness and good wear resistance.

The processes most commonly used to create the vapour species are as follows: thermal evaporation by resistance or electron beam heating of a melt; cathodic sputtering; and cathodic arc deposition. Regardless of the technique there is an energy requirement of the depositing atom that will have to be satisfied in order to ensure a high quality non-columnar coating. This is best summarised by Messier's refined grain growth model[12], see figure 1. The energy requirements will vary in relation to the melting point of the material, such that higher meting point materials require higher adatom energies than lower melting point materials, in order to form a non-columnar deposit. The deposition temperature will therefore strongly affect the deposition morphology, as can be seen in figure 1. An additional input of energy to the depositing atom can be supplied by simultaneous ion bombardment of the growing film. This is generated by applying negative potentials to the substrate in an ionised plasma (usually argon), the argon ions will then bombard the film and inpart energy to the depositing atoms.

The method of deposition applied in this study is based on the cathodic sputtering process. A negatively charged target (source material) is immersed in a glow discharge containing positively charged argon ions, the ions are accelerated towards the target, and as a result of impacts on the target surface, metal atoms are removed by momentum transfer.

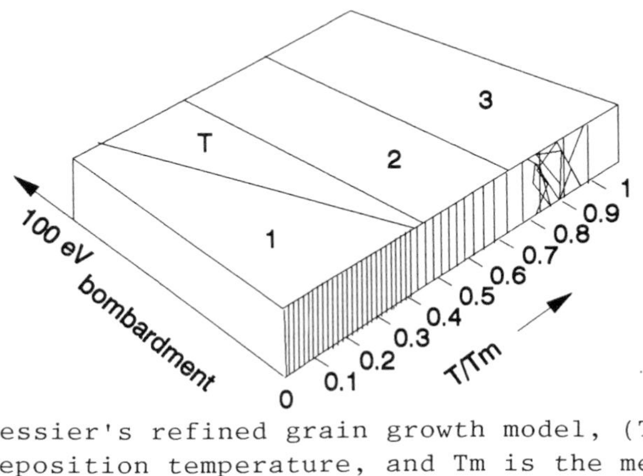

Figure 1 Messier's refined grain growth model, (T is the deposition temperature, and Tm is the melting point of the material)

The introduction of a radial magnetic field above the target further intensifies the discharge by magnetically confining the secondary electrons produced during the process. The combination of the cathode target and magnetic field is commonly known as the magnetron. The conventional magnetron has a tightly confined magnetic field configuration so that all the discharge (plasma) is strongly confined to the near target area, see figure 2a. Recently[13-19], it has been shown that by altering the standard magnetic arrangement to what is commonly termed an unbalanced configuration (see figure 2b), increasing levels of ion bombardment at the substrate can be achieved. These higher levels of bombardment are very useful in many applications where modification of the depositing films is necessary to ensure a high quality dense coating[20,21].

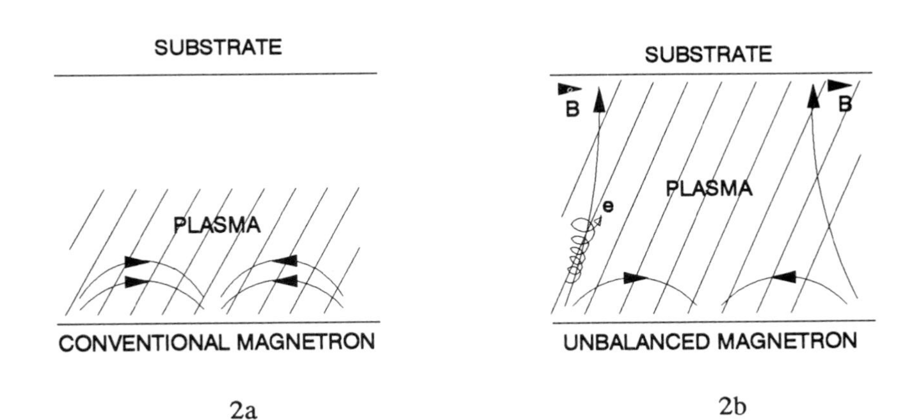

Figure 2 The magnetic field arrangement for a conventional and unbalanced magnetron

3 METASTABLE ALLOY DEPOSITION BY PVD TECHNIQUES

The use of modern PVD techniques to investigate metastable film deposition is fairly limited in the non-reactive field. Much of the documented research is mainly concerned with alloys based on the copper system for electrical switching applications, and the copper chromium system provides a useful comparison of techniques. By electron beam evaporation[22] substantial increases in solid solubility in the Cu/Mo, Cu/W, and Cu/Cr systems have been seen. In the case of the copper/chromium system the limit of chromium incorporation was 30% in copper. This can be compared to a prediction of 0.02% chromium incorporation in the FCC copper from the equilibrium phase diagram. Magnetron sputtering[23] of copper and chromium has been shown to exhibit single phase deposition across the whole alloy range, with a change from FCC copper to a BCC chromium structure at approximately 28% chromium. Similar results have been obtained by diode sputtering[24] where chromium incorporation between 20-30% could be achieved depending upon the deposition temperature. If the results are compared to RSP of copper/chromium then the advantages of vapour deposition can be seen. Splat cooling and melt spinning have produced supersaturations only as high as 1.8%[25], and pulsed laser irradiation has pushed solubility up to 4%[26]. The drawback with rapid quenching from the melt can be as a result of the limited liquid solubility of some alloy systems, dictating a maximum solubility which can never be exceeded.

Holleck[27] has drawn a schematic PVD phase field for the copper/chromium system based on kinetic and thermodynamic considerations, see figure 3. The deposited phase will be dependent upon both the deposition temperature and the alloy composition. The reliance on deposition temperature is important when considering the resulting morphology of the thin film. From figure 1, a decrease in deposition temperature will increase the likelihood of an open columnar structure. If, therefore, additional ion bombardment can lead to a low temperature, high quality film deposition, this will be beneficial during metastable alloy formation.

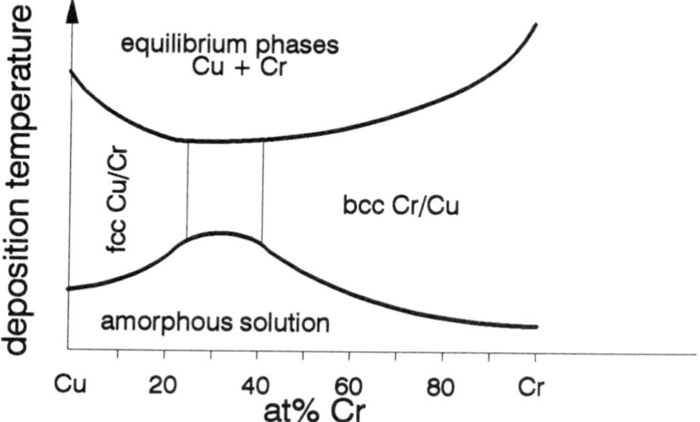

Figure 3 Hollecks PVD phase field for the Cu/Cr system

4 UNBALANCED MAGNETRON DEPOSITION OF METASTABLE ALLOYS

During various studies, various metastable solid solutions have been deposited by unbalanced magnetron sputtering, including: aluminium/magnesium for sacrificial protection of aircraft fasteners; aluminium/tantalum for oxidation investigations; aluminium/lead for bearing applications; stainless steel/silicon for grain boundary simulation; palladium/silver for selective diffusion barriers; copper/chromium for grain size refinement in bulk films. A variety of magnetron configurations has been employed depending upon the substrate and film requirements, see figures 4a-f. This paper will present only the results of the deposition of Al/Mg and Cu/Cr metastable alloy films.

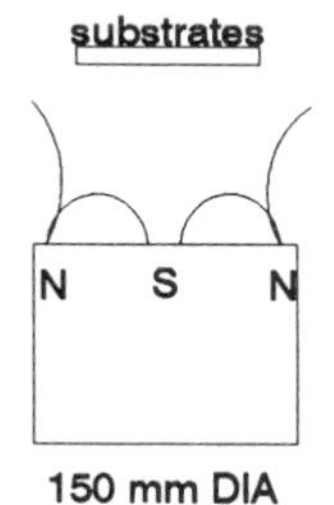

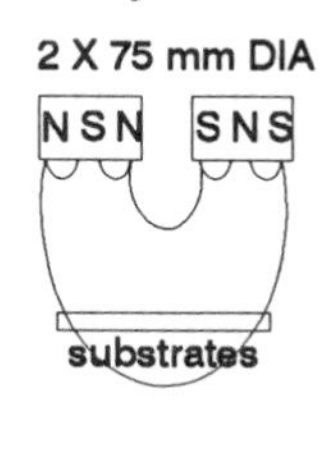

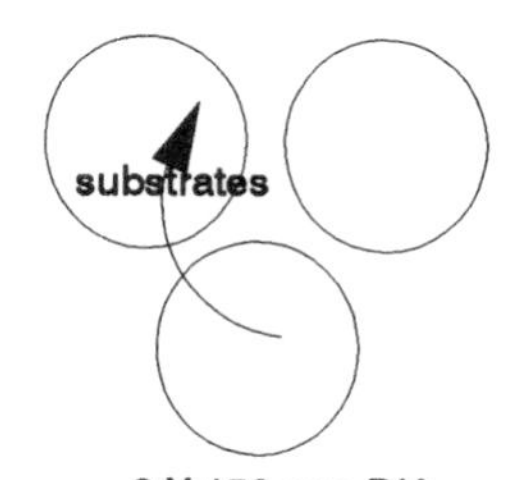

3 X 150 mm DIA

Single Planar Magnetron for Deposition from Alloy or Hybrid Targets

4a

Dual Planar Magnetrons for Co-Deposition

4b

Tri-Magnetron Arrangement for Multilayer or Alloy Deposition, and Very High Rate Bulk Film Deposition

4c

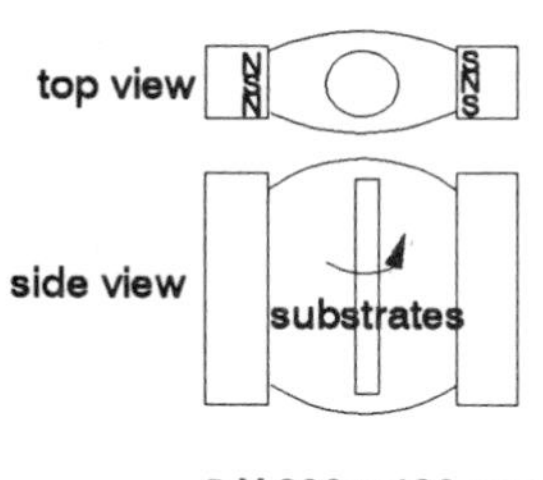

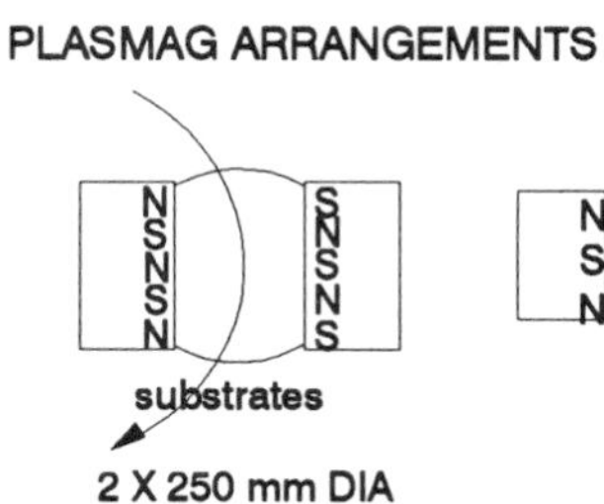

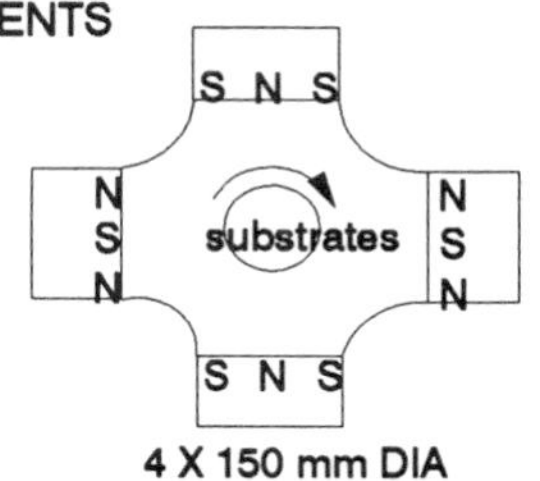

2 X 300 x 100 mm
Dual Closed Field Magnetrons for Alloy, Metal & Reactive Deposition onto Complex Shapes

2 X 250 mm DIA
Closed Field Twin Race-Track Magnetrons for Through-Put Deposition on Complex Shapes

4 X 150 mm DIA
4 Magnetron Arrangement for Exotic Alloys and Reactive Films

* PLASMAG is a Trademark of D.G.TEER COATING SERVICES, U.K.

4d 4e 4f

Figure 4 Various magnetron arrangements for alloy deposition

Deposition of Supersaturated Al/Mg Alloys

A range of aluminium/magnesium alloy films has been deposited from a 150mm diameter hybrid target (see figure 4a)[28], and by co-sputtering from two 75mm diameter targets (see figure 4b). The deposition rate from the hybrid target was approx. 0.2μm/min and the varying alloy contents were determined by the area of aluminium or magnesium inserts contained in the target. The resulting films covered the full alloy range and contained mostly single phase structures. The deposition conditions and results are cited in reference 28. Despite the films being deposited onto very rough 'application' surfaces they were of very high quality with a closed non-columnar morphology. Very high supersaturations were seen, and a change over from FCC aluminium to CPH magnesium took place at approximately 55% magnesium. From approximately 20% magnesium in solid solution, a gradual introduction of a glassy morphology was apparent in the fracture section of the films, see figure 5. It is not clear whether these areas are truly amorphous, or microcrystalline. Figure 6, shows the transmission electron microstructure of a glassy region of the alloy Al/22%Mg. Between 40% and 50% magnesium, only the glassy morphology was apparent. Within this glassy phase extra diffraction peaks appear which suggests some deviation from a single phase structure. However the peaks do not correspond to any published intermetallic structures and do not resemble a distorted magnesium lattice.

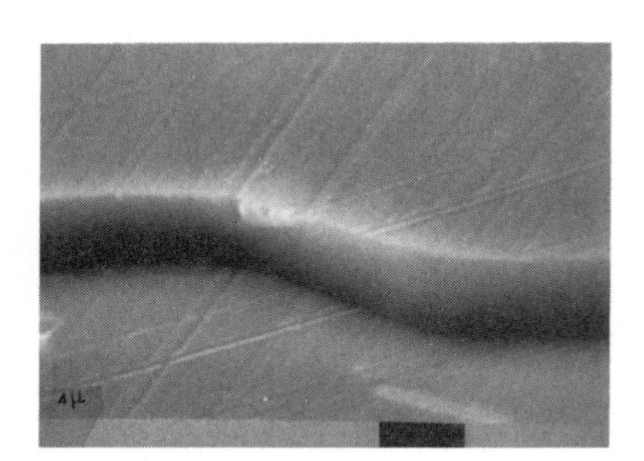

(layered appearance due to ion thinning)

300nm

Figure 5 SEM of fractured glassy phase

Figure 6 TEM of glassy phase

A second approach adopted involved co-sputtering from aluminium and magnesium targets. This offered more control over the alloy mixture across the full range. The films were deposited using a Taguchi L9 array[29], the variables being: source to substrate distance (70, 105, 140mm); Al magnetron power (0.5, 1, 1.5kW); Mg magnetron power (0.3, 0.6, 0.9kW); and substrate bias (approx. -35V self-bias, -50, and -100V). Substrates were placed in rows across the magnetrons so that during each deposition run a wide range of compositions are deposited. Again, all the films had a fully dense non-columnar structure, see figure 5. The alloy phases present can be summarised as follows:

1. In the range 0 to 20% magnesium - supersaturated magnesium in FCC aluminium with a strong (111) orientation.

2. In the range 20 to 45% magnesium - the introduction of a glassy phase which was the only phase present at 45% magnesium content.

3. At 50% magnesium the intermetallic $Al_{12}Mg_{17}$ appeared. Above this composition the amount of glassy phase decreased until at 60% only the intermetallic was present.

4. At 90%+ magnesium only a CPH solid solution of aluminium in magnesium was present, which had a strong (0002) preferred orientation.

Deposition of Supersaturated copper/chromium

The Al/Mg films were deposited by sputtering downwards. Hence, for the hybrid type target the alloying metal was made to form an insert in the target. For the copper/chromium deposition the films were deposited from a 150mm diameter magnetron by sputtering upwards. Two types of targets were used; a copper/1%chromium/0.2%zirconium oxygen free target and a hybrid type copper-chromium target. The hybrid target was made very simply by placing regular small areas of chromium on the surface of an OFHC copper target. The technique provides an easy way of varying the alloy contents of films, since the resulting alloy content is directly proportional to the area of alloying metal placed on the target, see figure 7. One disadvantage is that the poor cooling of the pieces of alloying metal can result in the pieces melting if too high a magnetron power input is applied.

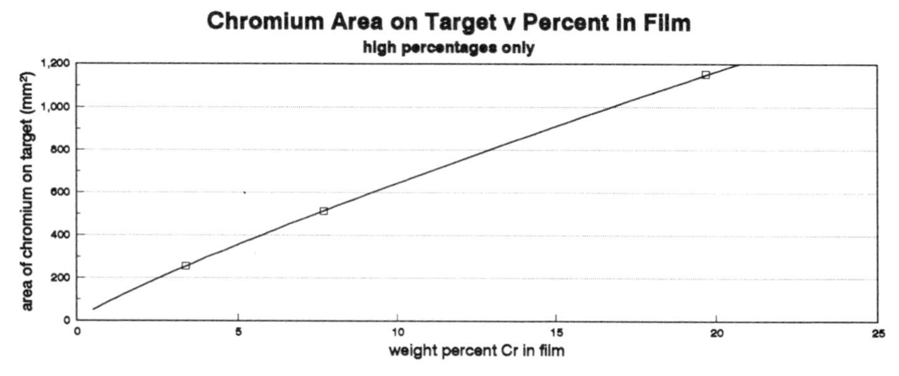

Figure 7 Change in film composition with area of chromium on the target

The addition of the 1%Cr and 0.2%Zr to a copper target was designed to inhibit grain growth that would otherwise occur during deposition of pure OFHC copper over extended deposition times. The aim is to create fully dense films of ultra-fine grained copper materials up to thicknesses in excess of 1mm. Comprehensive studies into the microstructure of copper films deposited under a variety of deposition conditions[20,21,30] have shown that at high deposition rates (1.5μm/min+) over extended deposition periods (8 h), grains can grow to diameters in excess of 1mm. Investigations into the grain size of copper films containing small alloy addition (1%) have shown that over deposition periods of 1h there is a considerable reduction in grain size at high deposition rates (at 2.5μm/min, typically 200nm for pure Cu and 40nm for a copper alloy)[31]. Over the extended deposition periods at high deposition rates the small alloying additions can restrict the grain size to well below the micron level, see figures 8a&b. XRD confirmed that all the alloy addition was in solid solution in the FCC copper.

Further alloy additions where made using the hybrid target, to produce films with up to a maximum of 20% chromium. Due to the power restriction imposed by using this type of hybrid target, the deposition rate was only approximately 0.25μm/min over the 1h period. X-ray diffraction analysis of the films showed that all the chromium additions were in solid solution. The films containing less than 1% Cr had very strong (111) preferred orientation and films with more than 1% Cr had a total (111) orientation. The grain size of the films with less than 3% Cr were greater than those of pure copper, but at higher percentages the grain size was less than that of pure copper. The initial increase is probably due to the extra heat input to the growing film caused by the 'glowing hot' chromium pieces on the magnetron target. Microhardness measurements on the films under a series of loads (to ensure no substrate effects), showed there to be a power law rise in the hardness with chromium incorporation, see figure 9. This is largely due to solid solution hardening as the grain size refinement at these low deposition rates is negligible.

____ 150nm

Figure 8(a) TEM of OFHC copper, 1.2μm/min deposition rate, 480 mins deposition time

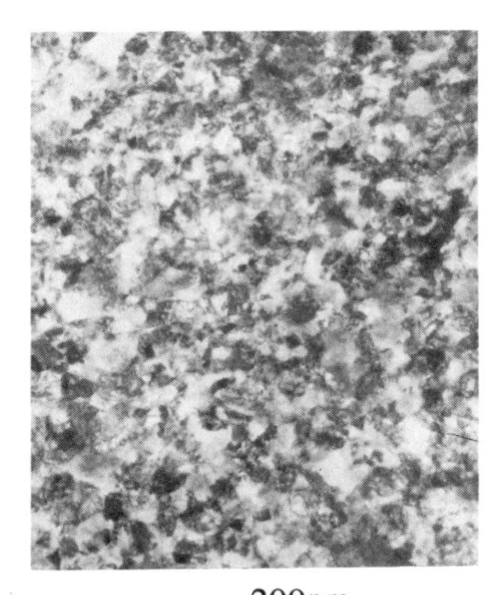

____ 200nm

Figure 8(b) TEM of Cu/1%Cr/0.2%Zr, 1.2μm/min deposition rate, 480 mins deposition time

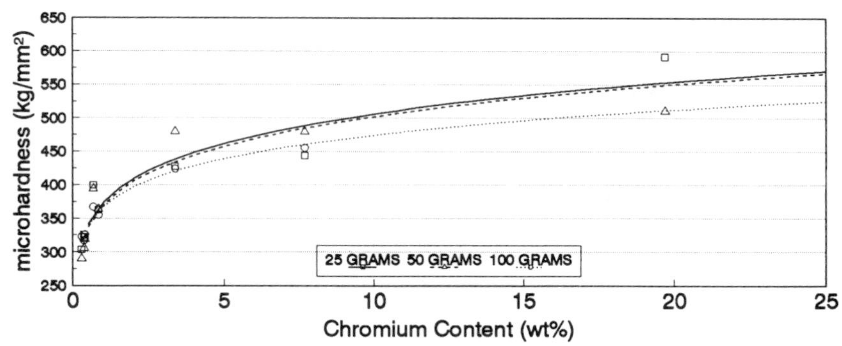

Figure 9 Change in microhardness with increasing chromium content

5 CONCLUSIONS

The formation of supersaturated solid solutions of alloy mixtures by PVD techniques provides the opportunity to study the properties of these materials, and also provides a process which can readily apply the material benefits to surface regions. The most apparent improvement in material properties offered by these solid solutions are very large increases in the hardness, and, subsequently, enhanced wear resistance. If the improved tribological properties are combined with other material properties such as corrosion resistance and electrical conductivity, then the alloys may meet the requirements for a number of applications.

The use of unbalanced magnetrons in the processing of thin films has increased the operating envelope for the deposition of fully dense structures. This is primarily due to enhanced ionisation at the substrates, increasing the mobility of the adatoms, leading to higher mobility conditions at lower temperatures. Evidence suggests that not only pure metal and alloy films benefit from this lower temperature deposition, but also the reactive deposition of ceramic coatings can be achieved at considerably lower temperatures than are normally required[32]. Hence, materials that are highly temperature sensitive can be coated with hard wear resistant coatings without losing their bulk properties.

The dependence of the solubility on the deposition temperature (see figure 3), means that the unbalanced magnetron lends itself as a most favourable technique in this respect. This can be shown by comparing the published deposition rates of the other PVD techniques, with the rates used during this study. Electron beam evaporation[22]; 0.12μm/min, room-temp. substrates, 1μm film thickness, to yield a 10% solubility gap between 30-40%Cr. RF co-sputtering[24]; 1μm/hour, 1.5-3μm film thickness, 23%Cr soluble at 55°C, and 8%Cr soluble at 110°C. Magnetron co-sputtering[23]; 0.036μm/min, gave total solid solubility with a phase transition at about 27%Cr. This study; 0.25 μm/min, temperature estimated at 70°C minimum, 15μm thickness, gave 20%Cr soluble in copper (higher percentages were not investigated).

ACKNOWLEDGEMENTS

The authors would like to thank the Defence Research Agency for funding the work, and Mr G.France for assistance with TEM preparation.

REFERENCES

1. H.Jones, in 'Ultrarapid Quenching of Liquid Alloys', ed H.Herman, Academic Press, New York, 1981.
2. P.Duwez, R.H.Willens, W.Klement, J.Appl.Phys, 1960, 31, p1136.
3. R.Pond, R.Maddin, Mater.Sci.Eng., 1969, 23, 87.
4. H.S.Chen, C.E.Miller, Rev.Sci.Instr., 1970, 41, 1237.
5. E.Babic et al, in Proc. 1st Int. Conf. on Metastable Metallic Alloys, Brela, 1970, Fizika(Yugoslavia).
6. Y.V.Murty, R.P.I.Adler, J.Mater.Sci, 1982, 17, 1945.
7. R.E.Maringer, C.E.Mobley, J.Vac.Sci.Technol., 1974, 11, 1067.
8. R.W.Cahn et al, Mater. Sci. Eng., 1976, 23, 83.
9. S.Mader, J.Vac.Sci.Technol, 1965, 2, 35.
10. S.D.Dalgren, in Proc. 3rd Int. Conf. on Rapidly Quenched Metals, Brighton 1978, ed B.Cantor, Metals Society, London, 1978.
11. R.W.Cahn, in Physical Metallurgy II, ed Cahn, Haasen, Elsevier Science Publishing, 1983.
12. R.Messier, A.P.Giri, R.A.Roy, J.Vac.Sci.Technol., 1984, A2(2), 500.
13. B.Window, N.Savvides, J.Vac.Sci.Technol., 1986, A4, 196.
14. B.Window, N.Savvides, J.Vac.Sci.Technol., 1986, A4, 453.
15. B.Window, N.Savvides, J.Vac.Sci.Technol., 1986, A4, 504.
16. S.Rossnagel, J.J.Cuomo, Vacuum, 1988, 38 2, 73.
17. R.P.Howson, A.G.Spencer, K.Oka, R.W.Lewin, J.Vac.Sci.Technol., 1989, A7(3), 1230.
18. S.Kadlec, J.Musil, in Research Report 073/89, Czechoslovak Academy of Science, 1989.
19. D.G.Teer, Surface & Coating Technology, 1988, 36, 901.
20. D.P.Monaghan, R.D.Arnell, in Proc. IPAT, Brussels, 1991, 316, CEP Consultants, Edinburgh.
21. D.P.Monaghan, R.D.Arnell, Surface and Coatings Technol., 1991, 49, 298.
22. A.G.Dirks, J.J. van den Broek, J.Vac.Sci.Technol., 1985, A3(6), 2618.
23. A.P.Payne, B.M.Clemens, Mat.Res.Symp.Proc.Vol.187, 1990, Materials Research Society, 39.
24. D.McIntyre, J-E.Sungren, J.Greene, J.Vac.Sci.Technol., 1988, sum. abst.
25. G.Falkenhagen, W.Hofman, Z.Metallkd. 1952, 43, 69.
26. J.F.M.Westendorp et al, J.Mater.Res., 1986, 1, 5.
27. H.Holleck, Surface and Coatings Technol., 1988, 36, 151.
28. R.D.Arnell, R.I.Bates, Vacuum, 1992, 43, 105.
29. R.Roy, in A Primer on Taguchi Methods, Van Nostrand Reinhold, New York, 1990.
30. D.P.Monaghan, R.D.Arnell, Vacuum, 1992, 43, 77.
31. D.P.Monaghan, R.D.Arnell, Proc. ICMCTF92, San Diego, 1992, in press.
32. D.G.Teer, *private communication*, to be presented at the IVC-12, The Hague, Oct 1992.

3.4.3
The Compound Layer Characteristics Resulting from Plasma Nitrocaburising with an Atmosphere Containing CO_2 Gas Additions

E. Haruman, T. Bell, and Y. Sun

WOLFSON INSTITUTE FOR SURFACE ENGINEERING, SCHOOL OF METALLURGY AND MATERIALS, UNIVERSITY OF BIRMINGHAM, UK

1 INTRODUCTION

Thermochemical surface treatments to impart superior surface properties to engineering components have been well documented[1,2]. Ferritic nitrocarburising is a thermochemical treatment during which nitrogen and carbon are simultaneously introduced into the surface of ferrous materials at temperatures between 550-580^{0}C. As a result, a thin compound layer consisting predominantly of ϵ iron carbonitride together with γ' iron carbonitride is formed, beneath which is a diffusion zone. Nitrocarburising is generally employed to upgrade cheaper materials such as plain carbon steels and low alloy steels. The principle improvements in mechanical properties associated with the process are increased fatigue and yield strength and improved wear resistance, and in certain situations improved corrosion resistance. The improved fatigue strength is due to the retention of nitrogen in solid solution when the component is quenched from the treatment temperature. Improvements in wear and corrosion resistance are related to the presence of a compound layer usually consisting of the mono phased ϵ iron carbonitride.

In principle, nitrocarburising can be carried out using solid, liquid, gaseous or glow discharge environments to effect the mass transfer of the alloying elements. Industrial applications of liquid and gaseous nitrocarburising processes have long been established[3,4,5], whereas the glow discharge technique is still facing major problems in obtaining a mono phased ϵ compound layer structure[6-10]. However, despite this limitation the advantages of the environmentally friendly glow discharge process are widely known[11-13], and these include reduced energy and treatment gas consumption, as well as accurate and reproducible control of the processing parameters. The principles of glow discharge or plasma processing have been explained elsewhere[11,14,15].

Most previous investigations [6,7,8,10,16] of plasma nitrocarburising have been carried out using CH_4-containing atmospheres. The results obtained indicated that it was difficult to produce mono phased ϵ compound layers, and that the cementite phase is formed even at very low CH_4 additions (~1%). The cementite + ϵ mixed compound layer has been proven to be brittle, as is the $\gamma'+\epsilon$ compound layer. Accordingly, attempts have been made in the present investigation[17] to plasma nitrocarburise ferrous materials using atmospheres containing CO_2. The purpose of the present paper is to characterise the compound layers produced in $N_2 + H_2 + CO_2$ gas mixtures.

2 EXPERIMENTAL METHODS

2.1 Materials and Plasma Nitrocarburising Procedures

The materials used in the present work are Armco pure iron, carbon steel (En 8 with 0.46% carbon), and low alloy steel (En 40B with 3.2% chromium). The plasma furnace used was a 20 kW GZ nitriding unit manufactured by Klockner Ionon GmbH, which had been modified to suit the nitrocarburising process (Figure 1). The unit was supplied with a cold-walled vacuum chamber, and hence could only be used for nitrocarburising followed by furnace cooling and not oil quenching.

The specimens were plasma nitrocarburised at 570^{0}C with 3 mbar pressure, and a total gas flow rate of 25 litres per hour. Gas atmospheres, consisting of $N_2 + H_2 + CO_2$, with various nitrogen contents between 50% and 87% and CO_2 contents between 0% and 15% were employed. Heating took approximately 75 minutes and the treatment duration was 3 hours. Table 1 shows the complete series of experiments used during the course of this work.

2.2 X-ray Diffraction Analysis and Metallography

X-ray diffraction analysis and metallographic studies were carried out to assess the structures of the compound

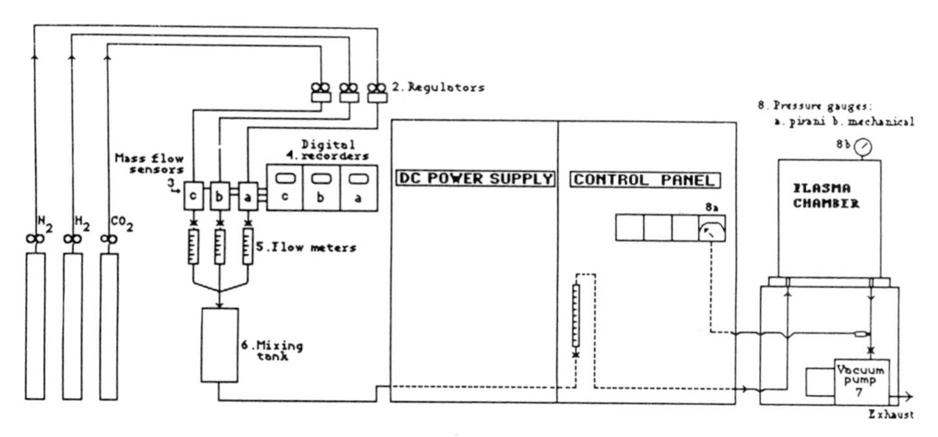

Figure 1 Schematic representation of a modified 20kW GZ plasma nitriding unit

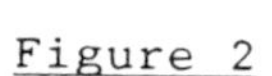

Figure 2

X-ray diffraction pattern of plasma nitrocarburised Armco iron treated with (a) 50% N_2, (b) 75% N_2, (c) 87% N_2 gas, and a constant CO_2 (1%) level

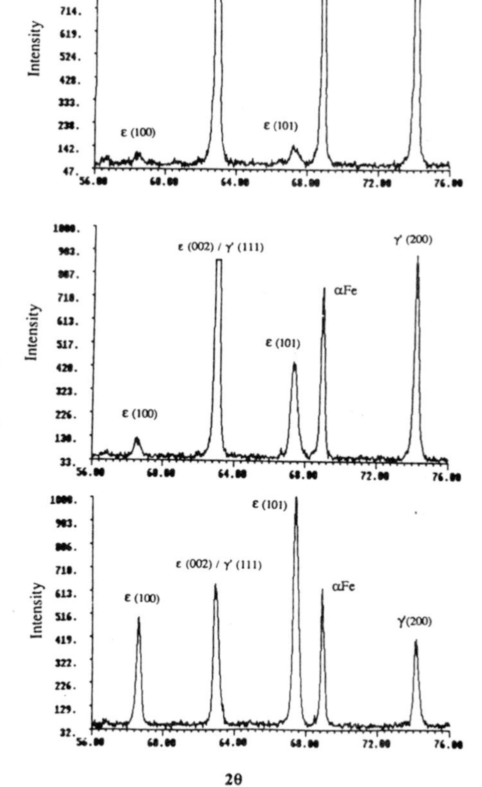

Table 1 Experimental series and the resulting compound layer structures

Gas Composition N_2 CO_2 H_2 (vol.%)			Treatment Time (hrs)	Armco Iron Structure	En 8 Structure	En 40 B Structure
				-	-	-
50	1.0	49	3	γ'^{*}/ε	γ'^{*}/ε	γ'^{*}/ε
75	1.0	24	3	γ'/ε	γ'/ε	γ'/ε
87	1.0	12	3	ε/γ'	ε/γ'	ε/γ'
87	2.0	11	3	ε/γ'	ε/γ'	ε^{*}/γ'
87	3.0	10	3	ε/γ'	ε/γ'	ε^{*}/γ'
87	4.0	9	3	ε/γ'	ε/γ'	ε / Fe3C
87	5.0	8	3	ε^{*}/γ'	ε^{*}/γ'	ε/Fe_3C
86	6.0	8	3	ε^{*}/γ'	ε^{*}/γ'	ε/Fe_3C
87	7.0	6	3	-	-	-
84	8.0	8	3	ε/γ'	ε/γ'	ε/Fe_3C
82	10.0	8	3	ε/γ'	$\varepsilon/\gamma'/Fe_3C$	ε/Fe_3C

Note: γ'/ε = gamma prime phase is more predominant in the compound layer.
ε/γ' = epsilon phase is more predominant in the compound layer.
γ'^{*}/ε = compound layer is composed predominantly of the gamma prime phase.
ε^{*}/γ' = compound layer is composed predominantly of the epsilon phase.

Table 2 The compound layer thickness

% N_2	% CO_2	Armco iron (μm)	En 8 (μm)	En 40B (μm)
50	1	8.2	7.9	4.6
75	1	10.6	11.3	7.1
87	0	8.9	10.0	7.4
87	1	11.3	13.2	8.2
87	2	12.8	14.9	7.3
87	3	13.5	15.9	6.8
87	4	14.1	16.8	1.7
87	5	14.4	17.2	1.2

layers. The analysis was performed using a Phillips PW 1050 diffractometer with CrK_α radiation to study the structure throughout the entire layer. This is possible because CrK_α has a penetration of ~14μm for iron carbonitride layers. In addition, an X-ray profiling technique was performed on specimens showing large amounts of ε phase. The technique involved the sequential removal of approximately 1-3μm from the surface layer using 1200 grade SiC abrasive. After each removal the specimens were re-analyzed with CuK_α diffractometer (Phillips 1020) that has a shallow penetration for iron carbonitrides (~4μm). These steps were repeated until the compound layer was completely removed. With this profiling procedure the variation in structural composition across the compound layer could be precisely determined.

Metallographic studies were performed using both optical techniques and scanning electron microscopy. Prior to examination, edge-mounted and polished specimens were first lightly etched with 5% nital for Armco iron and 2% nital for En 8 and En 40B steels. They were then further re-etched by reagent A (1 cc conc. HCl mixed with 10 cc ethanol, 1 cc mix + 99 cc of 5% nital) for Armco iron and reagent B (1 cc conc. HCl mixed with 15 cc ethanol, 1 cc mix + 99 cc of 2% nital) for En 8 and En 40B steels. The use of these reagents allows for sensitive discrimination between ε iron carbonitride and γ' iron nitride.

2.3 Light Elements Depth Profiling Analysis

Concentration profiles of nitrogen and carbon across the compound layers are a reflection of the phases coexisting in the layers. Nitrogen, carbon, and oxygen depth profiling analyses of selected specimens were investigated by Electron Probe Micro Analysis (EPMA) and Glow Discharge Optical Emission Spectrometry (GDOES).

The EPMA was performed with a JEOL JXA 800A instrument with an accelerating voltage of 20 keV, a beam current , 30 nA, a probe diameter ~1μm, and a beam take-off angle 40^0. The analysis was conducted by scanning the probe beam from the outermost layer down to the substrate with 1μm distance between each point, in an inclined direction. The contents of nitrogen, carbon and oxygen were quantified by measuring their K_α intensities with reference to standard samples of Si_3N_4 for nitrogen, $C_{(graphite)}$ for carbon, and SiO_2 for oxygen. The external standards Armco iron, $\gamma'Fe_4N$, and $\epsilon Fe_{2-3}(N,C)$ of known composition, were also used for the purpose of calibration, thus ensuring the accuracy of measurement.

Depth profiling by glow discharge spectrometry was performed with GDS-750 QDP, supplied by LECO Ltd. This relatively new method of surface analysis has become increasingly accepted for industrial application, due to its simple operation and rapid analysis. In principle, the analysis involves cathodic sputtering of the specimen surface by applying a controlled voltage at a controlled argon pressure across the specimen being analyzed. Surface atoms removed by this sputtering process diffuse into the argon plasma where excitation provides spectra with a linear relationship between element concentration and spectral intensity. The emitted spectra are then characterised by the spectrometer system, and a 386 IBM Compatible PC quantifies the values on the basis of a standard reference. In this investigation, measurements were performed at 61 mA and 704 volts.

3 RESULTS AND DISCUSSION

3.1 Compound Layer Structures

X-ray diffraction analysis (Table 1) identified that the compound layers produced by plasma nitrocarburising with an atmosphere containing CO_2 gas mainly consisted of the $\epsilon Fe_2(N,C)_{1-x}$ and $\gamma'Fe_4(N,C)$ phases, and were almost free of cementite. The results showed that the formation of the ϵ phase was enhanced by the use of a high nitrogen content in the atmosphere. For example, when a 1% CO_2 atmosphere was used, the specimens treated with 87% N_2 produced a greater amount of ϵ as compared with those treated using 50% and 75% N_2, as revealed by the X-ray diffraction patterns in Figure 2. Therefore, the following discussion will be focused on the 87% nitrogen atmosphere.

With the Armco iron specimens, the level of formation of the ϵ phase was enhanced by an increase in CO_2 content up to 5%. At 5% CO_2 the compound layer was predominantly composed of the epsilon phase with a thin γ' band adjacent to the substrate. An X-ray diffraction pattern, showing

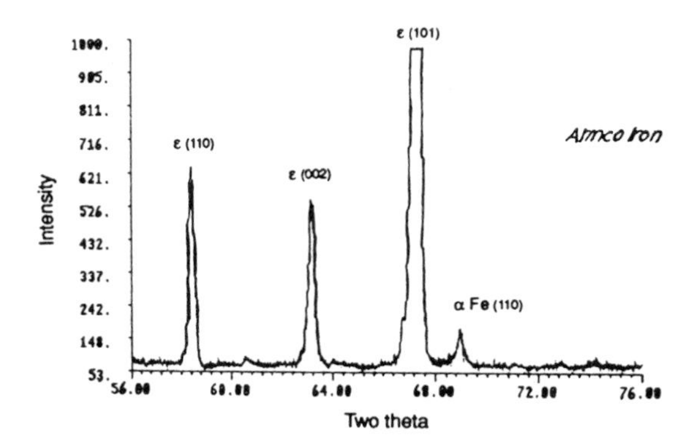

Figure 3 X-ray diffraction pattern of predominantly ϵ phased compound layer

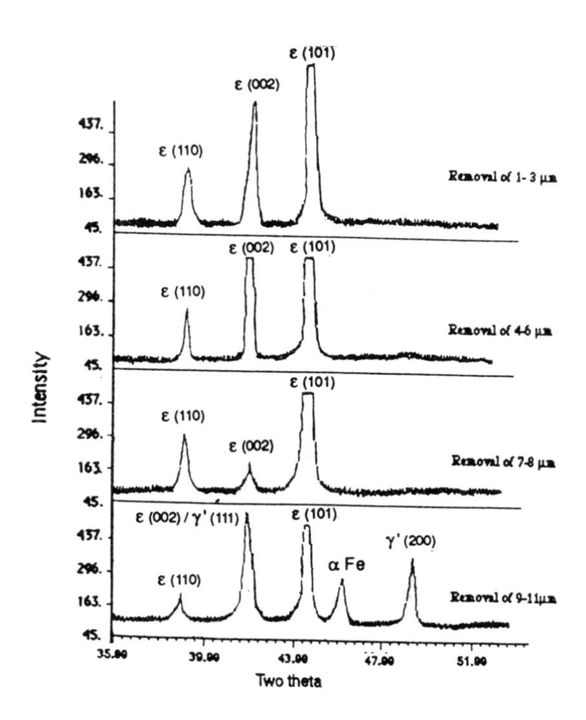

Figure 4 X-ray depth profiling patterns of predominantly ϵ phased compound layer

only the ε phase peaks obtained from the compound layer treated with this condition, is given in Figure 3. The use of CrK_α radiation did not detect the presence of the γ' phase after removal of ~10μm from the layer (Figure 4). The micrographs of the compound layers on Armco iron at 1%, 3% and 5% CO_2 compositions, are shown in Figure 5. It is clear from Figure 5 that the γ' phase almost disappears when the CO_2 content is increased to 5%. Previous work has shown that the formation of this fringe of γ' phase in the compound layer - substrate interface could be due to slow cooling in the furnace[10].

When the CO_2 content was increased to 7%, the glow discharge failed to raise the temperature above 520^{0}C, and consequently the formation of compound layer was hindered. This occurrence may be due to the reason that when the CO_2 content was fixed at 7%, the hydrogen level in the atmosphere became 6%, and as a consequence, the hydrogen level was insufficient to produce an adequate emission of secondary electrons and surface sputtering, which controls the specimen temperature. The presence of hydrogen in the atmosphere has been observed to provide the majority of secondary electron emissions and sputtering on a ferrous surface. A critical level of around 10% has been suggested as the minimum for a positive-nitriding capacity[18].

In the case of En 8 steel, the results followed a similar trend to those found with Armco iron. Unlike Armco iron where the ε and the γ' phases were clearly separated into two sub-layers, this material produced dual mixed phase ε and γ' compound layer structures. The γ' phase was randomly scattered in the layer but with the ε phase making some direct contact with the substrate. In the presence of carbon in the substrate, a lower nitrogen ε phase was formed in areas surrounding the pearlite grains (Figure 6). The formation of a low nitrogen ε phase is in good agreement with the Fe-N-C diagram (Figure 7) published by Slycke et al[19].

Attempts were made to produce a mono epsilon phased structure by increasing the CO_2 content up to 10% whilst maintaining the hydrogen content at 8%. The results however, still indicated the existence of the γ' phase in the compound layer, and furthermore a cementite phase was

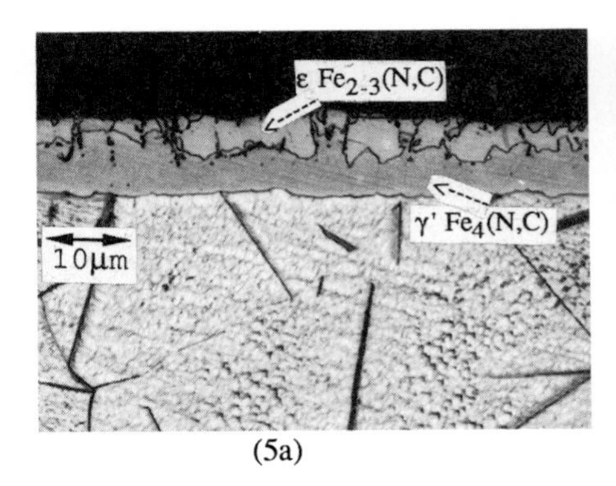

(5a)

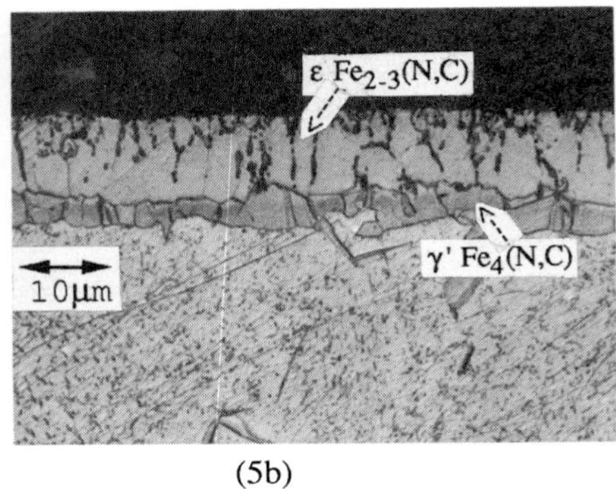

(5b)

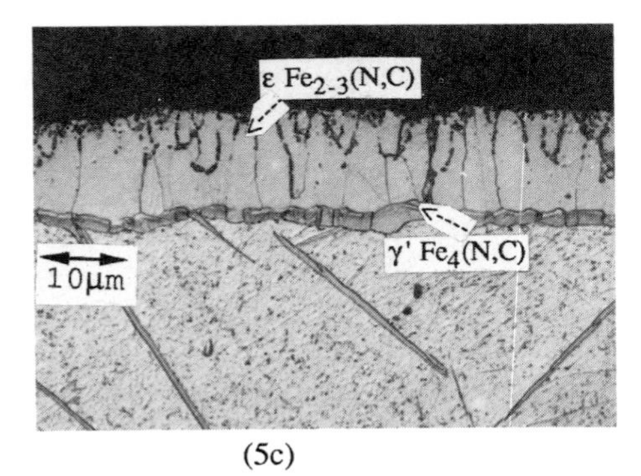

(5c)

Figure 5 Compound layers developed on Armco iron, treated with 87% N_2 and various CO_2 compositions, (a) 1% CO_2, (b) 3% CO_2, (c) 5% CO_2

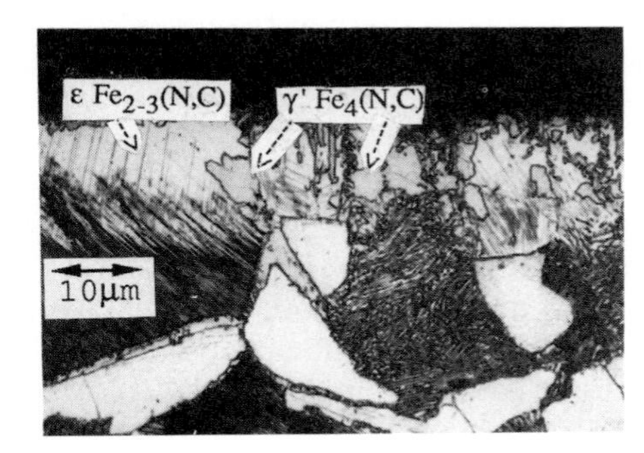

Figure 6
Dual mixed compound layer developed on En 8 steel, treated with 87% N_2 + 5% CO_2 + 8% H_2

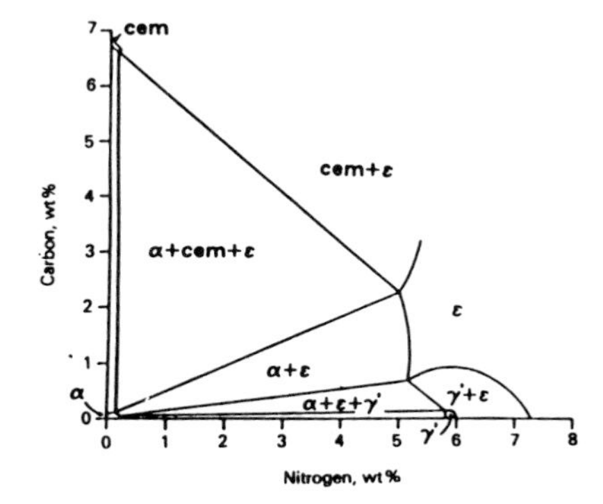

Figure 7 Fe-N-C phase diagram published by Slycke et al.[19]

found in the compound layer on En 8 steel, when the CO_2 addition was 10%.

In previous work using CH_4-containing atmospheres[6], the presence of cementite was detected at only a 1% CH_4 addition (on both Armco iron and En 8 steel) while in this work cementite formed only when the CO_2 level was increased to 10%. Clearly, the use of a CO_2-containing atmosphere gives a beneficial effect in the control of cementite formation. It is interesting to note that when the CO_2 content in the atmosphere was above 8%, a very thin Fe_3O_4 iron oxide (less than 1μm) layer developed on the top surface during furnace cooling. This phase was not distinguishable by the metallographic studies, but CuK_α X-ray analysis confirmed its presence (Figure 8).

The compound layer structures developed on the En 40B steel were mixed dual phases similar to those found on the En 8 steel. With this material, however, the predominantly epsilon phased structure occurred at only 2-3% CO_2 addition. At 4% CO_2, the presence of cementite was verified by X-ray diffraction analysis (Figure 9), thus indicating the strong influence of the alloying elements on the resulting compound layer structure. Electron microprobe analysis revealed that up to 4% chromium is dissolved in the compound layers. The dissolved chromium thus may hinder the diffusion of nitrogen and carbon through the layers. Micro hardness measurements clearly indicated that the compound layers on En 40B steel had a higher hardness than those on En 8 steel, which has no inherent chromium content. This suggests that chromium forms finely dispersed chromium nitrides in the compound layer.

Furthermore with the En 40B steel, the chromium carbides in the diffusion zone tend to react with the diffusing nitrogen to form chromium nitride, due to the higher affinity between chromium and nitrogen. The free carbon liberated from this reaction will therefore precipitate as cementite in the grain boundaries, some carbon from which will diffuse out due to a concentration gradient with respect to the surface. This together with the carbon transferred from the atmosphere, enriches the bottom-part of the compound layer with carbon. A further increase in the carbon content of the atmosphere allows cementite to develop in the compound layer region close to

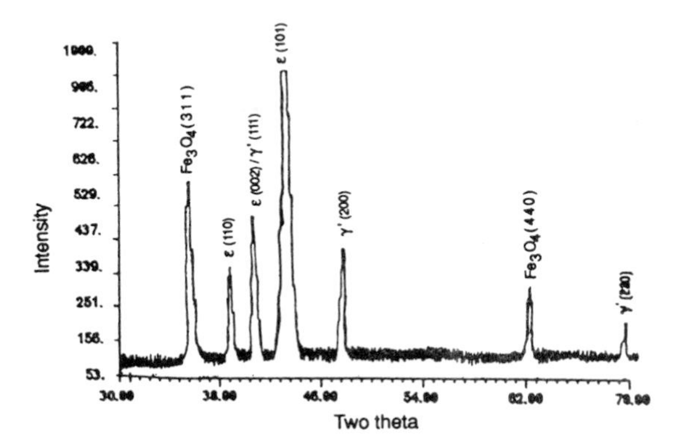

Figure 8 CuK$_\alpha$ X-ray diffraction patterns indicating the presence of Fe_3O_4 oxide on the compound layer

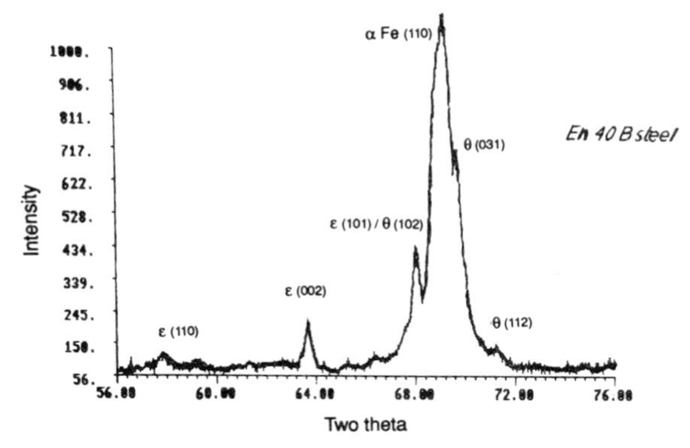

Figure 9 X-ray diffraction of plasma nitrocarburised En 40B steel, indicating the existence of cementite in the compound layer

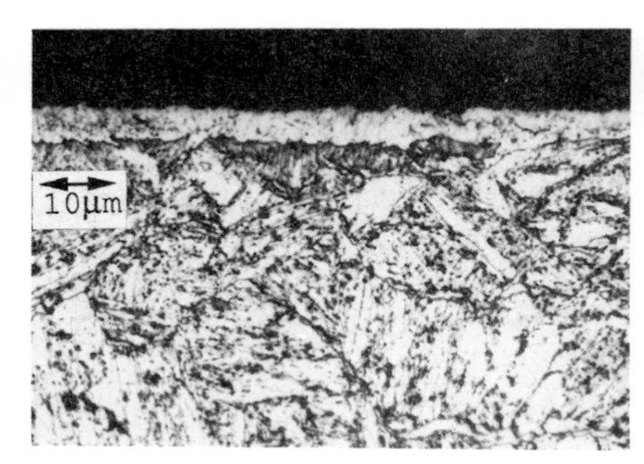

Figure 10
Micrograph of the compound layer developed on En 40B steel

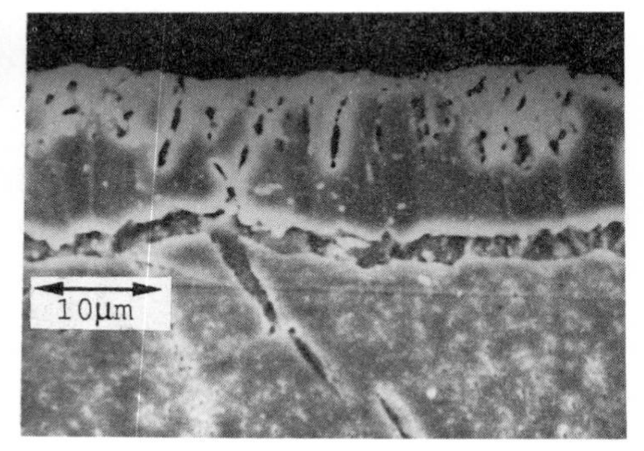

Figure 11 Scanning electron micrograph of plasma nitrocarburised Armco iron showing some pores in the compound layer

the substrate, as was observed with the specimen treated with 5% CO_2. Figure 10 shows a typical compound layer structure developed on the En 40B. With this material, the coexisting phases are compact and finely distributed, and optical metallographic examination could not distinguish between the phases.

3.2 Compound Layers Thickness

The thickness of compound layers produced at 570^{0}C with a 3 hour treatment, depends not only on the nitrogen and CO_2 levels in the treatment atmosphere, but also on the content of carbon and other alloying elements (Table 2). The results clearly indicate that increasing the nitrogen level in the atmosphere gives rise to an increased compound layer thickness on all the materials investigated. This tendency is due to the increased nitrogen "activity" in the plasma. Increasing CO_2 levels in the atmosphere, however, does not always increase the compound layer thickness. Depending on the chemical composition of the substrate, it may reduce the compound layer thickness. For example, the compound layer thickness on Armco iron and carbon steel (En 8) increased with an increase in CO_2 content, but the compound layer thickness on low alloy steel (En 40B) decreased with an increase in CO_2 content, thus indicating a strong influence of the substrate composition on compound layer growth. Under identical process conditions the compound layer on En 8 was the highest and the compound layer on En 40B was the lowest. With the En 8 steel, one can suggest that the carbon from the pearlite in the substrate could enhance the diffusional growth of the compound layer. With En 40B steel, however, cementite formation and chromium nitride precipitation impede the compound layer growth. At high CO_2 contents in the atmosphere (e.g. 5% CO_2), where an enriched carbon level in the compound layer is in equilibrium with respect to cementite, a complete suppression of compound layer growth was found to occur.

3.3 Porosity in the Compound Layers

The present experiments have revealed that the use of high nitrogen atmospheres produces a proportion of pores in the compound layers developed on Armco iron. This porosity was, however, not found in the case of En 8 and En 40B steels, due to lower nitrogen levels across the compound layer. A scanning electron micrograph of the Armco iron, indicating the presence of porosity in the compound layer, is given in Figure 11. It is obvious from this Figure that pores are preferentially located at grain boundaries in the ϵ sub-layer perpendicular to the surface.

The formation of porosity during nitrocarburising is due to the metastability of the epsilon phase with respect to nitrogen gas, which can lead to the precipitation of molecular nitrogen at the grain boundaries[20,21]. A recombination of molecular nitrogen creates pore networks which develop channels penetrating the epsilon grain boundaries in order to have contact with the open atmosphere. In the case of very high nitrogen activity in the layer, pores are also developed in the interior of the epsilon crystallites[21]. In the present plasma nitrocarburising experiments, this type of pore did not frequently occur.

Light element analysis in the region of the pores, by EPMA, indicated that some oxygen (2.6 wt.%) was dissolved in the surfaces of those pores which had a direct contact with the atmosphere. This could well be the reason why cementite did not develop in the compound layer on Armco iron, even when the CO_2 content was increased to 10%. In this case, the oxygen in the channels lowers the activity of carbon below the activity required to form cementite. With reference to gaseous processes[20,21], the formation of cementite during nitrocarburising is due to a 'denitriding phenomenon' during pore development, which leads to carbon absorption and accumulation in the area of the ϵ grains adjacent to the pore channels. In relation to the present work, this theory of cementite formation was not suitable for plasma nitrocarburising with an oxidising capacity. Cementite was found in the pore-free compound layers, as in the case of En 40B and En 8 steels. Hence, the description of cementite formation in Section 3.1 is more appropriate.

3.4 Light Element Analysis of Compound Layers

Electron microprobe analysis of nitrogen, carbon, and oxygen concentrations across the predominantly ϵ phased compound layers (Figure 12) exhibited that, of all the materials investigated, the compound layer on Armco iron

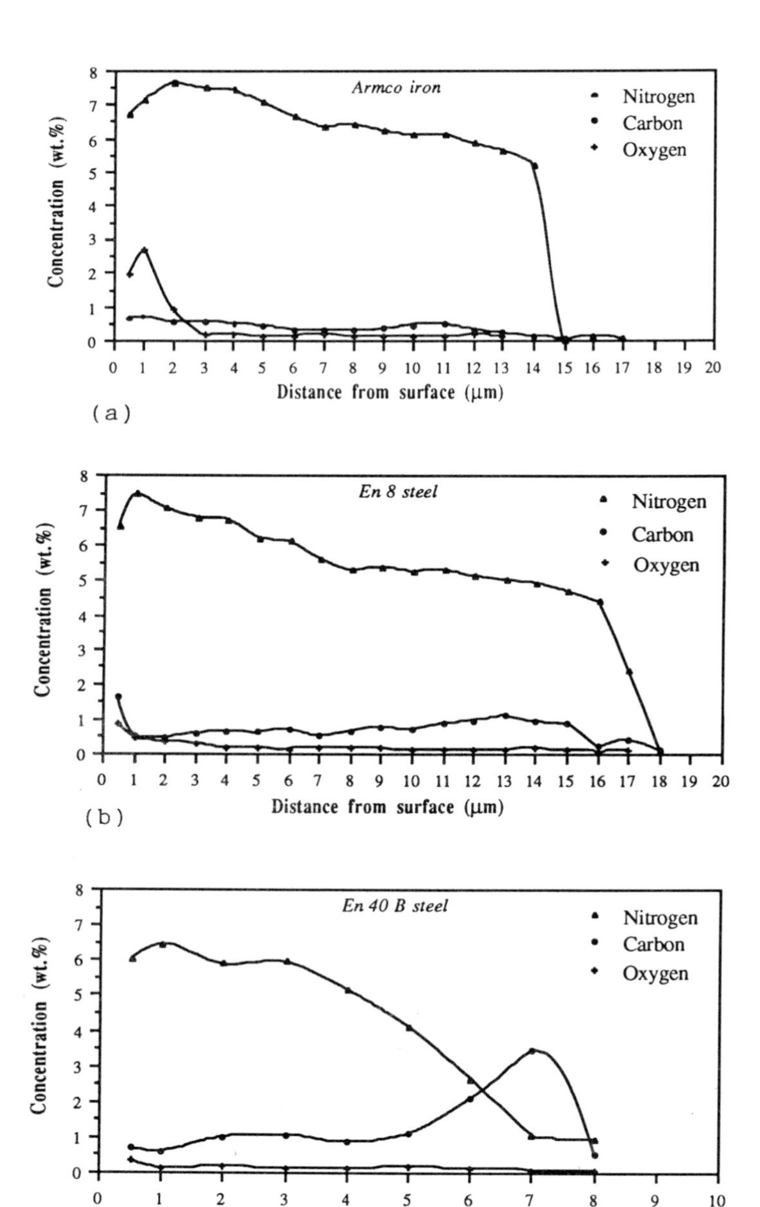

Figure 12 Light element concentration profiles (by EPMA) across the predominantly ε phased compound layers; (a) for Armco iron treated with 87% N_2 + 5% CO_2 + 8% H_2, (b) for En 8 steel treated with 87% N_2 + 5% CO_2 + 8% H_2, (c) En 40B steel treated with 87% N_2 + 3% CO_2 + 10% H_2

contained the highest nitrogen composition and the lowest carbon content, whilst the compound layer produced on En 40B steel contained the lowest nitrogen composition and the highest carbon content. This means that when the carbon composition in the compound layer is reasonably high, i.e., in the case of En 8 and En 40B steels, the nitrogen interstitial content in the ϵ phase is reduced to a lower value. The present investigation also revealed that compound layers dissolved only a negligible amount of oxygen. At the outermost layer the oxygen was reasonably high (1.6 - 2.6 wt.%), but abruptly decreased to well below 0.2 wt.% within the layer. The rôle of oxygen in plasma nitrocarburising thus lies in controlling the carbon activity in the region of the glow discharge - surface interface, similar to that argued for gaseous nitrocarburising[19]. Furthermore, the analysis also supported the argument[8-10] of a low nitrogen compound layer produced by plasma nitrocarburising. The average nitrogen compositions observed were below 6.2 wt.%, compared to 6.5 - 7 wt.% obtained by gaseous nitrocarburising with the endothermic gas-based atmospheres[20].

With Armco iron, the electron microprobe examination also revealed that the average carbon level in the compound layers increased with an increase in CO_2 level in the atmosphere (Figure 13). It is also evident that the carbon level is low in the outer part of the compound layer and steadily increases at some distance below the surface, before it drops when approaching the compound layer-substrate interface. This phenomenon of lowering the carbon content in the outer part of the compound layer is possibly due to the decarburisation of the surface during slow cooling in the furnace. The same phenomenon was also observed during plasma carburising with similar furnace operating conditions[22].

Depth profiling analysis by GDOES showed reasonable agreement with the analysis by the EPMA. The absolute concentrations of nitrogen and carbon obtained from this analysis were in close agreement with those of EPMA, whereas with oxygen, the interpretation relied on its relative concentration. Careful examination showed that the use of a reference standard for profile analysis is of prime importance and high vacuo conditions are also preferable for a high degree of accuracy. The nitrogen, carbon, and oxygen concentration profiles across the

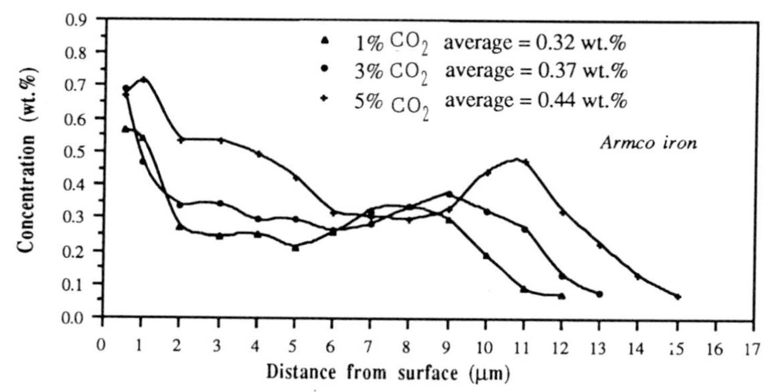

Figure 13 Carbon concentration profiles (by EPMA) across the compound layers produced from different CO_2 gas compositions, at a constant (87%) N_2 level

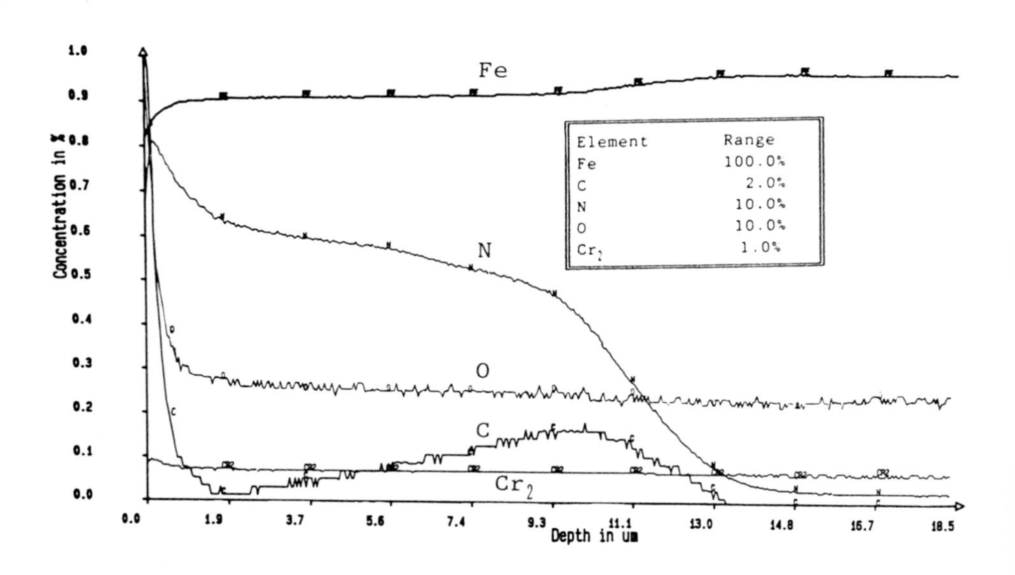

Figure 14 Depth concentration profiles of elements across the plasma nitrocarburised layer on Armco iron (treated with 87% N_2 + 5% CO_2 + 8% H_2), obtained from GDOES analysis

compound layer developed on Armco iron, En 8 steel, and En 40B steel, treated with 87% N_2 + 5% CO_2 + 8% H_2, are shown in Figures 14, 15 and 16. The same specimens for Armco iron and En 8 steel were also analysed by EPMA (Figure 12(a) and (b)).

It is clear from GDOES analysis that the nitrogen concentration profile on Armco iron is almost uniform before it drops considerably in the bottom region of the compound layer (Figure 14). With En 8 steel (Figure 15), the nitrogen concentration continuously decreases with distance from the surface. With En 40B steel (Figure 16), the trend is similar to that of En 8 steel, but with the overall nitrogen concentration being smaller than in Armco iron and En 8. Moreover, the tendency for high carbon peaks to be at some distance below the surface was clearly shown by the GDOES technique. The analysis also shows a higher carbon concentration and thinner compound layer on En 40B, compared to Armco iron and En 8 steel. Clearly, the carbon peak shown in Figure 16 corresponds to a zone of cementite formation. The EPMA depth profiling for this zone in the compound layer of ~1μm thickness was not possible, due to the limitation of beam size diameter (~1μm). Finally, taking into account the carbon profile in Figure 16, the GDOES method of surface analysis corroborates the mechanism of cementite formation previously described in Section 3.1, i.e., cementite is formed at the bottom part of compound layer.

4 CONCLUSIONS

On the basis of present studies, it is concluded that a high nitrogen content atmosphere in the glow discharge is required for the production of a compound layer with a predominant ϵ phased structure. The use of CO_2 gas in the plasma nitrocarburising process effectively delays the formation of cementite and, therefore, a wide range of CO_2 gas contents can be applied to control the compound layer structures. The low alloy steel (En 40B) has a great tendency to form cementite in the compound layer due to the influence of alloying elements, thereby indicating the strong influence of the substrate composition on the nature of the compound layer. Compound layers with a reasonable degree of porosity are produced by plasma nitrocarburising with high nitrogen additions. The pores are found in the grain boundaries of the ϵ phase

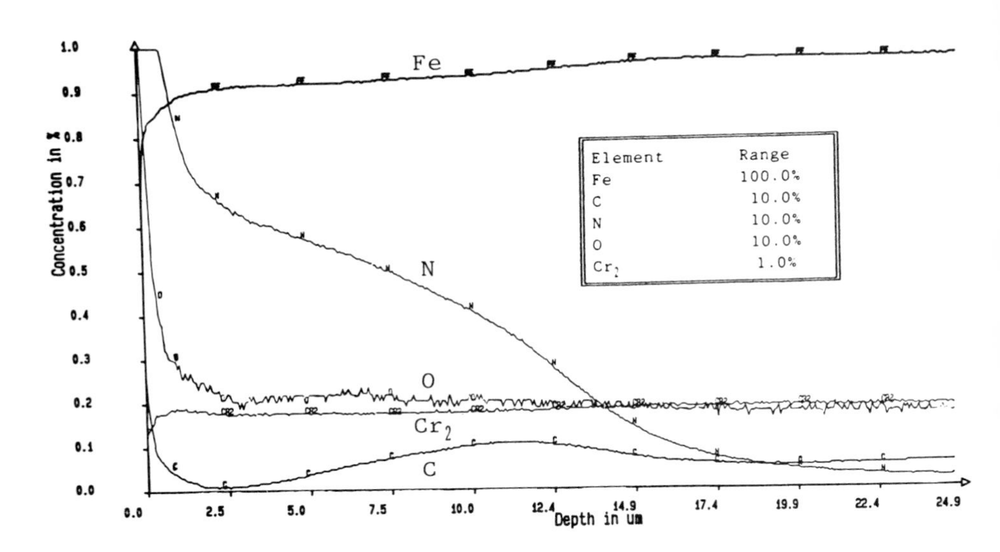

Figure 15 Depth concentration profiles of elements across the plasma nitrocarburised layer on En 8 steel (treated with 87% N_2 + 5% CO_2 + 8% H_2), obtained from GDOES analysis

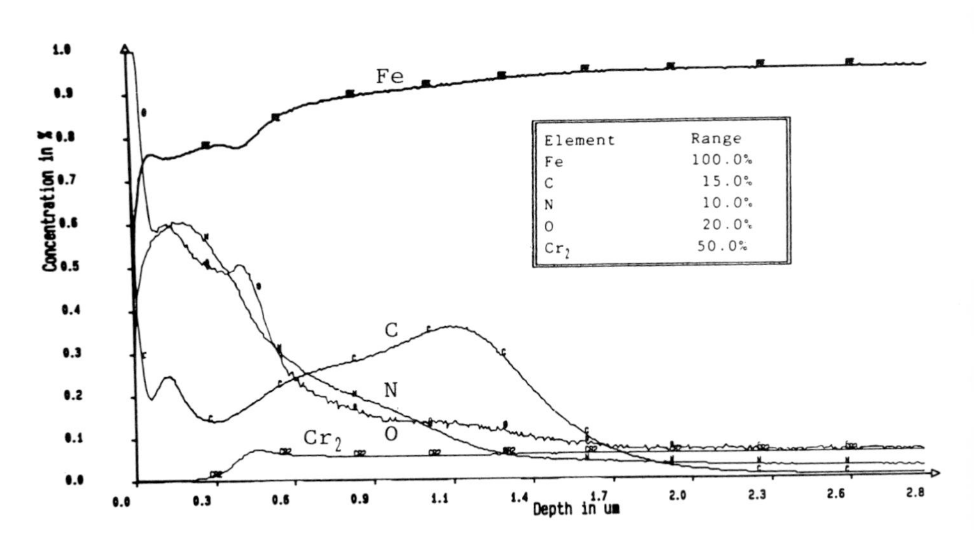

Figure 16 Depth concentration profiles of the elements across the plasma nitrocarburised layer on En 40B steel (treated with 87% N_2 + 5% CO_2 + 8% H_2), obtained from GDOES analysis

perpendicular to the surface, and oxygen may be dissolved in the pore channels that have direct access to the atmosphere. Also, the rôle of oxygen is particularly important in controlling the activity of carbon in the surrounding glow discharge. An increase in CO_2 gas level in the atmosphere generally increases the compound layer thickness, but in the presence of carbide and nitride-forming elements in the substrate the compound layer thickness is seen to decrease continuously as CO_2 is increased. Furthermore with plasma nitrocarburising, cementite formation is likely to initiate at the bottom region of the compound layer.

ACKNOWLEDGEMENTS

One of the authors (E. Haruman) is grateful to the Indonesian Government and the World Bank for financial support during the course of this work, and for the experimental assistance of Mr. J. Farmer with various aspects of the work.

REFERENCES

1. T. Bell, 'Survey of Heat Treatment of Engineering Components,' The Met. Soc., London, 1976.
2. H.C. Child, 'Surface Hardening of Steel,' The Oxford University Press, 1985.
3. G. Wahl and I.V. Etchells, Heat Treatment '81, The Met. Soc., London, 1983.
4. C. Dawes and D.F. Tranter, Heat Treatment of Metals, 1985, 3, 70.
5. C. Dawes, Heat Treatment of Metals, 1990, 1, 19.
6. J. Hadfield, MSc Thesis, University of Birmingham, 1986.
7. T. Sone, Trans. Japan Inst. Met., 1981, 22(4), 237.
8. A. Burdese, et al., Heat Treatment '84, The Met. Soc., London.
9. A.M. Staines, Heat Treatment of Metals, 1990, 4, 85.
10. T.H. Lampe, St. Eisenberg and G. Laudin, to be published 1992.
11. B. Edenhofer, Heat Treatment of Metals, 1974, 2, 59.
12. V. Korotchenko and T. Bell, Heat Treatment of Metals, 1978, 4, 88.
13. Hombeck, Heat Treatment '83 Shanghai, The Met. Soc., London, 1984.
14. A.M. Howatson, 'Introduction to Gas Discharge',

Pergamon Press, London, 1976.

15. A.C. Dexter , M.I. Lees and T. Farrell, Int. Conf. on Plasma Heat Treatment Sci. and Tech., PYC Edition, Paris 1987.
16. K.T. Rie and T.H. Lampe, Heat Treatment '84, The Met. Soc., London, 1984, p.33.1.
17. E. Haruman, PhD Thesis, University of Birmingham, 1992.
18. L. Petijean, et al., Proc. Conf. 4th Int. Colloquium on Plasmas and Cathodic Sputtering, Nice, 1982, p.183.
19. J. Slycke, et al., Scandinavian Journal of Metallurgy, 1988, 17, 122.
20. A. Wells, PhD Thesis, University of Liverpool, 1982.
21. M.A.J. Somers and E.J. Mittemeijer, Surface Engineering, 1987, 3(2).
22. A.M. Staines, T. Bell and H.W. Bergmann, Heat Treatment '84, The Met. Soc., London, p.48.1.

3.4.4
Plasma Spraying of Polymer Coatings

D. T. Gawne

DEPARTMENT OF MATERIALS TECHNOLOGY, BRUNEL, THE UNIVERSITY OF WEST LONDON, UXBRIDGE, MIDDLESEX, UK

1 INTRODUCTION

The polymer powder coatings industry in the UK currently has a market value of approximately £150 million per year and is one of the most rapidly expanding sectors of the engineering coatings industry. Powder coating involves applying a dry particulate polymer on to the component, usually by electrostatically spraying followed by stoving (∙180°C, 10 minutes) or fluidized bed dipping of preheated substrates. Applications include double glazing frames, motor components, pipes, bicycles, office equipment, refrigerators and cookers. Governmental legislation, health and safety regulations and consumer pressures are favouring powder coatings at the expense of paints. As well as avoiding the problems of emission caused by the solvents in wet paints, powder coatings also obviate the problem of waste disposal, particularly as the disposal of wet paint sludge is becoming increasingly expensive. Powder coating involves relatively little waste as stray powder or overspray can be collected and re-used.

Powder coatings are generally applied to low-cost, non-critical applications as the deposits contain significant porosity, limited adhesion and thickness control. This paper concerns the use of plasma spraying as an alternative method of applying powder coatings.

The technology of plasma spraying has been developed to a high degree of sophistication for metallic and ceramic coatings but very little attention has been given to polymers. The process has considerable potential for polymer deposition, however, because its high velocity, inert/reducing plasma is expected to reduce degradation and porosity while promoting adhesion. A major process advantage of plasma spraying is that it combines particle melting, quenching and consolidation into a single process. The work reported in this paper is directed at identifying the process conditions that enable polymer coatings to be deposited, comparing the behaviour of polymers in the plasma with that of ceramics and metals, and investigating the structure and mechanical properties of the resulting coatings.

2 PLASMA SPRAYING

A plasma is a state of matter where normally stable gaseous atoms are excited to form positive ions and free electrons. In plasma spraying, the plasma is formed by using a gas-constricted electric arc where temperatures of 10000°C to 15000°C are usually created. The high-temperature gas expands drastically and leaves the nozzle at velocities of 2000 to 3000 m s^{-1}. The coating material in powder form is injected, using a carrier gas into the plasma jet, where it is heated, accelerated and projected on to a substrate to form a coating. Depending upon torch design and energy input, the velocity of the particles will reach between 250 and 600 m s^{-1} in a high-energy plasma system. There are various designs of plasma-spray torch but they are all based on common principles and Figure 1 provides a schematic representation. The plasma gas is usually argon or nitrogen often with additions of hydrogen. The arc, which is initiated by a high-frequency, high-voltage discharge is struck between a water-cooled electrode (cathode) and a water-cooled copper nozzle (anode). Powder injection is usually provided by a single port external to the nozzle.

Plasma spraying is being used increasingly as a method of manufacturing metallic and ceramic coatings on engineering components. The coating thickness can be from 10µm to several millimetres and a high deposition rate (~10kg h^{-1}) is attainable with the process.

3 EXPERIMENTAL DETAILS

The coating powders used were nylon 11 supplied by Atochem UK Ltd. and polymethylmethacrylate supplied by ICI Paints plc. The substrate used was a plain carbon steel (080M40 grade) in the form of a plate of 6mm thickness. The steel was degreased and grit blasted with alumina grit (Metcolite C) under a blast pressure of 4 bar to give a surface roughness of 7µm R_a.

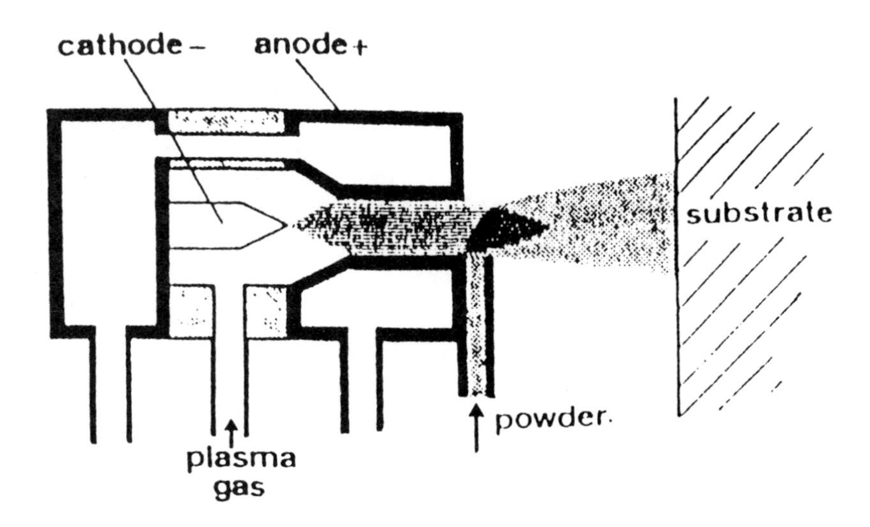

Figure 1 Schematic diagram of plasma torch.

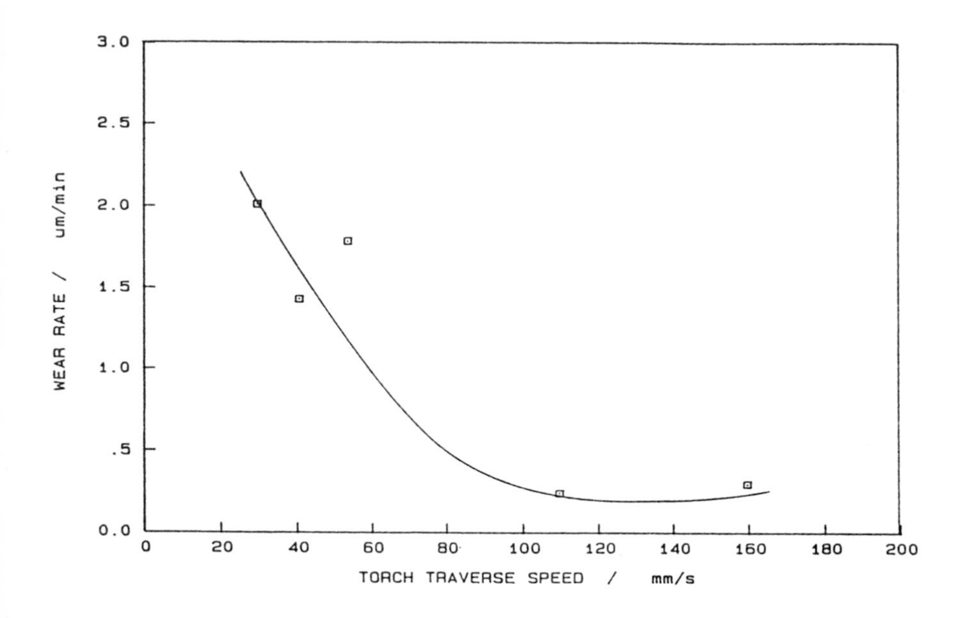

Figure 2 Effect of torch traverse speed on the wear of nylon 11 coatings in the ball-on-flat test.

Plasma spraying was undertaken using a Metco plasma spray system with an MBN torch, MCN control unit, 4MP powder feed unit and a fluidized hopper.

Wear performance was assessed using a reciprocating pin-on-flat machine with a stainless steel ball counterface of diameter 12.7mm. A load of 20N was applied to the steel ball, which slid at 50 cycles per minute over a track length of 30mm on the flat coated plate. Wear was evaluated by measuring the mean depth of the wear track on the polymer coating using a linear variable differential transducer.

4 DEPOSITION OF POLYMER COATINGS

A critical process parameter in the plasma spray deposition of polymer coatings is the speed at which the plasma torch moves across the substrate. The coating quality as measured by its wear rate deteriorates rapidly at slow traverse speeds as shown in Figure 2 for nylon 11. The slow traverse speeds produced deposits with relatively low melting temperatures, low enthalpies of fusion (Figure 3) and a relatively dark appearance, which indicate that significant degradation of the polymer had taken place. Similar results were obtained for PMMA coatings, where Figure 4 shows the effect of traverse speed on the mass loss of the coating.

The degradation is caused by overheating of the substrate since the slow traverse imparts much more thermal energy on a given area per pass of the plasma torch. The critical traverse speed above which no substantial degradation of the deposit takes place is a crucial parameter for plasma spraying of polymers. In this work, the critical speed was 100mm s^{-1} for nylon 11 and 200mm s^{-1} for PMMA indicating the lower thermal stability of the latter. However, the value will in general depend upon the spraying conditions used, such as the torch-to-substrate or stand-off distance.

The arc power, which is the product of the arc voltage and the arc current, is directly related to the thermal output of the plasma and is an important control parameter for the process. Figures 5 and 6 show the influence of arc power on the quality of nylon 11 coatings and PMMA coatings. Both materials show an optimal arc power: high wear rates are observed at both low and high power with a minimum in the range 21-28kW depending upon the polymer.

Examination of the coatings revealed major differences in macrostructure. Figure 7 shows a deposit

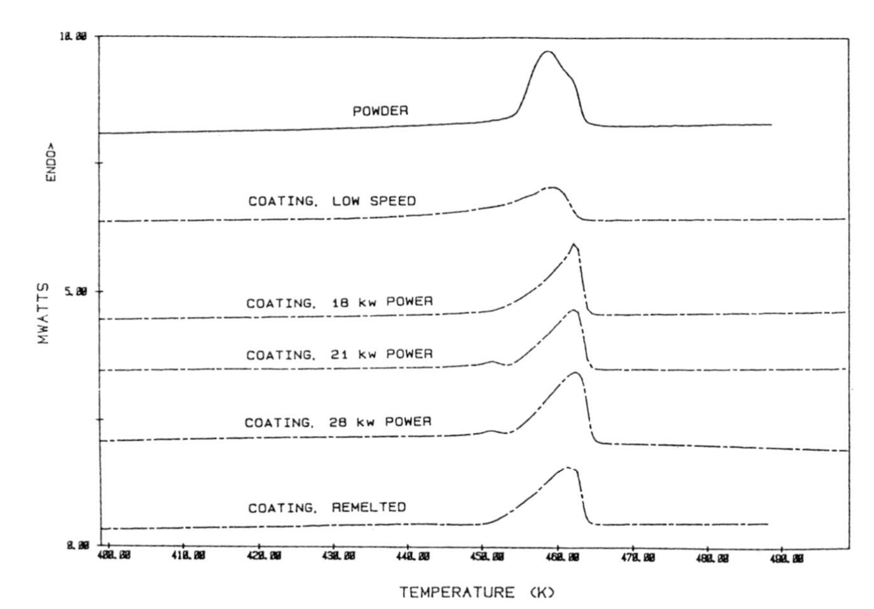

Figure 3 Differential scanning calorimetry traces for nylon 11.

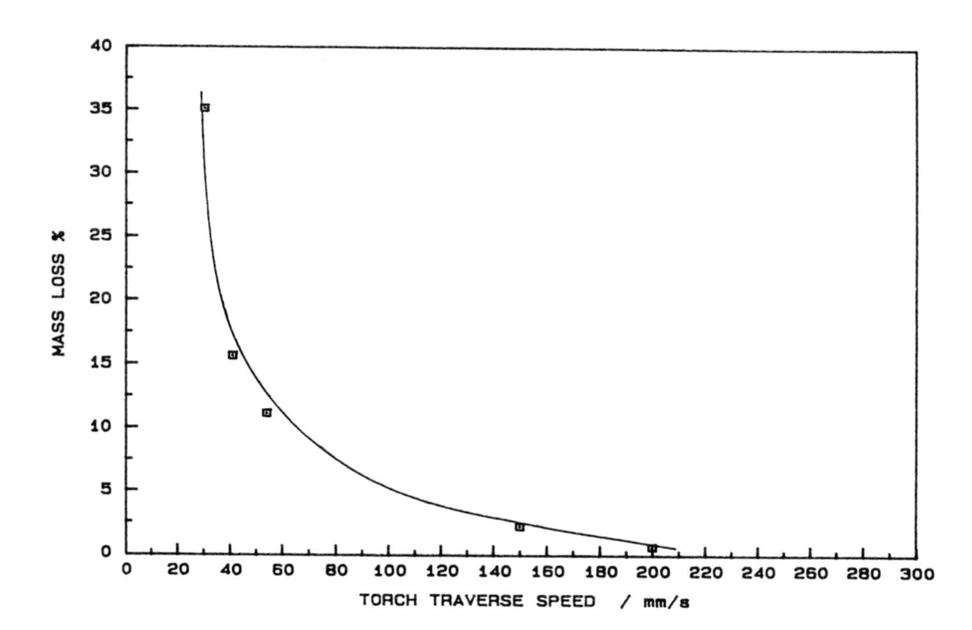

Figure 4 Effect of torch traverse speed on the mass loss of PMMA coatings.

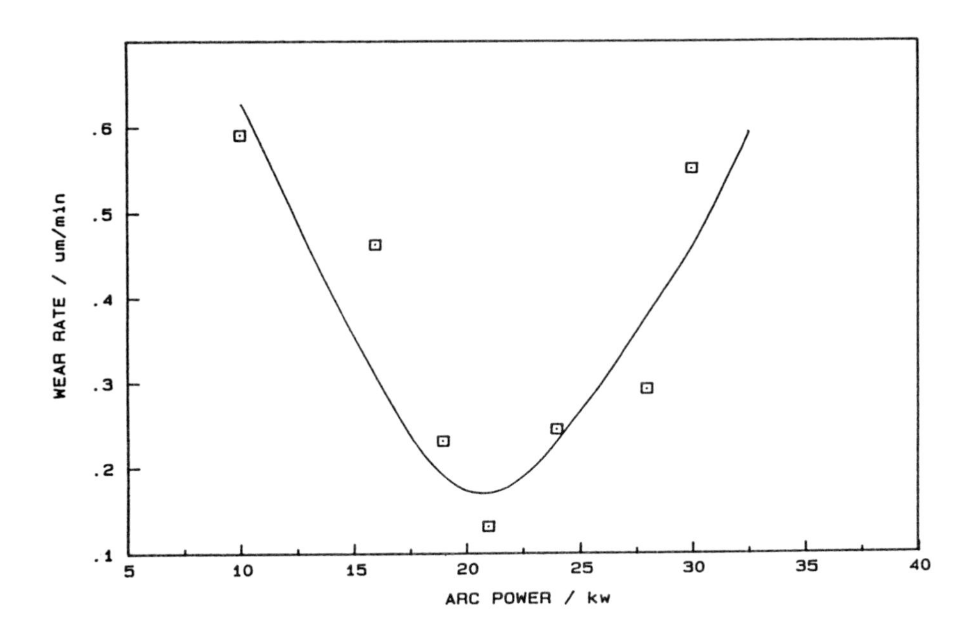

Figure 5 Effect of arc power on the wear rate of nylon 11 coatings in the ball-on-flat test.

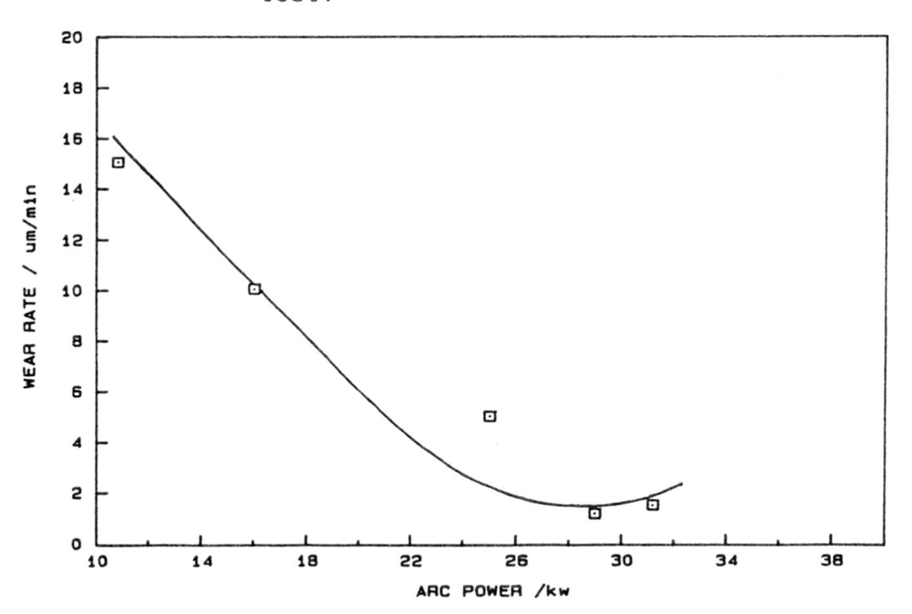

Figure 6 Effect of arc power on the wear rate of PMMA coatings in the ball-on-flat test.

produced under low (10kW) arc power: the consitituent particles are the same size and shape of the feed powder (Figure 8) indicating that only superficial melting of the particles has taken place. However, sufficient melting of the particle surfaces has occurred to enable them to bond together to form an aggregate, albeit with a high porosity due to the interstices between the particles.

Figure 9 shows through-thickness sections of nylon coatings sprayed at higher arc powers. Figure 9a is taken from a polished, etched cross-section and shows that the coating is composed of overlapping disc-shaped splats formed by the flow of the incoming spheroidal particles on impact with the substrate. Figure 9b gives a coating cross-section produced by fracturing in liquid nitrogen, which was deposited at 18kW, just below the optimal power. The extent of melting is now much greater but still incomplete. The particles labelled P on Figure 9 are only partially melted during deposition (as evidenced by their roughly equiaxed shape and similar size to the feed particles) while that labelled M denotes a fully melted particle which flowed on impact with the substrate to form a disc-like splat; the boundaries of the splat are indicated with arrows. The feature labelled I is an elongated void formed by entrapped gas in the interstices between the splats due to inadequate flow as the molten particles strike the substrate. The feature marked D is a spherical void possibly formed by gas evolution within the liquid particle.

A coating produced under the optimal arc power (21kW) is shown in Figure 10. The coating indicates a dense, coherent structure with much reduced incidence of voids or unmelted particles. Finally, Figure 11 is taken from a coating deposited under a high arc power (28kW). In this case, a high density of spherical (D type) voids are visible in a well-melted structure.

The arc power is related to the thermal energy output of the plasma jet. The higher the arc power, the greater the ability of the plasma to heat the injected powder feed. The sub-optimal power levels (<21kW) have insufficient thermal energy to melt completely the injected particles resulting in the formation of interstitial voids, as shown by I in Figure 9. The formation of these voids will be encouraged by low arc-power levels, since the associated low particle temperatures and high velocities will result in inadequate flow of the splats on impact with the substrate.

The formation of the spherical voids (e.g. D in Figure 9) is likely to have occurred in a molten, fluid liquid environment, where the observed minimum surface energy configuration can readily be achieved. Similar voids were observed in isolated splats generated by the

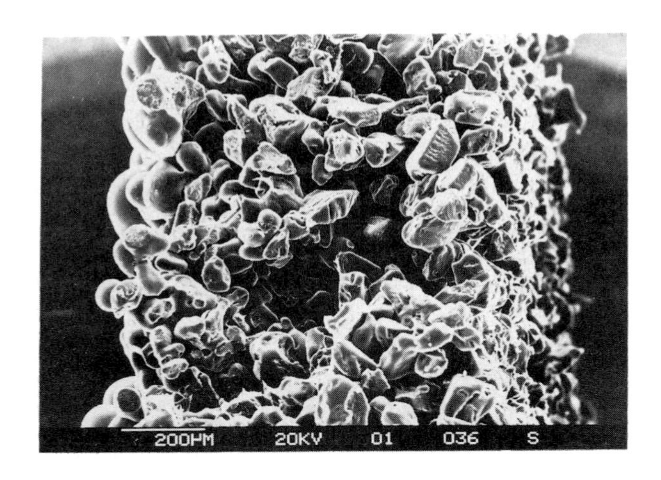

Figure 7 Scanning electron micrograph of a through-thickness section of nylon 11 coating produced under low arc power (10kW).

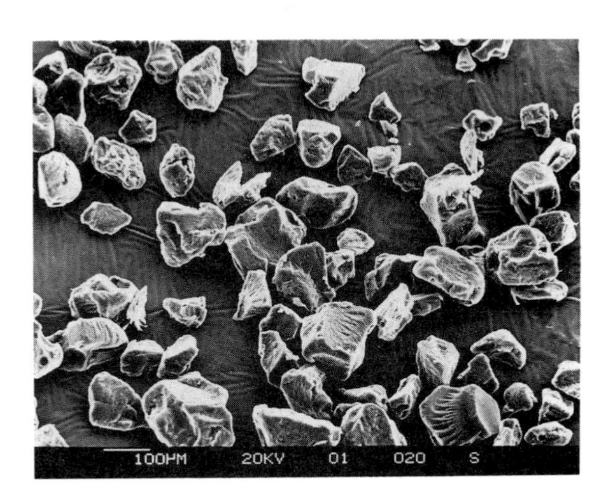

Figure 8 SEM micrograph of nylon 11 precursor powder.

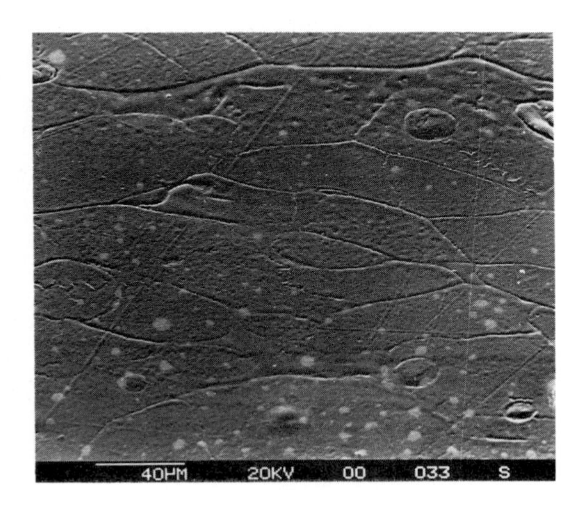

Figure 9(a) SEM micrograph of a polished through-thickness section of nylon 11 showing splat structure

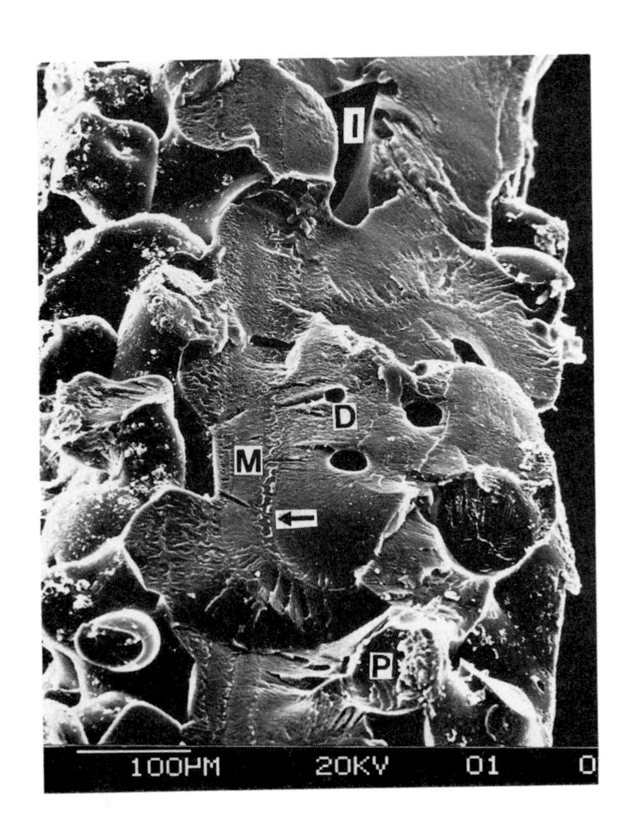

Figure 9(b) SEM micrograph of through-thickness section of nylon 11 coating produced under sub-optimal arc power (18kW). P denotes a partially melted particle, M a fully melted particle, I a void from entrapped gas in interstices between particles, D a spherical void in the interior of a melted particle.

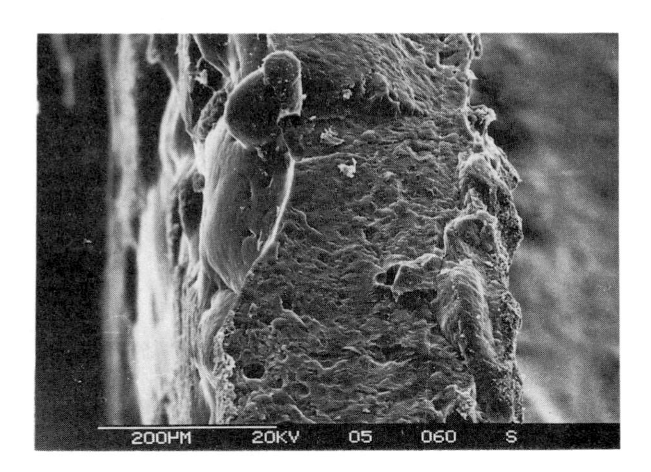

Figure 10 SEM micrograph of through-thickness section of nylon 11 coating produced under optimal arc power (21kW).

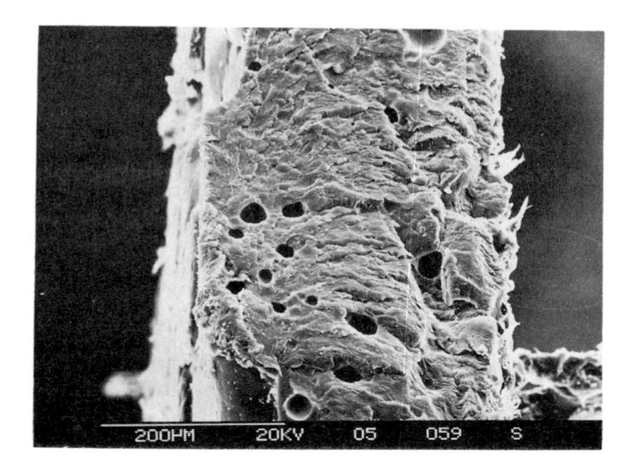

Figure 11 SEM micrograph of through-thickness section of nylon 11 coating produced under high arc power (28kW).

wipe test (Figure 12), in which the plasma torch is traversed so rapidly over a clean substrate that no overlapping occurs and separate splats are produced. This result and the fact that the voids in the coating are in the splat interior removed from the boundaries suggests that this type of void formation is due to nucleation within the polymer by thermal degradation.

The majority of the work reported in the literature on the degradation of nylon relates to nylon 6 and nylon 6.6 (1-4), and no systematic investigation of nylon 11 is available. Straus and Wall (2) have pointed out that the weakest link in the nylon chain is the C-N bond, since this has a significantly lower bond strength than the C-C bond (284 kJ mol^{-1} and 334 kJ mol^{-1} respectively). Degradation of nylon 6 and nylon 6.6 has thus been postulated (3,4) to take place by scission at the -$NH.CH_2$- group. It seems likely that a similar mechanism occurs in nylon 11:

$$-NH.CH_2(CH_2)_9CO-NH \,\vdots\, CH_2(CH_2)_9CO.NH-$$

with scission producing one fragment with a carbonamide end and another with an unsaturated hydrocarbon:

$$-NH.CH_2(CH_2)_9CO.NH_2 \ + \ CH_2.CH(CH_2)_8CO.NH-$$

The carbonamide group then splits off water to form a carbonitrile:

$$-NH.CH_2(CH_2)_9CO.NH_2 \rightarrow -NH.CH_2(CH_2)_9CN \ + \ H_2O$$

The water formed by this mechanism will be in the vapour state and capable of producing gas bubbles or voids in the coating. It can also hydrolyse the amide groups along the chain to give free amino and carboxyl end groups, which can interact in a number of ways to produce further gaseous products including carbon dioxide and water.

Similar behaviour was observed for the PMMA coatings: Figure 13a shows inadequate melting under low arc power and Figure 13b indicates the spherical voids generated by thermal degradation. In the case of PMMA, the gas bubbles or voids are formed mainly by thermal degradation by depolymerization leading to the gaseous monomer.

5 PARTICLE-PLASMA INTERACTIONS

The work clearly shows that polymers can be successfully plasma sprayed to produce sound coatings. It is not immediately clear, however, how such thermally sensitive materials can withstand plasma temperatures of 15000°C and why, for power levels below 21kW, these extremely high temperatures do not melt the polymer particles. It is therefore useful to consider the mechanisms by which heat is supplied to a particle and its consequent rate of

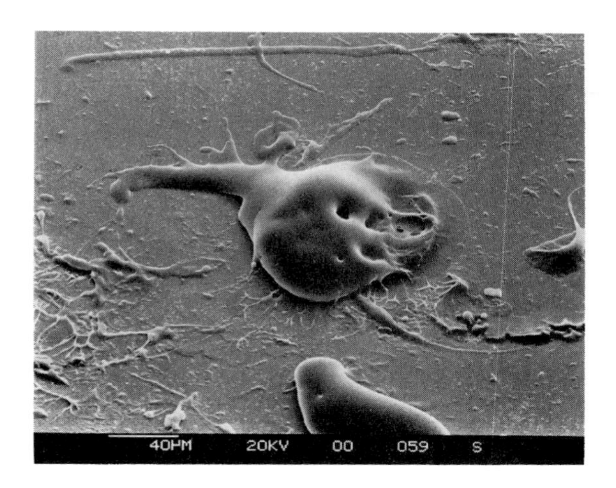

Figure 12 Voids within an individual splat of nylon 11 generated by the wipe test.

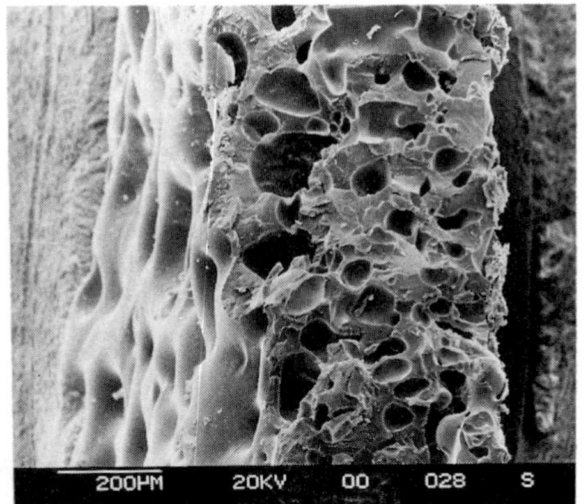

(a) (b)

Figure 13 SEM micrographs of through-thickness section of PMMA coatings (a) below optimal power (14kW), and above optimal arc power (31kW).

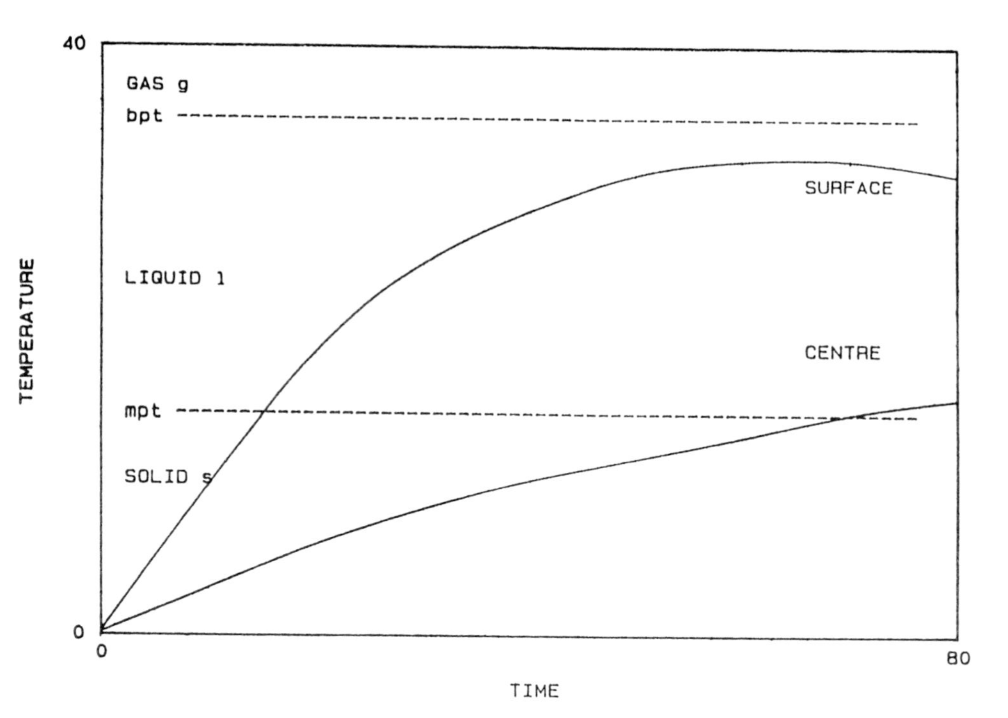

Figure 14 Schematic diagram showing the temperature rise of the surface and centre of a particle during flight in the plasma jet.

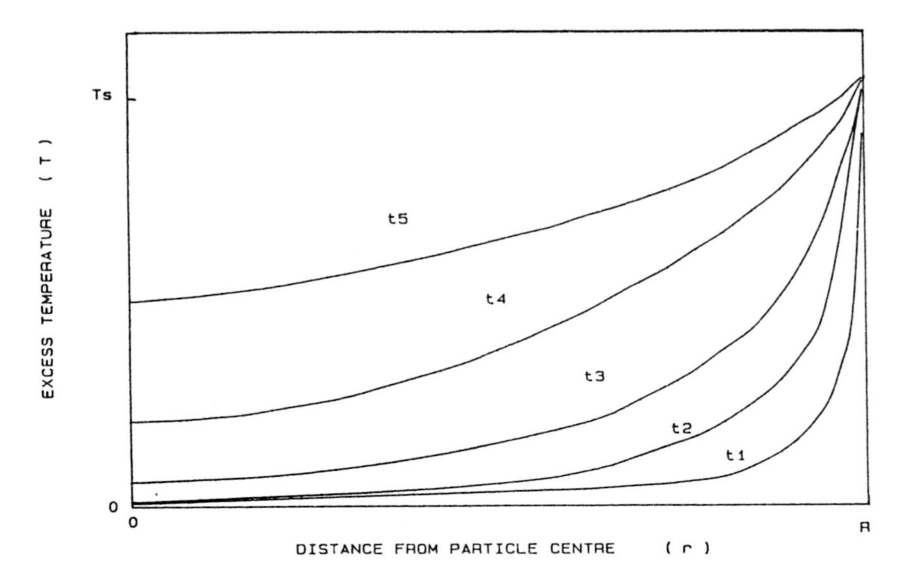

Figure 15 Schematic diagram showing the temperature rise within a spherical particle within a constant surface temperature T_s at r = R for a series of times increasing from t_1, to t_5.

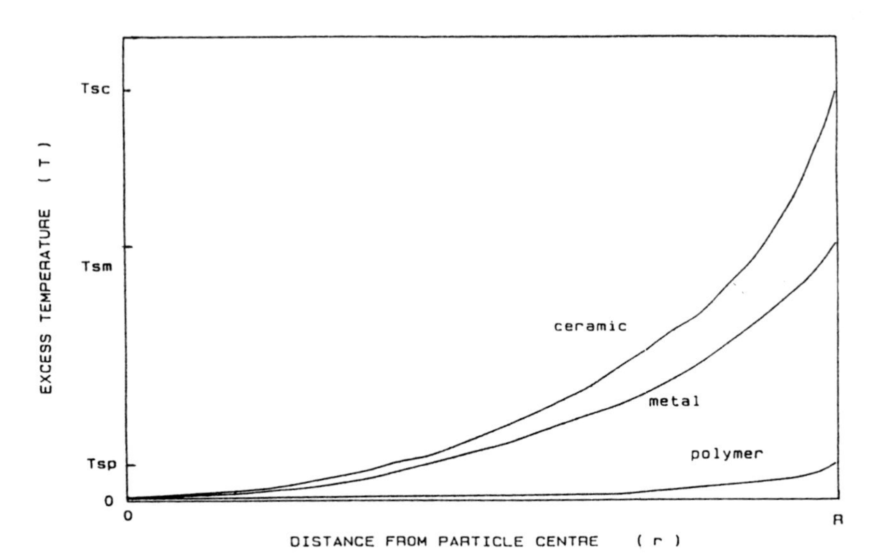

Figure 16 Schematic diagram showing the temperature profiles in polymer, metal and ceramic particles.

temperature rise.

A number of models have been developed to describe the temperature profile of a particle during its flight through the plasma jet. Most are based on heat conduction and transfer to a spherical particle. A recent model (5), for example, used the following basic heat transfer equation in the analysis:

$$\frac{\delta T}{\delta t} = \frac{1}{pCr^2} \frac{\delta}{\delta r} \left(K\, r^2 \frac{\delta T}{\delta r}\right) \qquad \text{........(1)}$$

where r is the radial distance from the centre of the particle and p, C and K are the density, specific heat and thermal conductivity of the particle respectively. T is the temperature at a time t. However, none of the models are able to predict particle temperatures from the basic process parameters such as arc current, gas composition and particle injection conditions, and are only able to predict trends in behaviour. Accordingly, a qualitative approach will be applied in this paper.

Figure 14 shows the temperature rise at the surface and centre of a spherical particle in the plasma. The temperature at the particle centre is limited by heat transfer and rises much more slowly than that at the surface. Melting is dominated by the rate of heat transfer into the interior of the particle. Figure 15 shows how the temperature profile changes with time within the particle with a constant surface temperature T_s. T_s has a major role in determining the temperature gradients within the particle and the resulting heat transfer rates. The value of T_s is related to the decomposition temperature of the material: approximately 440°C for nylon 11, 2700°C for nickel and 4500°C for alumina. As a result, the temperature gradients will increase in the same order as shown in Figure 16. The relatively low thermal conductivities of polymers (e.g. 0.29, 92 and 1.0 $Wm1^{-1}K^{-1}$ for nylon 11, nickel and alumina respectively) reinforce their low temperature gradients and produce low heat transfer rates. Thus, the extreme outer layers of the polymer particle are continually being removed by thermal degradation during its short flight time in the plasma jet but this acts to protect the interior of the particle and hence the material of the subsequent coating. Local cooling of the plasma gas takes place surrounding the polymer particles and a critical heat content equivalent to an arc power of 21kW is required to ensure melting of the particles. A high temperature is thus not the only condition and a sufficiently high enthalpy is also required.

6 COATING PROPERTIES

The properties of a plasma sprayed coating are dependent upon the state of the particles when they impinge onto the substrate. Ideally, all particles should be molten

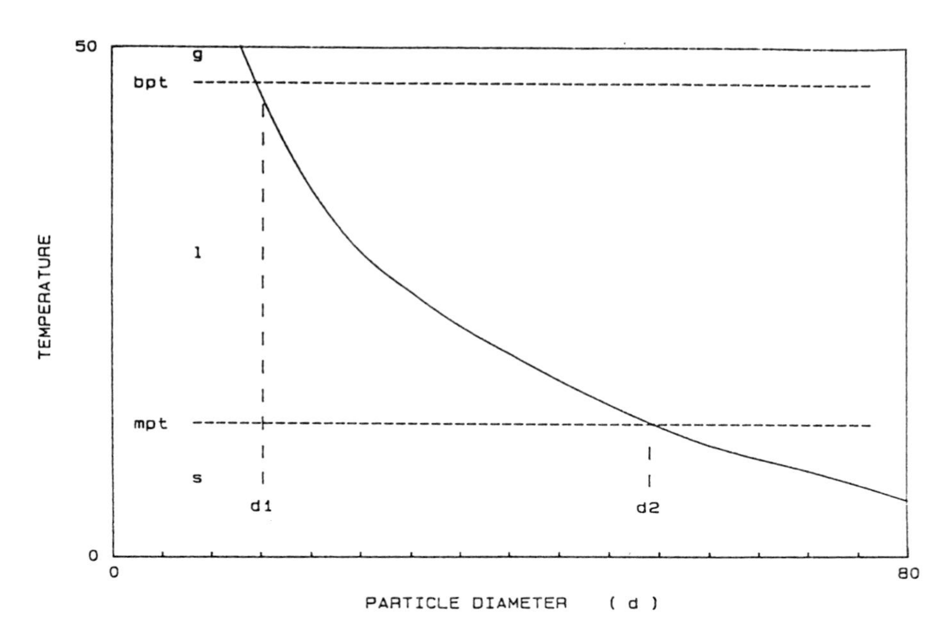

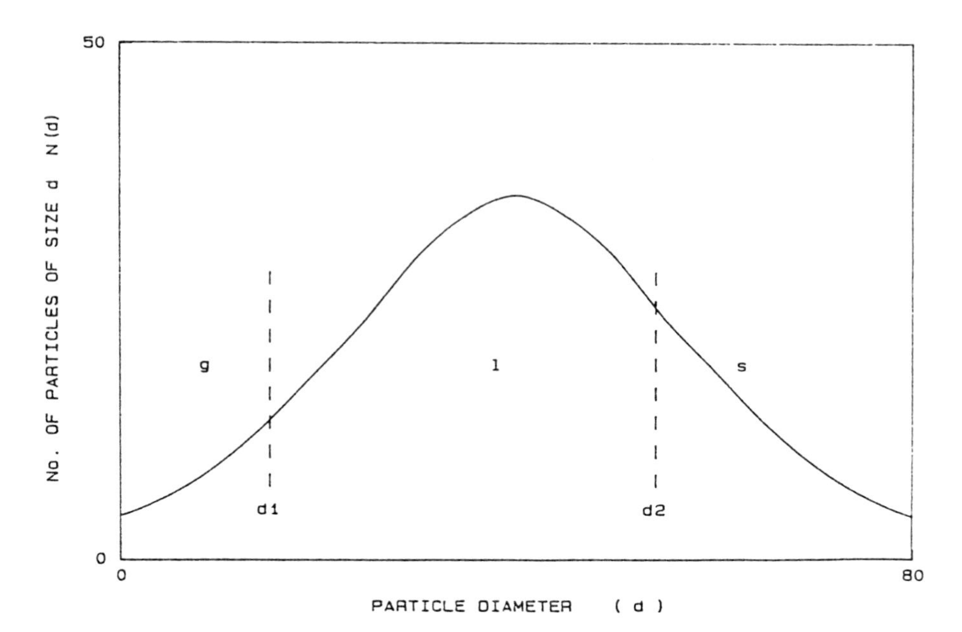

Figure 17 Schematic diagram showing the effect on the state of the particle of (a) particle diameter, and (b) particle size distribution.

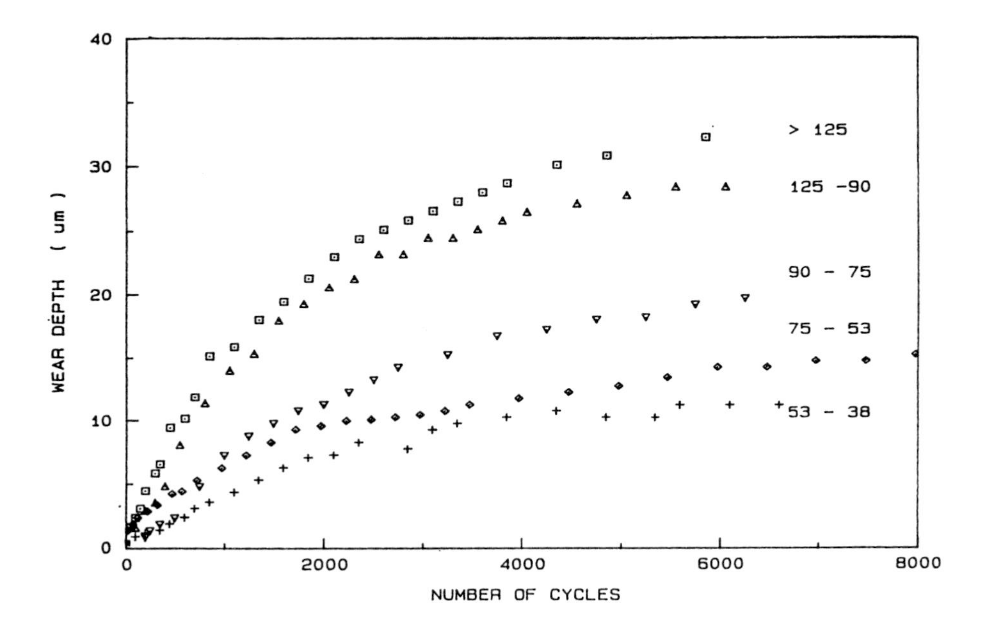

Figure 18 Effect of size classification on the wear of nylon 11 coatings.

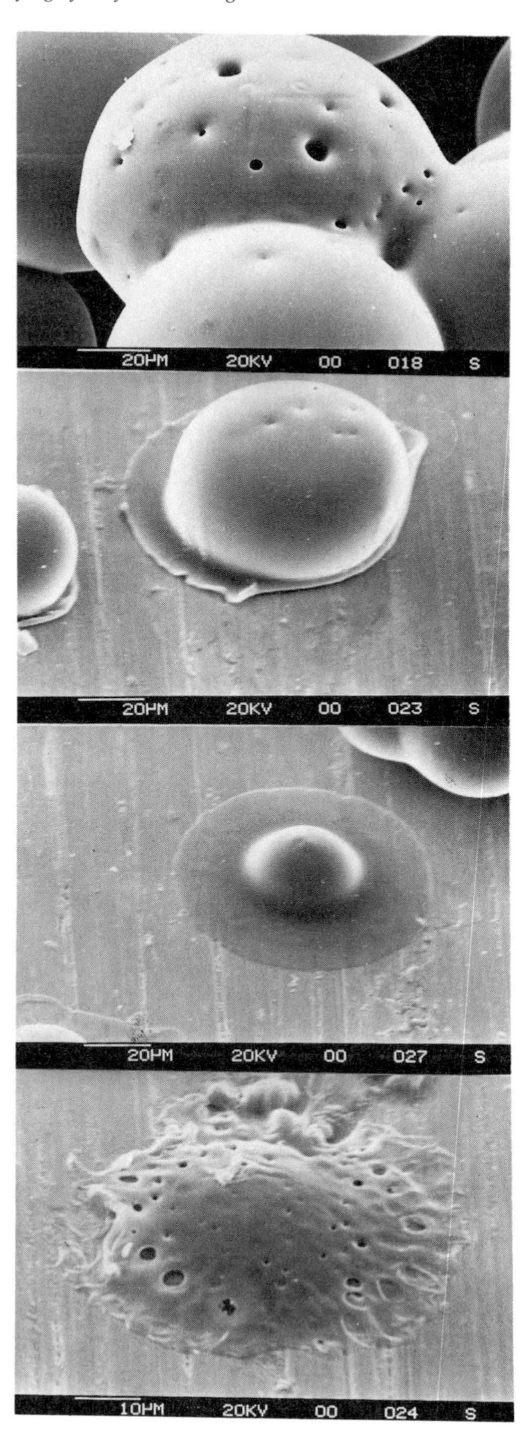

Figure 19 Effect of PMMA particle size on particle melting and impact flow in the wipe test.

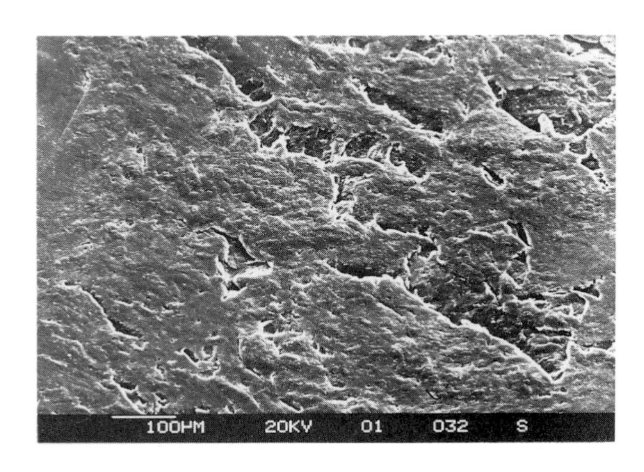

Figure 20 SEM micrograph showing nylon surface after wear against stainless steel counterface.

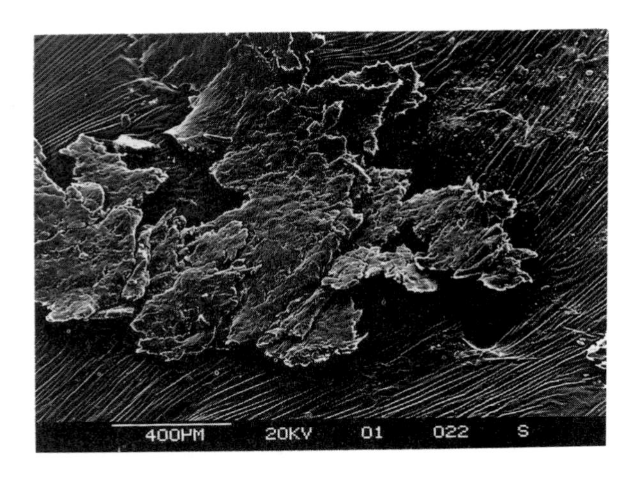

Figure 21 SEM micrograph showing debris particles generated by wear of nylon against stainless steel counterface.

on impact in order to maximize inter-particle contact and reduce porosity, although in practice this is rarely achieved. Figure 17a shows how the state of a particle just before impact is influenced by particle size and Figure 17b, the consequent tendency for a particle size distribution to generate a range of particle impact states. Figure 18 shows that size classification of the precursor powder to smaller sizes has a substantial benefit on deposit quality. A similar affect was observed with PMMA and Figure 19 indicates that the influence of particle size originates from the degree of particle melting and impact flow.

The wear behaviour of the coatings was followed using the steel ball-on-flat test. Transferred films of nylon were found on the stainless steel counterface indicating that adhesive wear is a major material-removal mechanism. Figure 20 shows that the worn nylon surface contains shallow pits from which material has been pulled out and transferred across the sliding interface. The debris particles collected during the wear test (Figure 21) consisted of flat platelets of polymer of similar dimensions to the splats making up the deposit. It appears likely that the nylon adheres to the sliding counterface and fractures along the underlying splat boundary to generate a debris particle. However, considerable plastic (irreversible) deformation of the polymer has also occurred and the appearance of grooves parallel to the sliding direction indicates that abrasive wear takes place: asperities from the harder steel counterface plough out the nylon surface.

The above wear mechanisms involve elastic-plastic deformation followed by fracture to produce a debris particle and generate a unit of wear. The ductility and resistance to fracture of the deposits will play a crucial role. The results in this work show that the macrostructure in terms of voids and unmelted particles has a major effect on deposit quality in addition to the microstructure, and that a degree of control can be achieved by manipulation of the plasma spraying process parameters.

7 CONCLUSIONS

1. Polymer coatings can be deposited by plasma spraying.

2. The coating quality deteriorates drastically for torch traverse speeds below 100-200 mm s^{-1} depending upon the polymer.

3. There is an optimal arc power for maximum deposit quality: lower power levels give incomplete melting and higher levels cause thermal degradation.

4. The macrostructure has a marked effect on the

coating properties, particularly unmelted particles and voids. The voids were caused by shadowing by unmelted particles, inadequate splat flow, gas entrapment or thermal degradation.

5. Size classification of the precursor polymer powder is shown to provide a major benefit to coating quality.

6. Adhesive transfer is a major wear mechanism in nylon 11 coatings with fracture occurring along splat boundaries.

7. Control over the structure and properties of nylon 11 and PMMA coatings can be achieved by manipulation of the plasma spraying process parameters.

ACKNOWLEDGEMENTS

The work has been carried out with the support of the Procurement Executive Ministry of Defence, the Science and Engineering Research Council and Metco Ltd. The authors would like to thank the above for permission to publish this paper.

REFERENCES

1. B.G. Achhammer, F.W. Reinhart and G.M. Kline, J. Res. Nat. Bureau Standards, 46 (1951) 391; J. Appl. Chem., 1 (1951) 301.

2. S. Strauss and L.A. Wall, J. Res. Nat. Bureau Standards, 63 (1959) 269.

3. B. Kamerbeek, G.H. Kroes and W. Grolle, in 'High Temperature Resistance and Thermal Degradation of Polymers', Soc. Chem. Indust. Monograph 13 (1961) p357.

4. G.W. Harding and B.J. McNulty in 'High Temperature Resistance and Thermal Degradation of Polymers', Soc. Chem. Indust. Monograph 13 (1961) p.392.

5. E.Pfender, Surf. Coating Technol., 34 (1988) 1.

3.4.5
Large Area Diamond-like Carbon Coatings by Ion Implantation

A. R. McCabe, G. Proctor, A. M. Jones, S. J. Bull, and D. J. Chivers

AEA INDUSTRIAL TECHNOLOGY, HARWELL LABORATORY, DIDCOT, OXFORDSHIRE OXI I ORA, UK

1 INTRODUCTION

Diamond-Like Carbon (DLC) is an attractive coating material because of its combination of low friction, high hardness (2000 - 3000 VHN) and chemical inertness. DLC coatings have been deposited on large complex geometry components in Harwell's Blue Tank ion implantation facility. A 40 keV nitrogen ion bucket ion source is used, both to modify the substrate surface, and to crack a low vapour pressure oil which is evaporated and condensed onto the substrate surface. Coating areas up to 1 metre diameter are common, with larger areas possible if coated using component manipulation. Coating of both large size individual components and large batches of small size components has been demonstrated.

Unlike the situation during many CVD / PVD techniques, the components never exceed 80 °C during the coating process, thus permitting substrates of a wide range of materials to be coated, including specialist tool steels and even certain high density polymers.

The use of various oil precursors has been investigated, these permitting the production of hard wear resistant coatings with extremely low coefficients of friction (0.02 - 0.15), along with a range of mechanical and electrical properties. The production and assessment of such coatings is presented, including measurements of their tribological performance.

Applications for wear resistance, corrosion protection and electrically conducting coatings are discussed, with examples from the engineering, electronic and biomedical industries.

The production of Diamond Like Carbon coatings has been demonstrated by a number of organisations, each producing coatings with certain interesting optical, electrical or mechanical properties. The techniques used to produce the coatings are practically as varied as the properties of the resultant coatings, all having their limitations. The deposition rate in direct ion beam deposition[1] and dual ion beam sputtering[2] are too low to be considered for industrial use, while it is hard to see laser ablation[3] being modified to permit large area deposition. The d.c.[4] and r.f.[5-7] plasma CVD methods involving the decomposition of a carbonaceous gas such as methane require substrate temperatures in the range 300 to 600°C. This clearly rules out their use for coating temperature sensitive materials such as polymers and low tempered steels.

The production of DLC at temperatures under 100°C has been developed and patented[8], and involves the cracking of a low vapour pressure oil precursor by a large area ion beam from a bucket ion source[9]. This method allows the coating of a wide range of substrate materials at commercially economic deposition rates and large surface areas. The coating area is merely limited by the size of the vacuum chamber and the manipulation in that chamber to move components through a 1 metre diameter processing zone. Further the mechanical, electrical and optical properties of the coatings can be modified by the selection of the process parameters and the use of a wide range of oil precursor additions.

2 THE PRINCIPLE

The principle of the Harwell ion implanted DLC process[10] is illustrated in the schematic, Figure 1.

The oil precursor, typically polyphenyl ether, is placed in an oil bath. This has a right-angle bend, which forms a wide directional nozzle through which the evaporated oil may escape. This nozzle is also heated electrically to prevent recondensation of the oil at this point.

At a bath temperature of around 140°C, and at a pressure of 5x10-6 torr, oil vapour emerges from the nozzle in a wide jet, and condenses onto any surface whose temperature is less than this. Consequently, a flux of oil vapour is continually condensing onto the substrates, mounted on a rotating table. Simultaneously, a 40-80 keV beam of mostly molecular nitrogen ions is impinging on the substrates.

As an energetic ion penetrates its way through the surface layer of condensed oil, it imparts much of its energy to the oil molecules, mostly by electronic excitation[11]. This results in a dramatic number of bond breakages, the release of volatile species such as oxygen and hydrogen and the solidification of the local area from the subsequent bonding of unsatisfied carbon atoms[12].

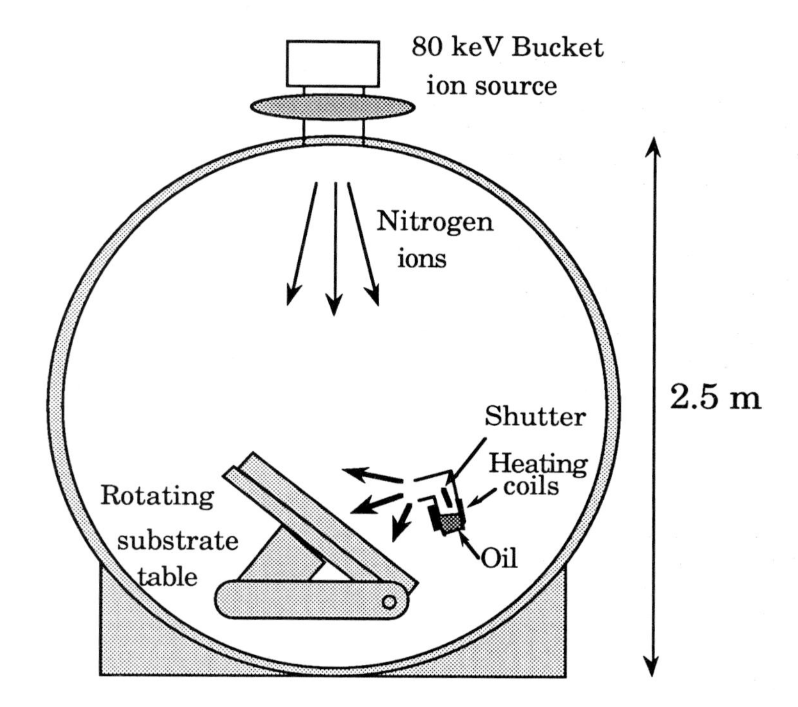

Figure 1 Harwell's Blue Tank Ion Implantation Facility

Adjusting the rate of oil deposition and ion bombardment allows control of the composition of the final material left deposited on the substrates, from a semi-solid hydrocarbon material to dense DLC, without the need for any substrate heating.

The use of a high energy ion beam in the DLC coating equipment has a number of subsidiary benefits:

(a) The ion beam can be used to sputter clean the substrates prior to coating;
(b) The substrates, often metals, can be implanted with a high dose of nitrogen to provide a hardened surface to deposit the coating onto;
(c) Early on in the deposition of the DLC coating, energetic ions are traversing the interface between coating and substrate. Ion beam mixing and ion beam stitching will occur, providing enhanced adhesion between the coating and substrate.

The DLC coating material can be deposited over the full 1 metre diameter of the ion beam at the substrate table, and replacing this table with specialist manipulation, larger areas can be coated by moving components through the coating area.

3 PRODUCTION OF A RANGE OF DLCs

The standard ion implanted DLC process has a number of variable process parameters with which to modify the nature of the coating deposited. In addition, the use of a range of low vapour pressure precursor oils has been studied.

Initial development concentrated on the use of polyphenyl ether, which has the empirical formula $[C_6H_4O]_n$ (where n=5 on average). The structure of this is illustrated in Figure 3. This form of DLC ('standard DLC') has been successfully applied to a range of substrates, including various polymers, glass, and a range of metals and alloys. Large area DLC coatings have been applied to polyimide sheet and steel process rollers.

The use of silicone precursor oils to produce silicated DLCs, and therefore potentially low friction and low wear rate coatings prompted trials of a number of precursor oils. All oils were initially assessed for vacuum compatibility down to 10^{-7} mbar, along with health and safety assessment of their decomposition products. A 'low' silicon containing precursor, pentaphenyl trisiloxane and a 'high' silicon precursor, octamethyl cyclotetrasiloxane (OMCTS) were selected for further trials. The chemical structure of these are given in Figures 2 and 4. A range of coatings have been produced using the trisiloxane, but the OMCTS was deemed unsuitable, and so trials were discontinued. The silicon content of the silicated DLC was successfully altered by modification of the ion bombardment, and hence the degree of cracking of the hydrocarbon chains. This enables varying silicon content DLCs to be produced from the one precursor. The effect of this method of control on other properties has subsequently been investigated.

The search for DLCs of lower friction coefficient and higher optical transmission and resistivity suggested that, although silicated DLCs were an improvement over the standard DLC, the use of fluorinated precursors should be investigated. Initial trials using a fluorinated precursor have produced such modified DLC coatings. Further work is required on these coatings before they can be regarded as properly characterised and their benefits fully assessed.

4 COATING ASSESSMENT

Microstructure. The microstructure of DLC has been studied by Raman and TEM. These techniques have confirmed the microstructure to be of fully disordered (amorphous) carbon, with no islands of graphitic or diamond structure. The density is virtually half way between that of graphite and diamond.

Composition. SIMS depth profiles have been obtained of both standard and silicated DLC coatings. These indicate

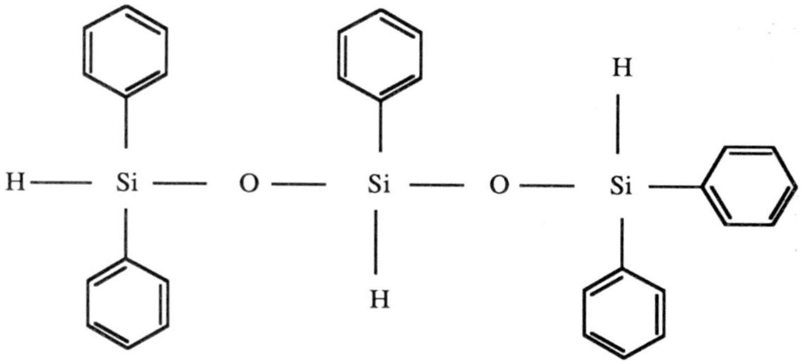

Figure 2 Pentaphenyl trisiloxane

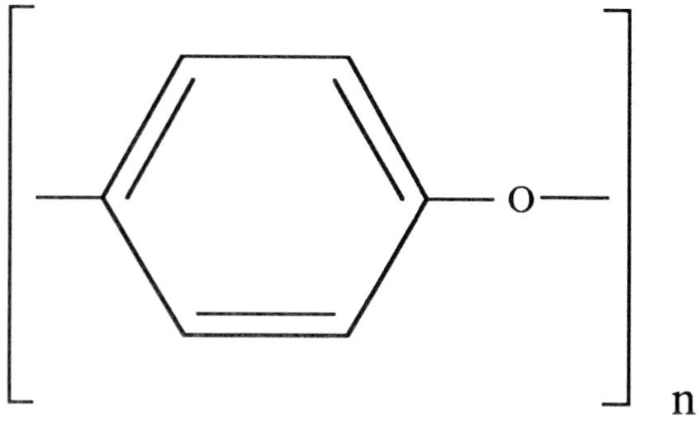

Figure 3 Polyphenyl ether (n~5)

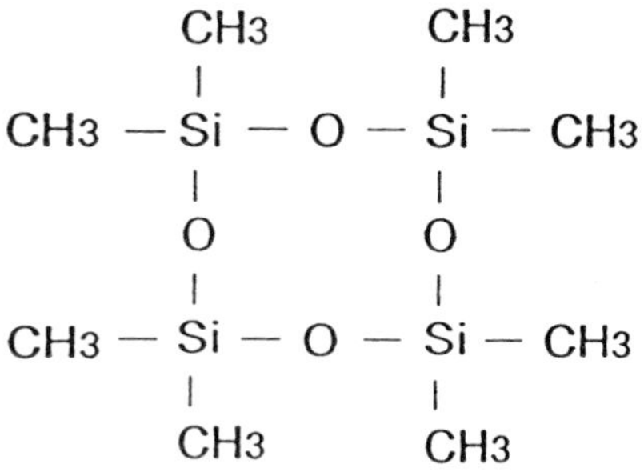

Figure 4 Octamethyl cyclotetrasiloxane (OMCTS)

relatively high hydrogen contents in each. The silicon content of the silicated DLC is approximately twice that of the oil precursor used - this increase being expected as a result of the sputtering of lighter constituents on cracking of the oil by ion bombardment.

Adhesion. The adhesion of DLC has been found to be partially substrate dependent, such that differing surface pretreatments are used to optimise the adhesion for any particular substrate. In some cases good adhesion can only be obtained after both a short substrate ion bombardment and a selected chemical treatment have been performed.

The adhesion to non-temperature sensitive substrates can be improved in some cases by preheating in situ to drive off weakly bound contaminants. This does not always give improvement, mainly due to the removal of these surface contaminants during a pre-coating ion bombardment. An example of this is illustrated in Figures 5a and 5b. Figure 5a shows the scratch test results for 1 μm standard DLC on tool steel. In this case a preheat does increase the adhesion. However in the case of 1 μm silicated DLC (again on tool steel), the adhesion is good anyway such that on preheating no improvement is observed.

Friction. The coefficient of friction between diamond and DLC is generally between 0.02 and 0.15, depending on the type of DLC and the load applied. Only for loads in excess of 4 kg is there a dramatic increase in friction, this being due to the destruction of the coating. This is illustrated in Figure 6.

Hardness. As the coatings are generally thinner than the indentation of the microhardness equipment, a computer model is used to extract the coating hardness from that of the coating/substrate composite[13]. Even after this compensation, the hardness is found to be substrate dependent, with a harder coating forming on a harder substrate. Typical values are 1000 VHN for DLC on mild and stainless steel, and 2000 VHN for DLC on tool steel. This in particular strengthens the case for ion bombarding softer materials prior to coating deposition. A slight increase in hardness with coating depth was also noticed, this being due to a change in the residual stress.

Residual stress. Residual stress is found to change with coating thickness, and indeed its nature can transform from tensile to compressive. This is illustrated in Figure 7. The residual stress can further be modified by altering the level of ion bombardment.

Resistivity. Resistivity is related to the structure of the coating. Generally the denser the coating, the higher the resistivity. A low density is more graphitic and so more conducting. The presence of silicon should also increase the resistivity. Four point probe resistivity measurements

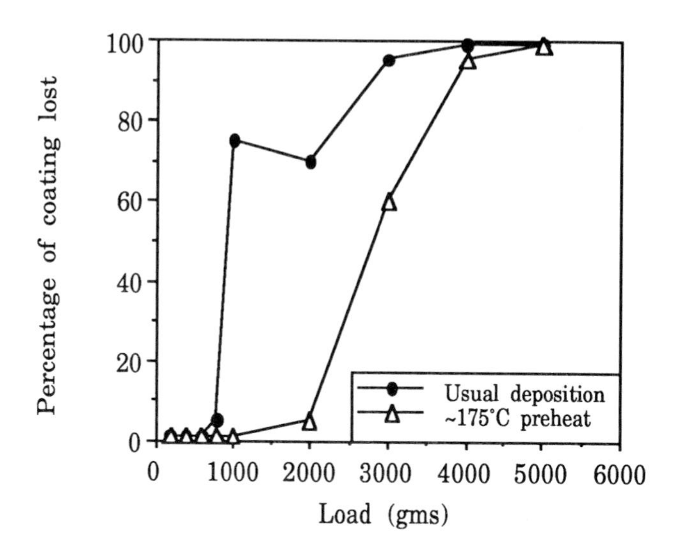

Figure 5(a) Scratch test results - 1 μm standard DLC on tool steel

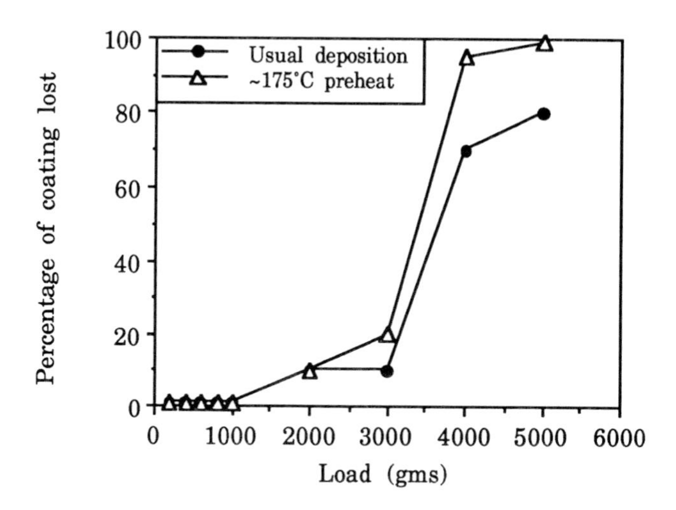

Figure 5(b) Scratch test results - 1 μm silicated DLC on tool steel

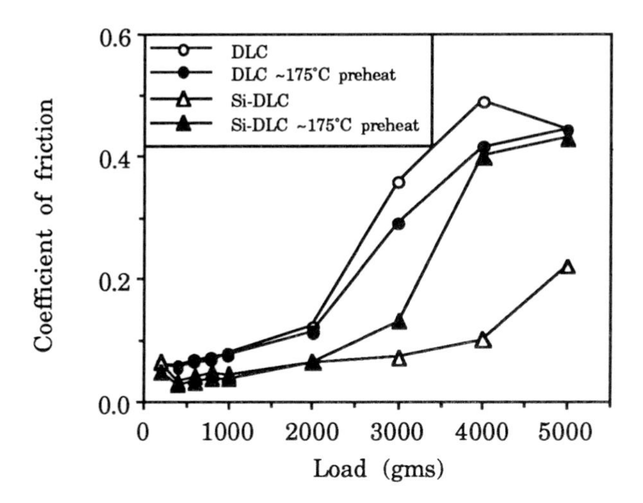

Figure 6 Friction of diamond on DLC coated tool steel (unlubricated)

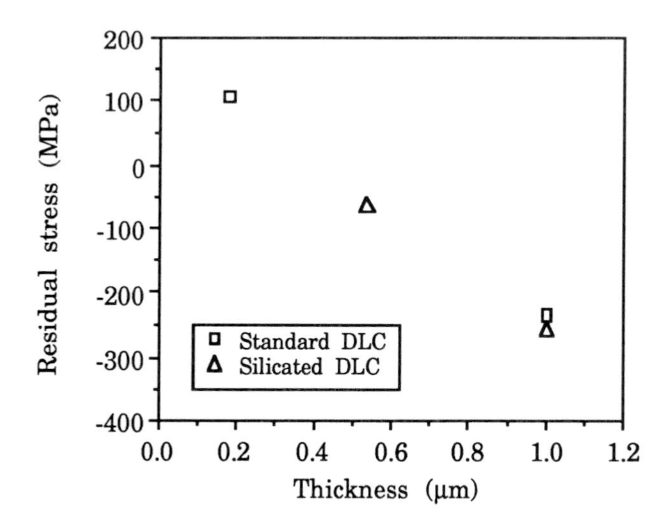

Figure 7 Residual stress vs. coating thickness of DLC on glass

show that this is the case:-

Standard DLC - resistivity	=	1.9	ohm-cm
Silicated DLC - "	=	110.0	"

Optical. Although ion bombarded DLC was initially produced in thick films for its wear properties, the use of the material for thin coatings is under investigation. It is clear that the use of silicated (and possibly fluorinated) precursors can modify the transmission of light of various wavelengths, possibly selectively. This is worthy of further investigation.

5 APPLICATIONS

DLC coatings have been applied to a range of components made from metallic, polymeric and glassy materials. The amorphous coatings can be used to protect polymers from solvent attack and show good resistance to aqueous corrosion.

Thick coatings (approx. 1 μm) are suitable for wear resistance and corrosion resistance, both in engineering and biomedical applications. The low coefficient of friction is of interest in engineering applications - either by itself or as an overcoat for TiN, while the biocompatibility of DLC is vital in its selection for trials for a number of orthopaedic applications. The largest component coated to date has been a cylindrical steel component 400 mm in diameter and 1.5 metres long and weighing just under two tons. This used the full dimensions of the Harwell facility (2.5 m diameter x 2.5 m long) in its manipulation. There is, however, no reason why larger components could not be coated given a larger vacuum chamber.

Thin coatings (<0.1 μm) are being assessed for their combination of electrical and optical properties and abrasion resistance. For electrical applications, the range of resistivity available from the various DLCs has the potential to allow the selection of coatings dependent on the coating specification. Depending on the oil precursors used, work is being considered to produce DLC which has a selective effect on the transmission of various frequencies from infra-red, through visible to ultra violet. The application of DLC to large areas of polymers such as polyimide has been demonstrated.

6 CONCLUSIONS

DLC coatings have been produced by an ion implantation technique. These have a number of advantages over DLC produced by other techniques. These include :-

* Deposition at substrate temperatures not exceeding 100°C, so permitting the coating of polymers and temperature

sensitive tool steels.

* The coating of large areas has been demonstrated, and the potential for scale-up is technically relatively simple to achieve.

* A range of mechanical, electrical and optical properties are possible both by modification of the process parameters, and the use of a range of precursors. The structure and properties of the DLC coatings are critically dependent on the level of ion bombardment.

* Applications are under assessment in the engineering, optical, biomedical and electronic industries.

ACKNOWLEDGEMENTS

This work is part of the long term Corporate Research Programme of AEA Technology.

REFERENCES

1. S. Aisenberg and R. Chabot, J. Appl. Phys., 1971, 42, 2953.
2. C. Weissmantel, K. Bewilogua, H-J. Erler and G. Reisse, Proc. Conf. Ion Plating & Allied Techniques, London, 1979 (CEP Consultants, Edinburgh, 1979) p. 272.
3. S.S. Wagal, E.M. Juengerman and C.B. Collins, J.Appl. Phys, 1988, 53, 187.
4. D.S. Whitmell and R. Williamson, Thin Solid Films, 1976, 35, 255.
5. J.C. Angus, H.A. Will and W.S. Stanko, J. Appl. Phys., 1968, 39, 2915.
6. L. Holland and S.M. Ojha, Thin Solid Films, 1979, 58, 107.
7. L.P. Anderson, S. Berg, H. Norstrom, R. Olaison and S. Tonita, Thin Solid Films, 1979, 63, 155.
8. P.D. Goode, W. Hughes and G.W. Proctor, 'Ion beam carbon layers', UK Patent No. GB 2122224 B (1986).
9. W.L. Stirling, P.M. Ryan, C.C. Tsai and K.N. Leung, Rev. Sci. Instrum., 1979, 50, 102.
10. C.J. Bedell, A.M. Jones and G. Dearnaley, 'Applications of Diamond Films & Related Materials', Elsevier, 1991, p.827-831.
11. J.F. Ziegler, J.P. Biersack and U. Littmark, 'The Stopping and Ranges of Ions in Solids', Pergamon, Oxford, 1985.
12. M.L. Kaplan, S.R. Forrest, P.H. Schmidt and T. Venkatesan, J. Appl. Phys., 1984, 55, 732.
13. S.J. Bull and D.S. Rickerby, Surf.Coat.Technol. 1990, 42, 149.

Part 4: Surface Analysis

Section 4.1 Corrosion Studies

4.1.1

Analysis of Non-conducting Coatings Using Glow Discharge Spectrometry (GDS) – Analysis of Surfaces and Interfaces – Technical Developments in Analysing Non-conducting Materials

J. L. Baudoin,[1] M. Chevrier,[1] P. Hunault,[2] and R. Passetemps[1]

[1] REGIE NATIONALE DES USINES RENAULT, B.P. 103, 8–10 AVENUE EMILE ZOLA, 92109 BOULOGNE BILLANCOURT, FRANCE

[2] INSTRUMENTS S.A., DIVISION JOBIN YVON, B.P. 118, 16–18 RUE DU CANAL, 91165 LONGJUMEAU CEDEX, FRANCE

1 INTRODUCTION

The field for GDS applications has until now been limited to the analysis of surfaces and coatings on electrically conducting materials.

A new glow discharge source developed and patented by RENAULT and marketed by ISA JOBIN YVON, has now made it possible to extend the field of applications to non-conducting materials. Typical application areas include:

Automobile Industry

- Analysis of prepainted steel sheet.
- Analysis of painted steel sheet (lacquer + sealer + cataphoresis + phosphatation + matrix).
- Analysis of glass and ceramics.

Building Industry

- Analysis of steel sheet that has been prepainted and prelacquered with a variety of lacquers (acrylic, polyester, etc.).

2 ANALYSIS OF NON-CONDUCTING MATERIALS

A new instrument configuration that includes modification of the GRIMM lamp and coupling of a high frequency generator, makes it possible to analyse non-conducting materials by GDS.

Comparison Between Classic GDS and High Frequency GDS on Thin Non-Conducting Materials

The chosen example is an organic coating (chromic passivation) on electrically zinc-plated sheet. The comparison was made at the same power level for both types of discharge, DC and HF (30 watts).

a. The use of the classic discharge (DC) produced instability at the beginning of the analysis, Figure 1(a), that appeared in the sample as dielectric micro-breakdown. The micro-breakdown produced poor depth resolution and caused elements from underlying layers to appear on the surface, for example Zn.

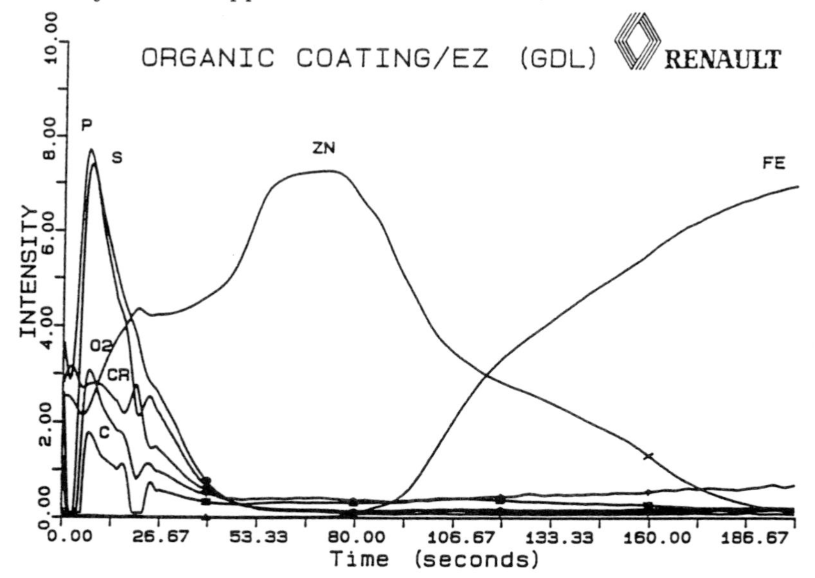

Figure 1(a) Organic coating/EZ (GDL)

b. On the other hand, with a high frequency discharge (GDS) there was no instability, Figure 1(b), and the depth resolution was correct. Observation of the analysed materials showed homogeneous erosion.

The new technique (GDS/HF) does away with the artifacts produced by the classic discharge on poorly conducting or thin (<5μm) non-conducting materials.

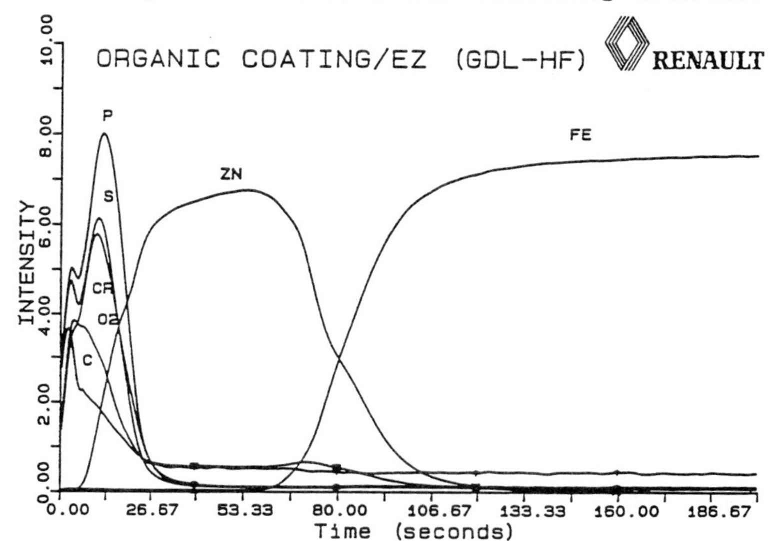

Figure 1(b) Organic coating/EZ (GDL-HF)

Non-Conducting Coating on a Conducting Substrate

Analysis of a Lacquer. Here we present a profile on a lacquer (Figure 2).

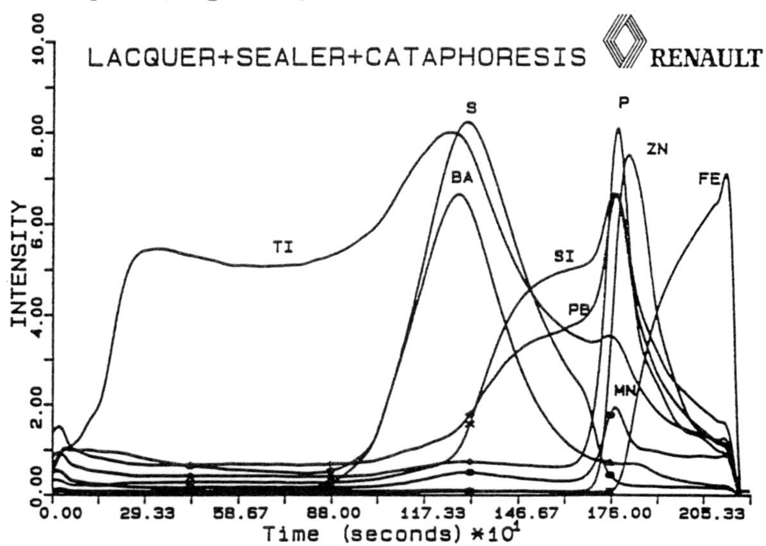

Figure 2 Lacquer + sealer + cataphoresis

The following treatments were applied to the sample which was initially electrically plated with zinc followed by:

- phosphatation,
- cataphoresis treatment,
- application of sealer,
- painting.

The succession of coatings shows up clearly in Figure 2 (continuous analysis). This example shows:

- Good in-depth resolution (separation of phosphate and Zn electrically plated layers after approximately 100μm erosion).
- Good sensitivity: here in the phosphate layer there is 0.5g/m^2 of phosphorus and 0.1 g/m^2 of manganese.

Furthermore, over several analyses the reproducibility was excellent. GDS/HF makes it possible to analyse multiple layers of non-conducting materials reliably, with good resolution and good sensitivity.

Analysis of Polymer Coating. The polymer studied was a polyester resin loaded with silica on an aluminium/zinc layer (Galvalum type), Figure 3.

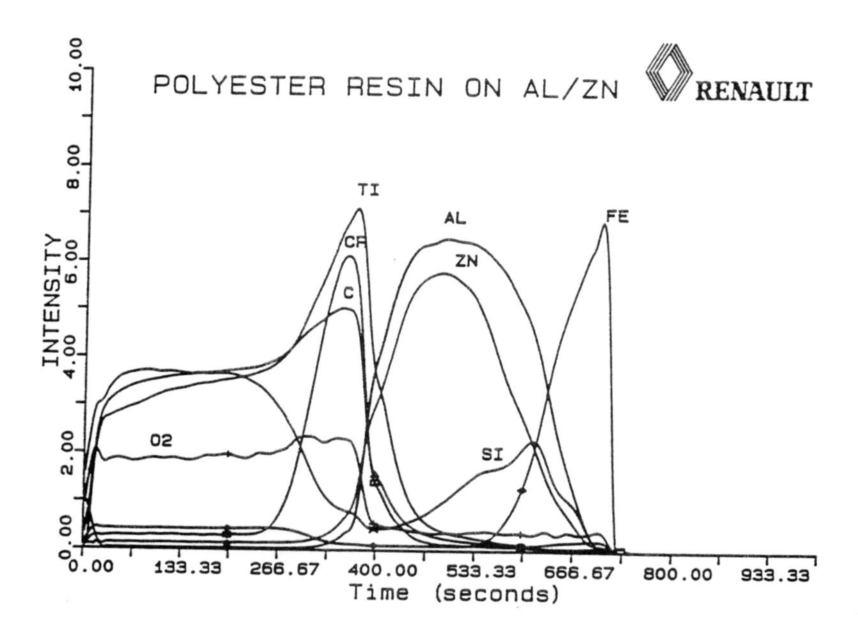

Figure 3 Polyester resin on Al/Zn

This coating is used in the building industry. Chromic passivation can be seen at the Al-Zn/polyester interface. Analysis of polyester or polyvinyl coatings can therefore be made with GDS/HF.

Analysis of Insulating Materials

Analysis of Glass. We present the analysis of a glass surface-treated with tin (Figure 4). The Si, O and Ca signals come from elements within the glass.

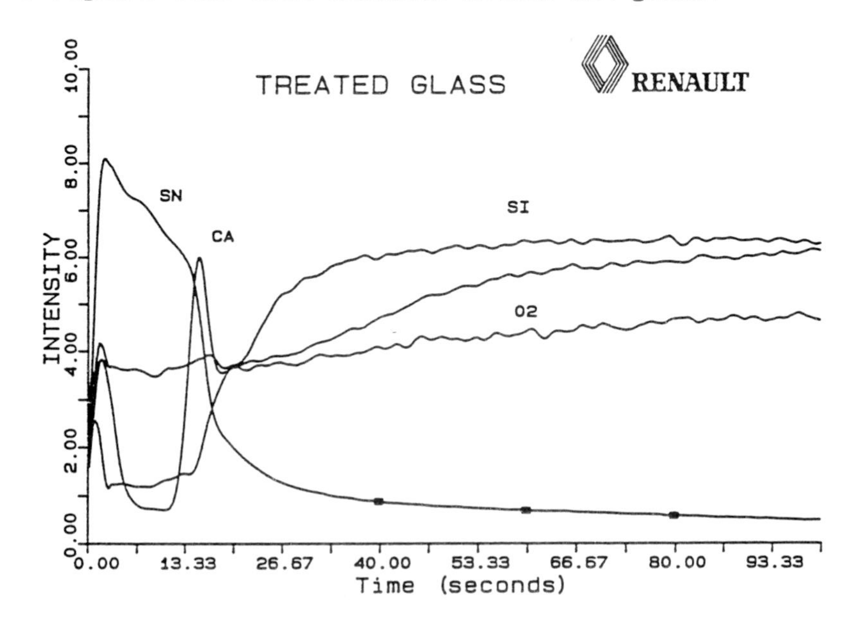

Figure 4 Treated glass

Analysis of a Ceramic. Analysis was made of an anti-wear ceramic coating from an engine compartment, Figure 5. We see the elements Zr and O, that are characteristic of ceramics. Ca, P, Fe, Si and C are polluting elements.

3 INSTRUMENTATION

This development made by RENAULT for the analysis of non-conductive material is now included in a new ISA Jobin Yvon GD Spectrometer (JY 50 S-GDS) using a simultaneous/ sequential spectrometer combination, and equipped with a rapid signal acquisition system.

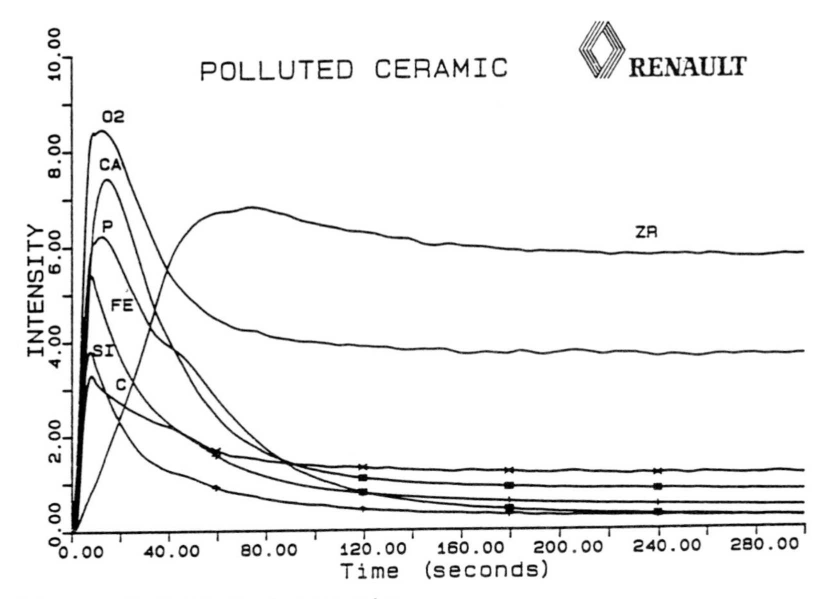

Figure 5 Polluted ceramic

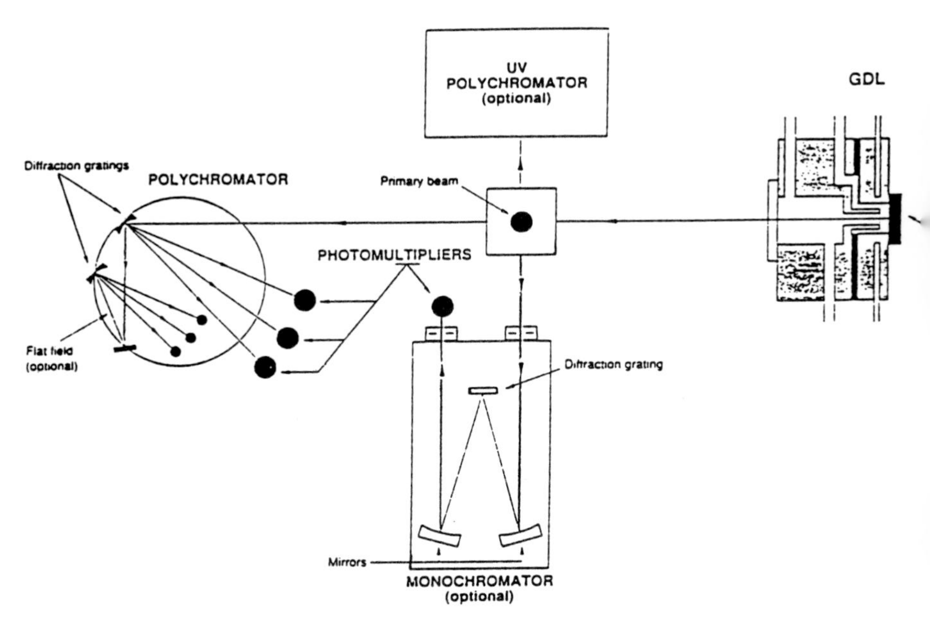

Figure 6 Schematic configuration of the JY 50 S-GDS combined (simultaneous and sequential) spectrometers for elemental and surface analysis

Here, the sequential part (monochromator) has the advantage of free selection of analytes, facilitating single element optimisation, while the simultaneous part (polychromator) has the advantage of high samples throughput. Figure 6 is the optical configuration of the JY 50 S-GDS.

We can note that a special spectrometer has been adapted for each spectral range, using the associated optical components. The main polychromator (simultaneous) will be used in the UV and visible ranges. A flat field polychromator (simultaneous) can be fitted within the main polychromator permitting analysis using lines in the near-IR. The VUV polychromator (simultaneous) is used between 110nm and 160nm, mainly for the detection of gases (O, H, N, Cl). The monochromator can work between 200nm and 500nm (or between 200nm and 800nm) depending on the choice of optical components.

All these spectrometers work at the same time, and due to a special optical interface can analyse the same area of the sample.

4 CONCLUSIONS

This new technique does away with the artifacts produced by the classic GDS in thin non-conducting coatings and on poorly conducting materials. It makes it possible to analyse a broad range of non-conducting coatings and materials directly, which is impossible for classic GDS. Typical applications include:

- Painted steel sheet,
- Prepainted steel sheet,
- Glass (windshields),
- Enamels,
- Ceramics.

The characteristics of GDS comprise:

- Good resolution,
- Good sensitivity,
- Fast analysis,
- Continuous analyses from the surface to approximately 200µm (variable depending on the material) are preserved even on these non-conducting sample types.

4.1.2
ESCA Evaluation of Zinc Phosphate Coatings

U. B. Nair and M. Subbaiyan

DEPARTMENT OF ANALYTICAL CHEMISTRY, UNIVERSITY OF MADRAS, GUINDY CAMPUS, MADRAS 600 025, INDIA

1 INTRODUCTION

The modification of metal surfaces with phosphate pretreatment is a well known method of improving their corrosion resistance, workability and paint-base properties. Although these conversion coatings in themselves provide little or no structural strength or corrosion protection, they play a significant role in the elimination of costly service failures by controlling underfilm corrosion when used in conjuntion with other finishes.[1,2]

Conventional phosphating formulations often need to be modified in order to suit the end use to which the coating is put to. To suitably modify a phosphating formulation it is essential to have a good understanding of the chemistry of the phosphating process and the complex phenomena involved in the nucleation and growth of phosphate crystals. This has led to the widespread use of electrochemical[3-5] and analytical techniques[6,7] to characterize the features and properties of phosphate coatings. Virtually every spectroscopic technique including infra-red spectroscopy,[9] X-ray fluorescence,[10] X-ray diffraction (XRD),[11] electron probe microanalysis (EPMA),[12] scanning electron microscopy (SEM),[13] X-ray absorption near edge structure (XANES) and extended X-ray absorption fine structure (EXAFS),[14] have been used to study the diverse characteristics of phosphate coatings such as their crystal size, crystal type and orientation, crystal density, microtopography, morphology and composition as well as the steps involved in their nucleation and growth.

More recently, surface microanalytical techniques such as electron spectroscopy for chemical analysis (ESCA) and auger electron spectroscopy (AES) have emerged as powerful aids in assisting the study of the mechanisms of aqueous corrosion processes[15] and in providing valuable information on interfacial interactions that govern failures of polymeric finishes on pretreated metal surfaces[16]. In combination with depth profiling, these techniques have also been used to evaluate post-treatment on phosphated steels,[17] interaction of polymeric additives on zinc phosphate coatings,[18] changes in morphological and chemical properties of phosphate crystals obtained on different types of steel substrates,[19] etc. In the present study the influence of a flotation surfactant viz., 1-Octadecanethiol (1-ODT) on phosphate coatings obtained from a calcium modified cold zinc phosphating formulation was evaluated using ESCA, with a view to obtaining information on the nature of its interactions with phosphate coatings on an atomic scale. Besides the determination of surface chemical composition, it was proposed to ascertain through depth profiling studies, the degree and depth to which this additive integrated itself in the phosphate coatings obtained in its presence, in order to correlate this information to its proposed mode of action in the formulated phosphating bath.

2 EXPERIMENTAL

A cold phosphating formulation of the following composition was used in the present study:

Zinc oxide	: 9 g l^{-1}
o-Phosphoric acid (85%)	: 20 ml l^{-1}
Calcium carbonate	: 1 g l^{-1}
Sodium nitrite	: 2 g l^{-1}

Immersion phosphating of hot rolled mild steel panels (IS:1079) of 96 cm^2 surface area were processed for 30 minutes at room temperature (27°C) in a processing sequence which included degreasing and pickling prior to phosphating. An optimum concentration of 75 mg l^{-1} of 1-ODT[20] was added to the formulated bath and the coatings obtained from the modified bath and the reference bath (containing no additive) were evaluated for their physical properties and corrosion performance.

1 cm x 1 cm sized samples of steel phosphated in the reference bath and 1-ODT containing bath were subjected to ESCA studies in a VG Scientific X-ray photoelectron spectrometer (model ESCALAB MK II), using a Mg K_α - X-ray source operating at a vacuum of about 10^{-8} to 10^{-9} torr. Survey and high resolution spectra of the phosphate layers present at the surface and at different strata of the coatings were obtained by using the argon ion sputtering technique. Qualitative and quantitative information regarding each layer of the phosphate coating was obtained from the precise determination of binding energies and integrated peak areas on the intensity versus binding energy plots obtained using a dedicated Apple II computer interfaced with the spectrometer.

3 RESULTS AND DISCUSSION

The physical properties and corrosion performance of phosphate coatings obtained in the reference bath and 1-ODT containing bath are compared in Table 1. It is evident from the data that panels phosphated in presence of 1-ODT possessed significantly lower coating weight as compared to the reference bath. However, their low absorption value[21] and weight loss due to corrosion in 3% NaCl indicated that coatings formed in the additive - containing baths were of low

Table 1 Physical properties and corrosion performance

Property	Bath Used	
	Reference bath	Additive-containing bath
Appearance	Greyish-white and uniform	Greyish-white, having scattered reddish-brown spots ,
Average Coating Weight (g m^{-2})	9.26	6.72
Absorption value (g m^{-2})	11.77	12.07
Weight loss in 3% NaCl (g m^{-2})	8.13	8.28

porosity and their corrosion performance was comparable to that of the panels coated using the reference bath.

Fig.1 represents the survey spectra obtained at the unetched surfaces of phosphate coatings formed using the reference bath and the 1-ODT containing bath. The survey spectrum of the uncoated steel surface is included for effective comparison. While carbon, oxygen, phosphorous, iron and zinc are the detectable elements for the sample coated in the reference bath, that coated in presence of 1-ODT showed additional peaks corresponding to sulfur. It was also evident that peaks corresponding to iron had disappeared completely on the surface layers of sample coated in presence of 1-ODT. In both the cases calcium was undetectable.

The detection of sulfur on the unetched surface of sample coated in the 1-ODT containing bath indicated its incorporation into the phosphate coating obtained from this bath and evidenced the ability of this surface-active agent to form films on the surface. The effectiveness of the additive film in sealing the pores of the phosphate coating is evident from the complete absence of iron at the surface layers of the coated sample, despite the fact that coating weight obtained in the presence of 1-ODT was about 27% less than that of the reference bath.

As mentioned in our earlier studies[20], the reddish-brown spots seen on the samples coated in the presence of 1-ODT were identified to be due to octadecyl thionitrite, a reaction product of the added thiol with nitrous acid generated in situ from the phosphating bath during processing.

$$C_{18}H_{37}SH + HONO \longrightarrow \underset{\text{(Octadecyl thionitrite)}}{C_{18}H_{37}SNO} + H_2O$$

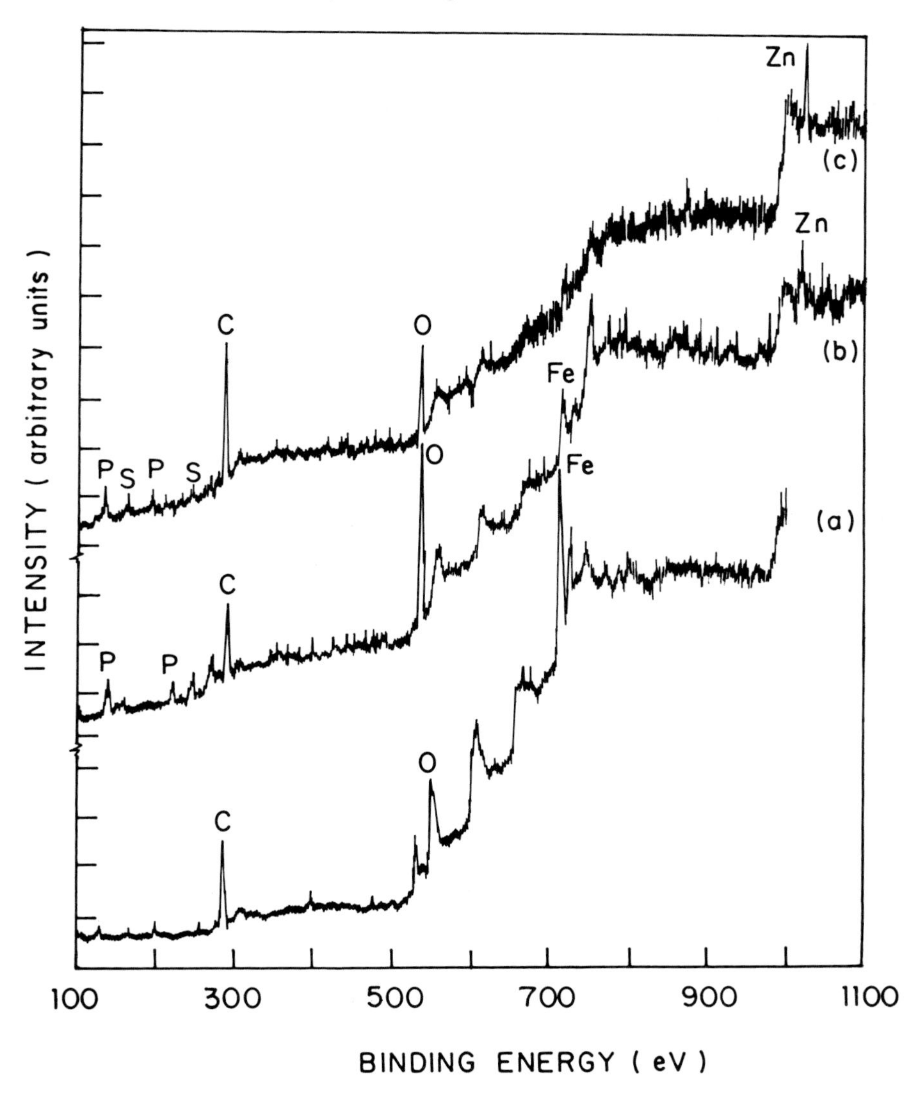

Figure 1 Plot of intensity vs binding energy for (a) uncoated sample, (b) sample coated in reference bath, (c) sample coated using 1-ODT

However, ESCA could not provide conclusive evidence for the formation of this derivative, as peaks corresponding to the binding energy of nitrogen were undetectable, probably due to the low concentration of the derivative formed. In order to ascertain whether the influence of 1-ODT on the phosphate coating formed was restricted to the formation of surface films or whether this surfactant integrated itself into the coating at every stage of coating formation, depth profiling studies were carried out.

The variations in the concentration of the constituent elements of coatings obtained in presence of 1-ODT, as related to sputter time are shown in Fig.2. It is evident that the surface of the specimen coated in presence of 1-ODT (sputter

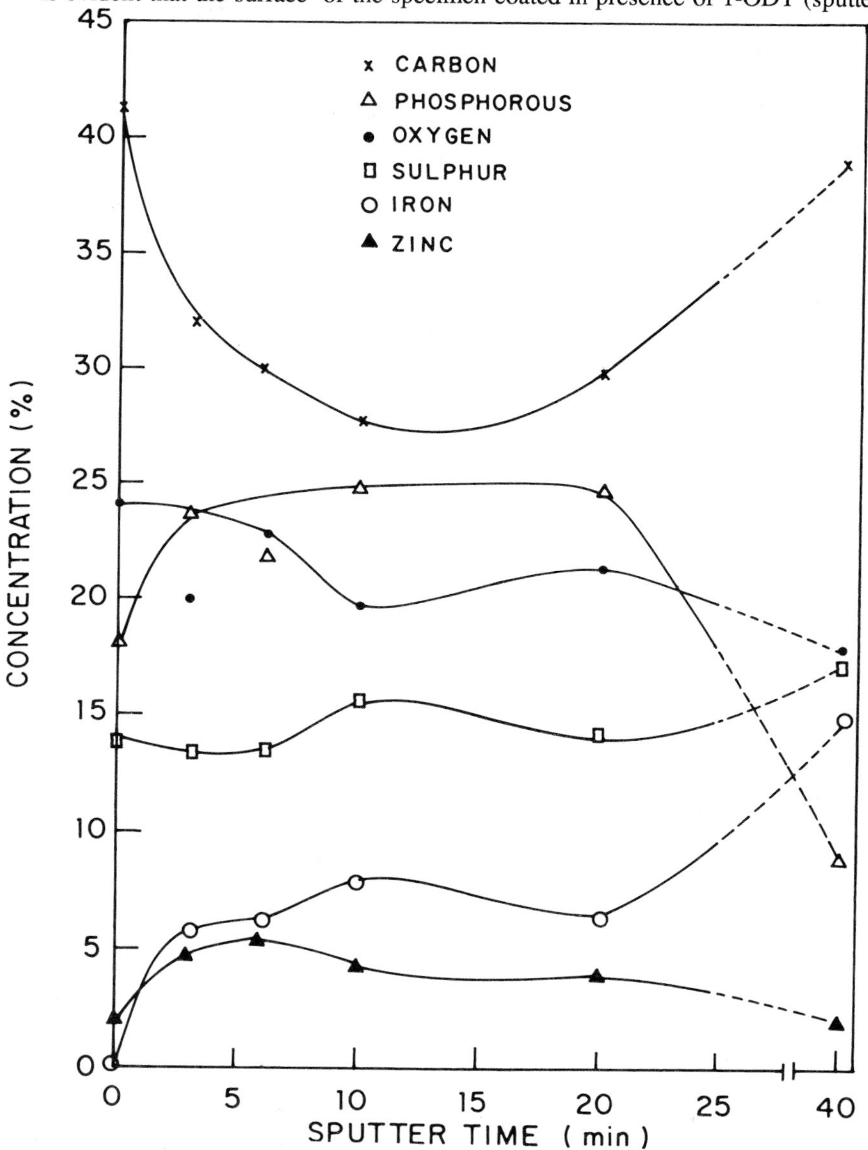

Figure 2 Variation of elemental concentration with sputter time

time of zero minutes) was rich in carbon, oxygen, phosphorous and sulfur while it contained relatively low concentrations of zinc and practically no iron. After a sputter time of 10 minutes, there was a significant decrease in the carbon and oxygen content in the coating while there was a remarkable increase in phosphorous. However, the phosphorous content decreased sharply as the substrate was approached (sputter time of 40 minutes). It is also evident that the layer closest to the metal surface was rich in carbon, sulfur and iron though it was depleted of zinc.

The trends observed in the changes in the concentrations of the constituent elements of the coating at different depths support the mechanism suggested earlier[20] to explain the action of 1-ODT in the formulated bath. The high concentrations of alkyl carbons (binding energy of 284.8 eV) at the surface as compared to the intermediate layers evidenced the adsorption of 1-ODT on the panel being phosphated. It may be considered that the adsorbed additive existed as the thiol (underivatized form) when adsorption occurred, as the high resolution spectra of sulfur at the layer close to the substrate showed a peak at 162.25 eV which correlated well with that of 1-ODT.[22] This adsorption phenomenon accounts for the decrease in coating weight as compared to that of the reference bath. The in situ derivatization of $C_{18}H_{37}SH$ into the thionitrite, though visually observed and analytically confirmed, could not be evidenced from the ESCA studies. However, the evident incorporation of sulfur into the growing layers of the phosphate coating indicate that much of the thiol added existed in the unconverted form until the completion of the phosphating process. The presence of sulfur (and hence the thiol) at all strata of the coating lead to the expectation that the crystals formed in the coating obtained from the 1-ODT containing bath would be of fine-grained nature as the thiol like other surface-active agents, would favour the controlled growth of crystals through their adsorption[23]. The increased concentration of carbon at the surface layers of these phosphate coatings also indicate its ability to form surface layers which would not only help in effectively clogging the pores of the coating but also prevent moisture ingress due to its hydrophobic nature. The combined effects of fine-grained, compact coatings of low porosity and considerable hydrophobicity account for the good corrosion resistance shown by these coatings in corrosive environments (Table 1), even in the absence of the conventional chromic acid post rinse.

4 SUMMARY

ESCA has been used to evaluate phosphate coatings obtained from a calcium-modified cold zinc phosphating bath obtained in presence of 1-ODT. The study reveals the effectiveness of ESCA to provide conclusive evidence of the integration of the additive at all stages of coating deposition. It is also suggested that good corrosion performance of the coatings obtained in the additive containing bath may be related to the formation of fine-grained and compact coatings of low porosity.

ACKNOWLEDGEMENTS

The authors express their thanks to **Mr.R.Rajendran** and **Mr.C.J. Gunasekaran** of the Regional Sophisticated Instrumentation Centre, Indian Institute of Technology, Madras, India, for their help in carrying out the ESCA studies. The financial support by the Council of Scientific and Industrial Research (CSIR), India, is also gratefully acknowledged.

REFERENCES

1. G.Lorin, "Phosphating of Metals", Finishing Publications Ltd., London, 1974.
2. D.B.Freeman, "Phosphating and Metal Pretreatment", Woodhead-Faulkner, London, 1986.
3. E.L.Ghali and R.J.A.Potvin, Corros.Sci., 1972, 12, 583.
4. K.Kiss and M.Coll-palagos, Corrosion, 1987, 43, 8.
5. R.W.Zurilla and V.Hospadaruk, Trans.SAE 780187, 1978, 87, 762.
6. J.P.Servais, B.Schmitz and V.Leroy, Mater.Perform., 1988, 27, 56.
7. D.D.Davidson, M.L.Stephens, L.E. Soreide and R.J.Shaffer, Trans.SAE 862006, 1986, 95, 1045.
8. T.Sugama, L.E.Kukacka and N.Carcicllo, J.Mater.Sci., 1984, 19, 4045.
9. N.Sato, K.Watanabe and T.Minami, J.Mater.Sci., 1991, 26, 1383.
10. N.Sato, T.Minami and H.Kono, Surf.Coat.Technol., 1989, 37, 23.
11. M.O.W.Richardson and D.B.Freeman, Trans.IMF, 1986, 64, 16.
12. G.Rudolph and H.Hansen, Trans.IMF, 1972, 50, 33.
13. H.W.K.Ong, L.M.Gan and T.L. Tan, J.Adhes., 1985, 18, 227.
14. N.Sato and T.Minami, J.Mater.Sci., 1989, 24, 4419.
15. N.S.Mc Intyre, in "Applied Electron Spectroscopy for Chemical Analysis", Eds.H.Windawi and F.F.L.Ho, Chemical Analysis Monograph series, John Wiley and Sons, Inc., New York, 1982, Vol.63, p.89.
16. W.J.van Ooij, in "Organic Coatings: Science and Technology " Eds.G.D. Parfitt and A.V.Patsis, Marcel Dekker Inc., New York, 1984, Vol.6, p.277.
17. L.Fedrizzi and F.Marchetti, J.Mater.Sci., 1991, 26, 1931.
18. T.Sugana, L.E.Kukacka, N.Carciello and J.B.Warren, J.App.Polym.Sci., 1985, 30, 4357.
19. N.Sato, J.Metal Finish.Soc.Japan, 1987, 38, 30.
20. U.B.Nair and M.Subbaiyan, Trans.Metal Finish.Assoc.India, 1992, 1, 9.
21. R.St.J.Preston, R.H.Settle and J.B.L.Worthington, J.Iron Steel Inst., 1952, 170, 19.
22. R.V.Duevel and R.M.Corn, Anal.Chem., 1992, 64, 337.
23. Z.Amjad, Langmuir, 1991, 7, 600.

Section 4.2 Wear Studies

4.2.1
STM and SIMS Analysis of the Interfaces and Intermediate Layers of TiN/Ti/Substrate Systems

K. A. Pischow, A. S. Korhonen, and E. O. Ristolainen

LABORATORY OF PROCESSING AND HEAT TREATMENT OF MATERIALS, HELSINKI UNIVERSITY OF TECHNOLOGY, VUORIMIEHENTIE 2 A, 02150, ESPOO, FINLAND

1 INTRODUCTION

The corrosion behaviour and wear properties are the main criteria to be considered when the usefulness of TiN coated stainless steel is to be assessed. However, the porosity of the PVD TiN-coatings causes severe pit corrosion attack of the substrate. There seem to be two major ways to improve the situation. Firstly, to improve pit corrosion resistance of the substrate and, secondly, to produce a coating without pores.

Freller and Lorenz[1] have made electrochemical porosity measurements of magnetron sputtered TiN films. Measurements show that the difference in the porosity must be the result of the difference in starting temperature. The films deposited at lower substrate temperatures show a porous Thornton zone I structure, whereas with substrate temperatures above 400^0C at the onset of deposition the film growth starts from a reduced number of growth centres which grow together in the course of the film growth. The densest film growth was measured close to a temperature of 300^0C.

The corrosion performance of PVD TiN coatings is strongly affected by porosity, which can cause local and rapid corrosion of the base material. Aromaa et al.[2] have used corrosion current to calculate porosity during investigations of three commercial TiN coatings but it seems to be difficult to find any simple correlation between the porosity and corrosion behaviour. The best corrosion behaviour was shown by a (Ti,Al)N + AlN coating having an insulating layer on the top of the coating, which increases the polarization resistance and decreases the corrosion current density.

Bardwaj et al.[3] have made corrosion tests with different intermediate Ti layer thicknesses and showed that a thin Ti layer tended to lower the critical current density. However, thicker Ti intermediate coatings tended to lower the passive region and in some cases actually accelerate pitting corrosion. The reason for this behaviour is not clear; however, very thin layers of titanium may be totally consumed to form TiN or TiO_2, whereas in the case of thick titanium layers, local galvanic cells may have been established between Ti and the other metals. They have also tested the affects of an intermediate sputtering on the corrosion behaviour of the coating and the results achieved suggested that propagating defects in the coatings were minimized by this intermediate sputtering process. Laser annealing of sputtered TiN films was pursued to eliminate point defects and reduce porosity in the TiN coatings. Laser annealing of the TiN coatings at low power levels reduced the critical current density and extended the passivation region relative to unannealed coatings.

Gröning et al.[4] have used X-ray photoelectron spectroscopy (XPS) to analyse the surface composition of austenitic AISI 316L and martensitic AISI 440C stainless steels after 1 h annealing in high vacuum at temperatures between 25 and 1000^{0}C. On both steels they observed strong chromium, manganese and silicon diffusion from the bulk to the surface in the temperature range between 350 and 850^{0}C. The composition of the forming surface layer (thickness about 50Å) was strongly temperature-dependent and completely different from the bulk. Chromium was found to be completely oxidized as Cr_2O_3 on the surface except for those temperatures at which high silicon was measured. It seems that Si reduces Cr_2O_3 to Cr and forms SiO_2. One reason for the well known differences in the quality of the adhesion of hard PVD coatings on stainless steels can be related to the surface composition of the steel substrate.

Shot peening is one of the cost-effective ways to compressively stress a surface to be coated. Shot peening before chromium plating of steel has been shown to improve fatigue life significantly[5,6]. Rickerby and Wood[7] have investigated the effect of peening on coating microstructure and subsequent oxidation behaviour of MCrAlY coatings. According to them, the primary benefits of glass bead peening can be summarized as the generation of an even compressive stress pattern in the surface layers and the elimination of leader defects from, and densification of, the coating when used after the deposition. Computer controlled shot peening has been used by Eckersley and Kleppe[8] and they divided its application into three main areas: firstly, surface preparation prior to coating; secondly, offsetting the negative metallurgical effects of coatings; and thirdly, adding beneficial properties of coatings. As a pretreatment for PVD TiN coating, the possible value of shot peening lies in the fact that the compressive layer most probably is thick enough to survive the sputter cleaning process. On the contrary the effects of the shot peening on the TiN coated surface are not so obvious because of the brittle nature of TiN, which might lead to severe microcracking rather than to densification and closure of leaders.

Our preliminary experiments have clearly shown that there are two kinds of pores in PVD TiN-coatings deposited on stainless steel substrates. Firstly, type T1 which are due to the orientation misfit between different columnar TiN grains. Secondly, type T2 which grow due to the different nucleation conditions on the sputtered SS-substrate surface. The hypothesis derived from the above is schematically illustrated in Figure 1. If we are using a Ti intermediate layer, very strong pitting corrosion can be expected in any corrosive environments in T2 type pores. In T1 type pores the formation of TiO_2 is believed to inhibit the corrosion reactions; however, solutions containing Cl^- ions are disastrous for these pores too. If the native oxide layer remains on the SS surface to be deposited with a Ti interlayer this is believed to lead to the reduction of Cr according to the equation $2Cr_2O_3 + 3TiO_2 \rightleftharpoons 3TiO_2 + 4Cr + 3O_2$. We also believe that the unsputtered oxide surface leads to less pores of the type T2 which would be beneficial for the corrosion behaviour. In our previous study[9] it was shown that it is possible to deposit TiN-coating using a Ti interlayer on the amorphous oxide layer of the stainless steel. An amorphous oxide layer should of course be an excellent corrosion barrier between stainless steel and TiN coating.

The aim of this study was, firstly to verify those aforementioned hypotheses by a set of experiments analysed with STM and SIMS, and, secondly, to study the effects of different passive layers and chromium layers produced by high temperature surface diffusion on the corrosion behaviour of the TiN coating-substrate system. Intermediate shot peening was also used in order to prevent growing pinholes penetrating through the whole coating thickness. Very rough preliminary corrosion tests were made to indicate the possible differences in corrosion behaviour.

2 EXPERIMENTAL

STM

The scanning tunnelling microscope used was a Struers Tunnelscope 2400 with a short scanning tube for scanning within a 6 µm x 6 µm area with a lateral resolution of up to 0.1 Å. The bias voltage was adjustable from 10 mV to 5 V in steps of 1 mV. The tunnel current could be varied from 0.01 nA to 10 nA in steps of 10 pA. During the scanning the bias voltage and the tunnelling current were low and kept constant at 0.1 V and 1 nA, respectively.

PVD TiN Coatings and Pretreatments

TiN films were deposited by an industrial size, reactive triode ion-plating apparatus, which has been described elsewhere[10]. Substrates were argon sputter cleaned as given in Table 1. The TiN interlayer used was about 50 nm thick.

Table 1 Coating parameters and pretreatments

SAMPLE	Coating and pretreatment sputtering time min	TiN coating thickness nm	Ti interlayer thickness nm	intermediate shoot peening
1.	60	2950	50	-
2.	60	2950	50	-
3.	60	950 + 3000	50	yes
5.	-	2950	50	-
6.	-	2950	50	-
7.	-	950 + 3000	50	yes
8. *	-	2950	50	-

* Sample 8 was annealed for 1h at 800^0C in vacuum before coating

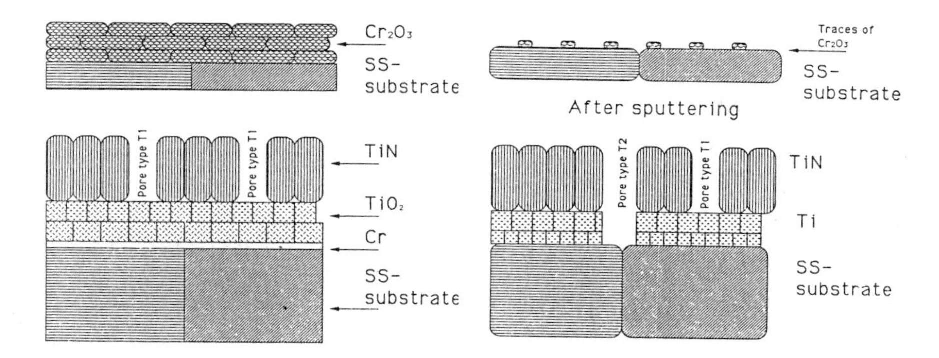

Figure 1 A schematic drawing of the different pore types in PVD TiN coatings

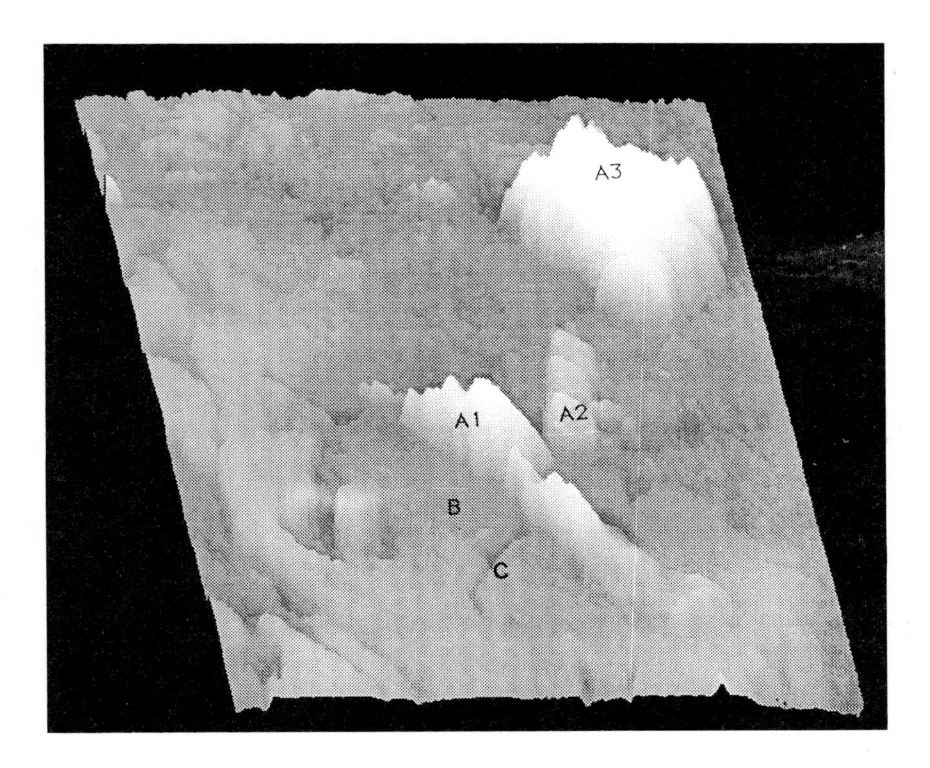

Figure 2 An STM image of the initial stage of TiN deposition on a sputtered stainless steel surface. Clusters of big TiN grains are marked A1, A2 and A3; a typical smooth area covered with small crystals of equal size is marked B; and an area with no TiN coverage is marked C

SIMS Analysis

Depth profiling was performed on a VG IX70S double focusing magnetic sector SIMS using an O_2^+ primary beam at 4 keV impact energy. A primary ion current of 50 nA was used with a spot size of 1000 nm and raster scanned over an area of 250 x 150 μm. A mass resolution of 1000 was used in this study. Profiling with the IX70S was performed by detecting positive $^{14}N^+$, $^{50}Ti^+$, $^{52}Cr^+$, $^{56}Fe^+$ and $^{64}TiO^+$ ions.

3 RESULTS AND DISCUSSION

STM Analysis

The initial stage of TiN deposition on a sputtered SS surface is shown on the STM picture in Figure 2, where three different growth modes of TiN can be identified: firstly area B covered with small equisized TiN grains, secondly, area A covered with large TiN grains being probably a consequence of the sputtering action during deposition and, thirdly, area C where no deposition of TiN can be seen. On places like C, a type T2 pore will most predictably be created.

Figure 3 shows STM pictures of fracture surfaces with three different interlayers: firstly, Figure 3(a) shows a typical fine-grained Ti-interlayer, secondly, Figure 3(b) shows the effect of sputtering on the carbon interlayer and on the upper left part of the picture the Ti-interlayer, Figure 3(c) shows details of the sputtered carbon layer and, finally, Figure 3(d) shows the sputtered oxide surface on the stainless steel substrate.

SEM Analysis

The SEM micrograph in Figure 5(a) shows the departed coating side after the adhesion test; a number of pinholes are clearly visible and the pit shown boxed in Figure 5(a) is magnified in Figure 5(b). Figure 5(c) was taken from the substrate side showing the bottom of the pinhole, which can be seen magnified in Figure 5(d). On the surface of this sample a thin carbon layer of 10 nm was vaporized before coating with a Ti interlayer and TiN. There is only a slight sputtering effect on the carbon layer.

The SEM micrographs illustrate the influence of a passive intermediate layer, which effectively protects sample 6 in Figure 6(a) when compared with sample 2 in Figure 6(b) and respectively sample 7 in Figure 6(c) and sample 3 in Figure 6(d).

The corrosion results for samples 2,3 and 7 are very similar; however, there are big differences in the appearances of the corroded surfaces. On sample 2 there are 27 large deep pits, on sample 3, 39 smaller pits, and on sample 7, 51 even smaller pits.

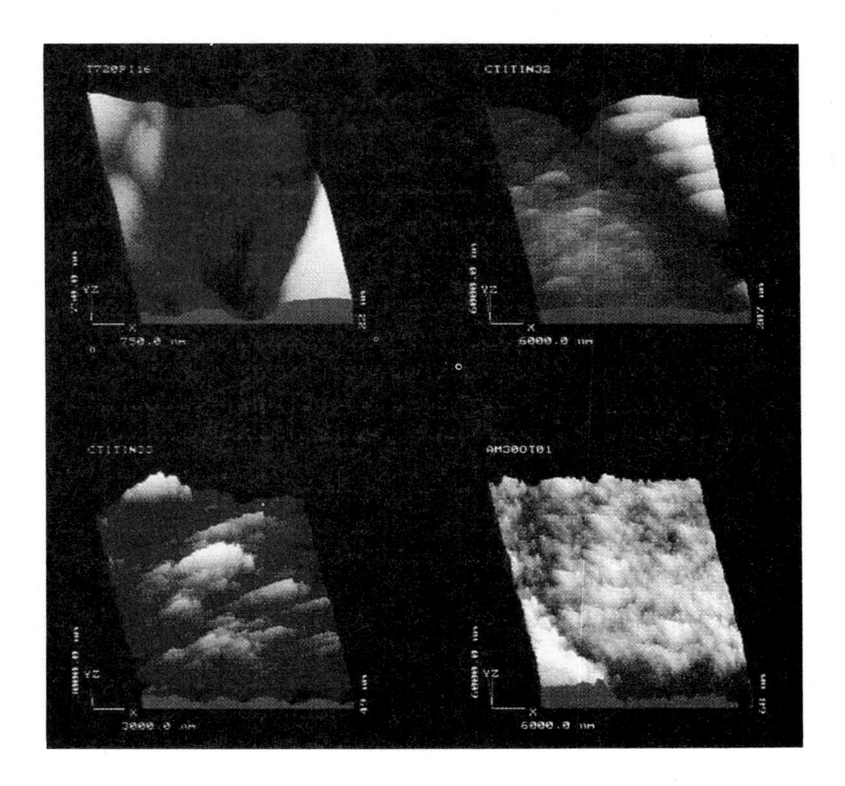

Figure 3 Shows STM scans of fracture surfaces with three different interlayers: (a) a typical structure of the fine grain size Ti-interlayer close to the TiN- coating interface, (b) sputtering effect on the carbon interlayer and on the upper left part of the picture the Ti-interlayer, (c) details of the sputtered carbon layer, (d) sputtered oxide surface on stainless steel substrate

SIMS Analysis

SIMS depth profiles of the elements($^{14}N^+$, $^{50}Ti^+$, $^{52}Cr^+$, $^{56}Fe^+$ and $^{28}Si^+$) of sample 6 are shown in Figure 7. The concentration of the elements changes gently at the interface. At the beginning of the sputtering, after the effect of the contaminants stops, only $^{14}N^+$ and $^{50}Ti^+$ can be seen; then deep in the coating, the chromium content starts to rise and finally just before reaching the coating-substrate interface iron will also be detected. Hardly any diffusion of silicon was measured. When the substrate was sputter cleaned and the passive layer was removed chromium and iron were detected only very close to the interface. Figure 8 shows SIMS depth profiles of sample 8, which was vacuum annealed for one hour at 800^{0}C. Here the behaviour of chromium and silicon is completely different. The rise of those elements is extremely sharp and there is a plateau on both curves, which might be caused from differences in binding energies. This

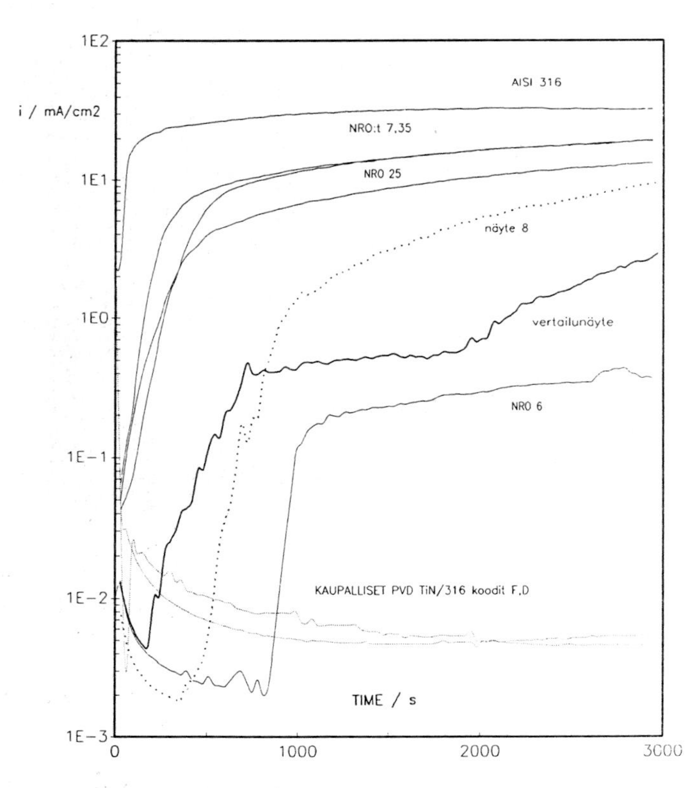

Figure 4 Potentiostatic curves in 0.1 N H_2SO_4 + 0.05 M NaCl

difference could be explained as a change from Cr_2O_3 to Cr, which is in agreement with the results achieved by Gröning et al.[11]; the opposite situation occurs with silicon in which the change from SiO_2 to Si is not easy to explain. One more important difference in these two figures must be pointed out and it is the behaviour of nitrogen, which in Figure 7 starts to decrease only at the interface, but in Figure 8 the nitrogen concentration starts to decrease deep in the coating at the same level where chromium starts its sharp rise.

Corrosion Tests

For samples 2,3,6 and 7 an on/off type potentiostatic corrosion test was used, which shows categorically whether

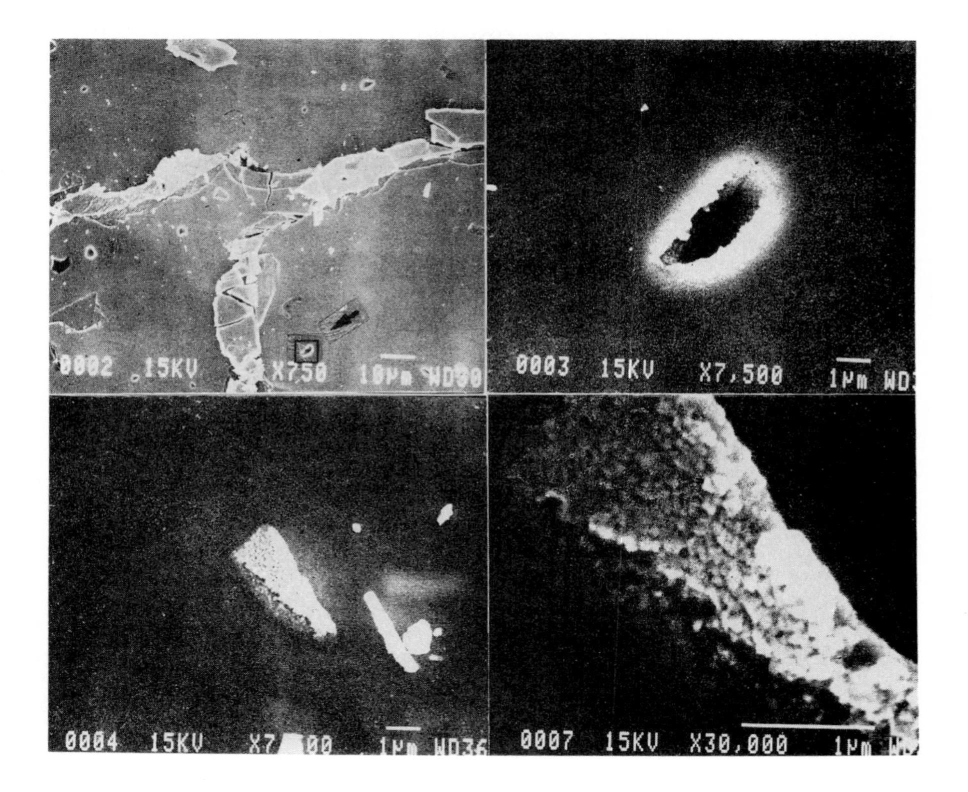

Figure 5 SEM micrographs of pinholes in TiN coating: (a) coating side showing several pinholes, boxed pit magnified in (b), (c) the bottom of the same pinhole on the substrate, (d) microstructure of the pinhole bottom

the coating will survive or not. The results are shown in Figure 4. As can be seen from Table 1, the only difference between samples 2 and 6 is in the sputter precleaning, which is not done for sample 6 at all leaving the passive layer unaffected. The corrosion behaviour of these two samples is completely different, in favour of sample 6, giving strong support to our hypotheses stated in the Introduction. The intermediate shot peening decreases the corrosion resistance of sample 3 and 7 to the same low level and at this stage the sputter cleaning seems to have no effect whatsoever.

Sample 8, which was annealed in vacuum in order to form a protective layer composed of Cr, SiO_2 annnd Cr_2O_3,

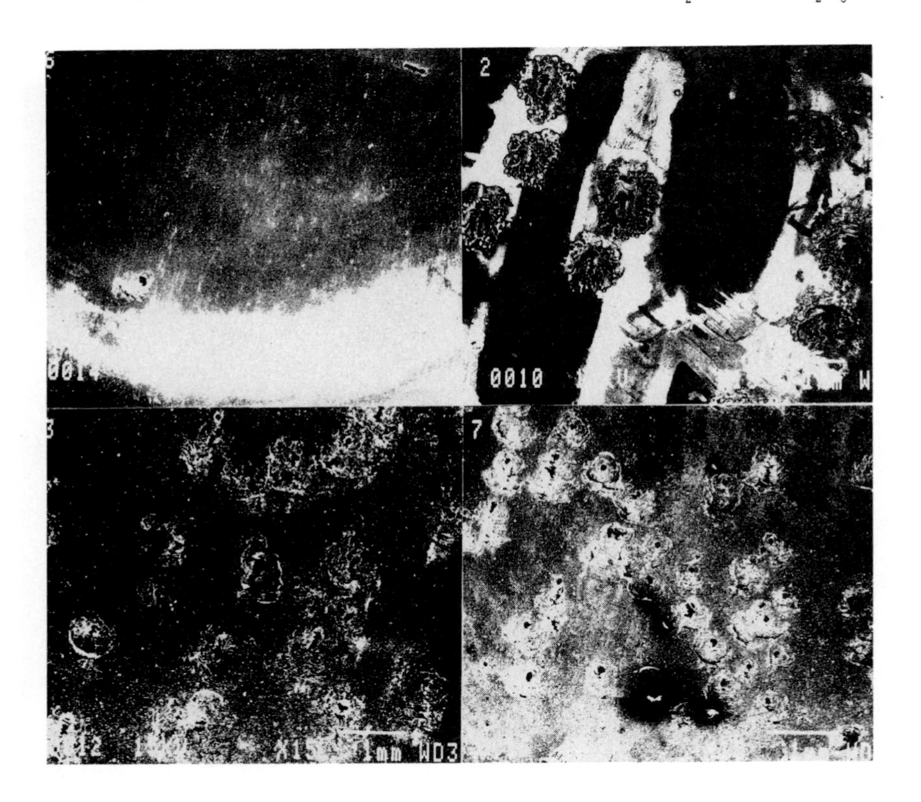

Figure 6 SEM micrographs of the sample surfaces after corrosion tests showing different degrees of pit corrosion attack on the surface: (a) sample 6 without sputtering, (b) sample 2 sputter cleaned 60 min., (c) sample 7 without sputtering with intermediate shot peening, (d) sample 3 sputter cleaning 60 min. with intermediate shot peening

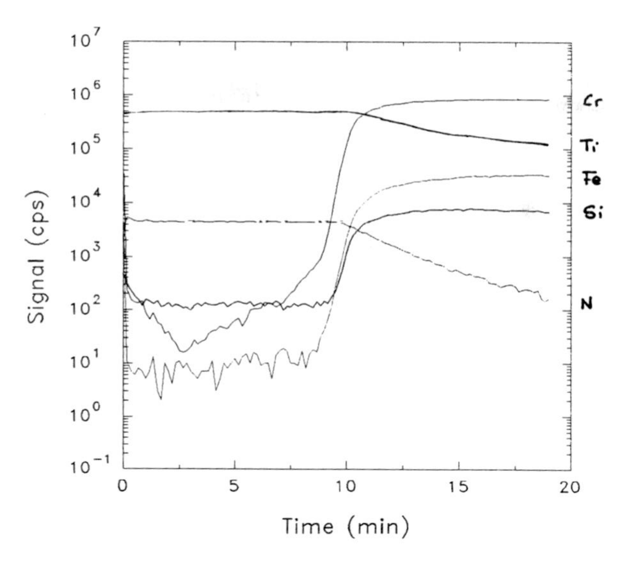

Figure 7 SIMS depth profiles from sample 6 with TiN coating without sputter cleaning

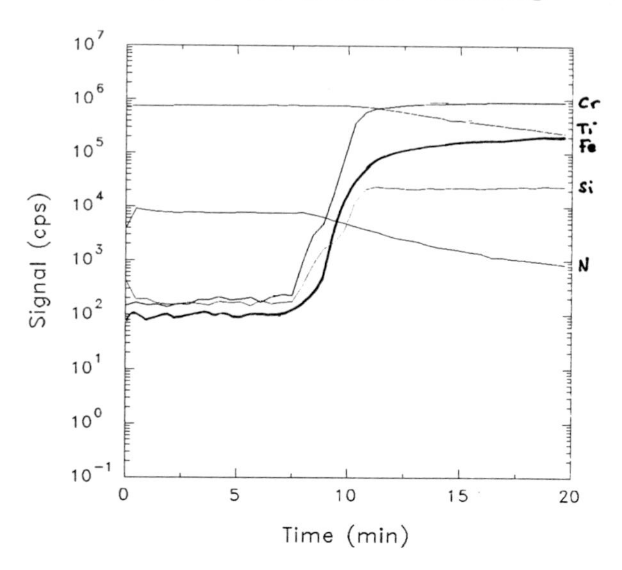

Figure 8 SIMS depth profiles from sample 8 annealed in vacuum one hour at 800°C

showed the same kind of corrosion behaviour as sample 6; at the beginning of the test it is markedly better than sample 6, but after approximately the same time the passive layer fails in one place.

4 CONCLUSIONS

There seem to be two principle ways to improve the corrosion resistance of the PVD TiN coated substrate system. Firstly, by producing a TiN coating without any type T2 pores (or deactivating them by filling) and, secondly, by raising the ability of the substrate materials to resist pitting corrosion. The latter will include noble intermediate layers, which must not be produced by a PVD process in order to avoid forming T2 type voids even in this layer. The effect of these kinds of intermediate layers seems to be decreased by the strong reorganization of the surface layer caused by the high surface diffusion rates due to the heavy sputtering.

The preliminary corrosion tests showed that corrosion behaviour was markedly improved when the coated surface was not sputter cleaned leaving the protective passive layer on the substrate surface. SIMS depth profiling of chromium showed that there was much chromium deep in the TiN coating, which is believed to be one of the main reasons for improved corrosion resistance. However, the reason for the final failure of the coating at one singular point is not yet fully understood.

The effects of vacuum annealing of the substrate were shown to increase the amount of chromium in the coating, leading to an even better start of the corrosion curve; as was the case when the deposition occurred directly on the native passive layer; however, the same kind of failure occurred even in this case.

The intermediate shot peening process was shown to deteriorate the corrosion behaviour of the tested samples. This is a consequence of the lack of plasticity of the TiN coating leading to severe cracking.

One reason for the sudden failure of the coating in the aforementioned cases might be the Ti-interlayer. We believe that a very careful optimization is needed in order to get full benefit from the passive layer. The thickness of the Ti intermediate layer must be monitored to a thickness sufficiently thick in order to partly reduce the Cr_2O_3 and to give the necessary chemical gettering between the substrate and TiN coating; and thin enough not to form a mechanically weak Ti layer, which is also believed to be detrimental to the corrosion behaviour.

The schematic mechanism stated above is of course a highly simplified explanation of a very complicated group of phenomena but is still believed to give guidance for producing coatings with better corrosion behaviour.

ACKNOWLEDGEMENTS

The authors wish to thank R. Suominen for taking the SEM pictures, M Turkia for the corrosion tests and E. Harju for making the coatings.

REFERENCES

1. H. Feller and H.P. Lorenz, J. Vac. Sci. Technol., 1986, A4(6), 2691-2694.
2. J. Aromaa, H. Ronkainen, A. Mahiout, S.-P. Hannula, A. Leyland, A. Matthews, B. Matthes and E. Broszeit, Mat. Sci. Eng., 1991, A140, 722-726.
3. P. Bardwaj et al., Appl. Surf. Sci., 1991, 48/49, 555-556.
4. P. Gröning, S. Nowak and L. Schlapbach, Appl. Surf. Sci., 1991, 52, 333-337.
5. T.J. Meister, Heat Treat., June 1991, 22-24.
6. "Shot Peening" in Metals Handbook, 5th edition, Part 5, pp. 138-149.
7. D.S. Rickerby and M.I. Wood, J. Vac. Sci. Technol., 1986, A4(6), 2557-2564.
8. J.S. Eckersley and R. Kleppe, Surf. Coat. Technol., 1987, 33, 443-451.
9. K.A. Pischow, L. Eriksson, E. Harju, A.S. Korhonen and E.O. Ristolainen, presented ICMCTF92.
10. J.M. Molarius, Helsinki University of Technology, Dissertation Thesis, Espoo, Finland, 1987.
11. P. Gröning, S. Nowak and L. Schlapbach, Appl. Surf. Sci., 1991, 52, 333-337.

4.2.2
Surface Analysis of Plasma Nitrided Layers on Titanium

H. J. Brading,[1] P. H. Morton,[1] T. Bell,[1] and L. G. Earwaker[2]

[1] SCHOOL OF METALLURGY AND MATERIALS, THE UNIVERSITY OF BIRMINGHAM, UK

[2] SCHOOL OF PHYSICS AND SPACE RESEARCH, THE UNIVERSITY OF BIRMINGHAM, UK

1 INTRODUCTION

In recent years a variety of surface engineering techniques has been used to improve titanium's tribological properties (characterised by high coefficient of friction and poor wear resistance) by producing a hard surface coating[1,2]. These protective coatings along with titanium's inherent advantages of high strength and low modulus, have led to the increased use of titanium for many industrial components - for example cutting tools and aeroengine turbine blades. Plasma or 'Glow Discharge' nitriding is a technique for producing a hard surface coating of titanium nitride on titanium which has been successfully used to improve the wear resistance of titanium under low load-bearing conditions.

The plasma nitriding process[3] makes use of an abnormal glow discharge[4], which is associated with high current and charge densities. The components to be nitrided are electrically isolated and placed or suspended in a vacuum furnace which is evacuated and back-filled with the treatment gas. A dc voltage is then applied between the components (cathode) and the furnace walls (anode) and the potential difference ionises the treatment gas producing the glow discharge. Positive ions in the treatment gas are accelerated towards the negatively connected components and hit the surface with high kinetic energy giving rise to sputtering of the surface, and heating of the components. In the plasma nitriding of titanium[5,6], nitrogen ions and/or accelerated neutral nitrogen atoms impinge on the surface and react to form a nitrogen-rich film which results in both the formation of a compound nitride layer on the surface (δTiN and ϵTi_2N phases) and the diffusion of nitrogen in the substrate. This has the effect of providing the titanium with a hard surface coating which adheres well to the substrate.

The nitriding process depends on a large number of parameters, all of which can affect the properties of the resulting coating. It is therefore important to be able to characterise the coatings produced in terms of their structure and composition. The depth profile of nitrogen through the coating, the oxygen contamination and the phase structure of the coating are important.

A range of complementary surface analysis techniques has been used to study plasma nitrided titanium. They allowed the analysis of both light and heavy elements and enabled surface compositions to be investigated. There are a large number of surface analysis techniques capable of analysing heavy elements, however many of these are not suitable for the analysis of light elements, and there are particular problems associated with the analysis of nitrogen and oxygen in the presence of titanium (peak overlaps and matrix effects). Partially because of these problems, no one surface analysis technique can provide an overall picture of the structure and composition of plasma nitrided titanium. The use of several complementary techniques, however, can provide a variety of information which could be combined to produce an overall picture of the coatings.

The analysis techniques included Rutherford backscattering and nuclear reaction analysis, complemented by electronprobe microanalysis, X-ray photoelectron spectroscopy and X-ray diffraction analysis. In this paper the advantages and limitations of each of these analysis techniques are highlighted to review their applicability to analysing titanium nitride coatings. In order to do this typical results from the various analysis techniques will be presented and compared.

2 SAMPLE PREPARATION

All the samples investigated were produced by glow discharge plasma nitriding of 'commercially pure' titanium or IMI 318 (Ti-6Al-4V), the nitriding being carried out using a 40kW (GZM 40) Klockner Ionon unit, designed specifically for the treatment of titanium components.

The specimen support and current lead-in were made of titanium to reduce contamination of the atmosphere and specimen surface by sputtering from these components. The specimens were first heated to the treatment temperature in the treatment gas mixture or in an inert argon atmosphere, all the heat being supplied by the glow discharge, and once the treatment temperature was achieved the argon was replaced by the treatment gas mixture at the treatment pressure. The plasma unit did not allow direct control of the voltage and current of the glow discharge; V and I varied according to the temperature and pressure setting. The treatment conditions used to produce the samples are listed in Table 1.

Table 1 Process parameters used to produce the samples investigated

Sample	Temperature	Pressure	Heat up gas	Gas mixture	Time	Alloy
Sample A	700°C	3 mbar	Argon	$90\%N_2/10\%Ar$	10 Hours	CP Ti
Sample B	700°C	5 mbar	$25\%N_2/75\%H_2$	$25\%N_2/75\%H_2$	20 Hours	IMI 318
Sample C	850°C	5 mbar	$25\%N_2/75\%H_2$	$25\%N_2/75\%H_2$	20 Hours	IMI 318
Sample D	850°C	2 mbar	$25\%N_2/75\%H_2$	$25\%N_2/75\%H_2$	24 Hours	IMI 318

3 EXPERIMENTAL PROCEDURES

Non-destructive NRA and RBS Measurements

RBS and NRA analyses were carried out using the 3 MV dynamitron accelerator at the School of Physics and Space Research in the University of Birmingham. Alpha particle backscattering was used to obtain high resolution (~0.01μm) depth profiles of nitrogen in the first micron of the coating and the $^{14}N(d,p)^{15}N$ nuclear reaction was used to investigate nitrogen to a depth of 4 to 5 microns, which included the nitrogen diffusion zone. RBS measurements were carried out using 2.0 MeV normally incident α-particles together with a 100μm depletion-depth silicon surface barrier detector (collimated to $2.5x10^{-3}$sr) placed at 150^0 to the incident beam direction. Data were accumulated for a total incident charge of 4 μC. The NRA measurements[7] involved the use of 1.1 MeV normally incident deuterons together with a similar detection system placed at 150^0 to the incident beam direction (data were accumulated for a total incident charge of 150 μC). Scattered deuterons were eliminated from the detector by the use of a 12μm aluminium absorber foil. The NRA experimental conditions were chosen so as to allow detection of protons from the $^{16}O(d,p0)^{17}O$ reaction as well as from the nitrogen (d,p) and (d,α) reactions. The experimental data were analysed using computer codes developed at the School of Physics and Space Research[8] and cross-sections taken from the literature[9]. The $^{16}O(d,p)^{17}O$ reaction was also used to obtain an upper limit of the oxygen concentration (averaged over the first 4 to 5 microns of the coating) by comparing the oxygen P_0 peak height with the nitrogen $P_{1,2}$ peak height and adjusting for the difference in reaction cross-section for these peaks.

EPMA on a Cross-section

EPMA measurements were carried out on a JEOL-JXA-840A (Model 10000 ZAF/4) electronprobe microanalyser linked to a computer (Link Systems Series 2 model 500 system running SPECTA ZAF-4/FLS software). The electron microprobe analyser was run in wavelength dispersive mode (WDX) with an LDE_1 diffracting crystal[10,11].

The specimen was prepared by putting a protective nickel layer on top of the nitrided layers and polishing the sample in cross-section. WDX spectra for titanium, aluminium and vanadium standards were measured along with a spectrum for unnitrided IMI 318 (Ti-6Al-4V). Full spectra for the region of the X-ray spectrum containing the nitrogen and titanium peaks (60^0 to 110^0 in 0.3^0 steps), were measured at 1μm intervals over the cross-section of the nitrided layers in order to obtain a nitrogen concentration depth profile. Peak heights and backgrounds were measured for all the peaks. As the only detectable nitrogen peak overlapped a titanium peak, it was assumed that the ratio of the heights of the two titanium peaks measured in the standard, would be the same for all positions on the sample, although there was no way to verify this. This allowed the titanium contribution to be subtracted from the overlapping nitrogen/titanium peak, leaving the nitrogen contribution. A similar assumption was used on the aluminium and vanadium peaks. The peak heights for each element were calculated as a percentage of the standard for that element.

The microprobe was set up to scan for oxygen in a series of 1µm steps for the first 15µm of the cross-section away from the surface of the nitrided layers and then scan in 10µm intervals for another 10 steps. Oxygen concentration was also measured on the nitrided surface (beam penetration depth = 1µm) of the coating rather than on a cross-section. Counts were measured for the central channel of the oxygen Kα X-ray peak for 100 seconds and for 50 seconds on positions either side to account for background continuum radiation. These were compared with similar measurements made on a standard SiO_2 sample. The beam spot was about 1µm in diameter and had a 1µm penetration depth. A probe voltage of 10 keV and a current of $1x10^{-7}$A were used.

XPS and In-situ Ion Etching

Two sets of XPS experiments were carried out, the first on a KRATOS XSAM 800 instrument at the University of Aston, which was coupled with a DEC PDP microcomputer. The second on a similar instrument coupled with a PDP 11/23 computer at the Hungarian Academy of Sciences, Budapest. These instruments used unmonocromated Mg Kα radiation (15KeV, 20mA), with a background pressure of typically $1.5x10^{-9}$ mbars[12,13].

The samples were prepared by slightly polishing the surface to produce a smooth finish. This was followed by a water detergent wash, a rinse in acetone and finally an ultrasonic rinse in n-heptane. Once in-situ in the spectrometer the sample was etched using Ar^+ ions in rastering mode ($P=5x10^{-6}$ mbar, 5keV, 15µA, 20 to 40 minutes), before the acquisition of the spectra. This procedure is believed to remove a minimum of 50 atomic layers. Room temperature oxidation of TiN is estimated to extend only to about the first 5-10 atomic layers (1-2µm)[14], and should therefore be removed.

Wide scan spectra were recorded for general characterization of the samples. Scans were made of the narrow characteristic lines of titanium, nitrogen, carbon and oxygen. On sample IB850, which was initially unetched, in-situ ion etch depth profiling was carried out ($P=5x10^{-6}$ mbar, 5keV, 15μA, 4 x 5 minutes).

X-ray Diffraction Measurements

XRD analysis, which was carried out on a Philips PW 1050 X-ray spectrometer, linked to a BBC Master computer, using unmonocromated Co Kα radiation, was used to obtain information about the nitride phases present in coatings. The specimens were scanned for 2θ angles between 30^0 and 100^0 and the resulting X-ray diffraction peaks were identified using JCPDS power index files[15].

4 RESULTS AND DISCUSSION

Non-destructive RBS and NRA Measurements

RBS could be used to analyse nitrogen and titanium and gave in-situ, non-destructive depth profiles to a depth of 1μm with a resolution of 0.01μm. As an example of this technique, Figure 1 shows two RBS spectra, one for a piece of CP Ti which was used as a standard and the other showing a spectrum for sample A (titanium nitride on titanium - see Table 1). The CP Ti spectrum shows an edge due to titanium at channel number 397. The spectrum for sample A has a titanium edge at the same position as the CP Ti spectrum and also shows an edge for nitrogen in channel 178. The contribution of titanium to the height of this spectrum is much greater than the nitrogen contribution because of titanium's higher atomic number, the Rutherford scattering cross-section being much larger for higher Z elements.

The number density of titanium nitride is much lower than in CP Ti due to the presence of nitrogen in the matrix. This appears on the spectrum as a lowering in the overall height of the profile for sample A, which depends on the number density for titanium in the target. Superimposed on this lowered titanium profile is the nitrogen contribution to the spectrum.

Two small peaks can be seen on sample A at channel numbers 471 and 430 which correspond to a small amount of surface contamination (to a depth of 0.015μm) from iron (0.15 atomic %) and copper (0.08 atomic %), which originates from the plasma unit. These peaks demonstrate the

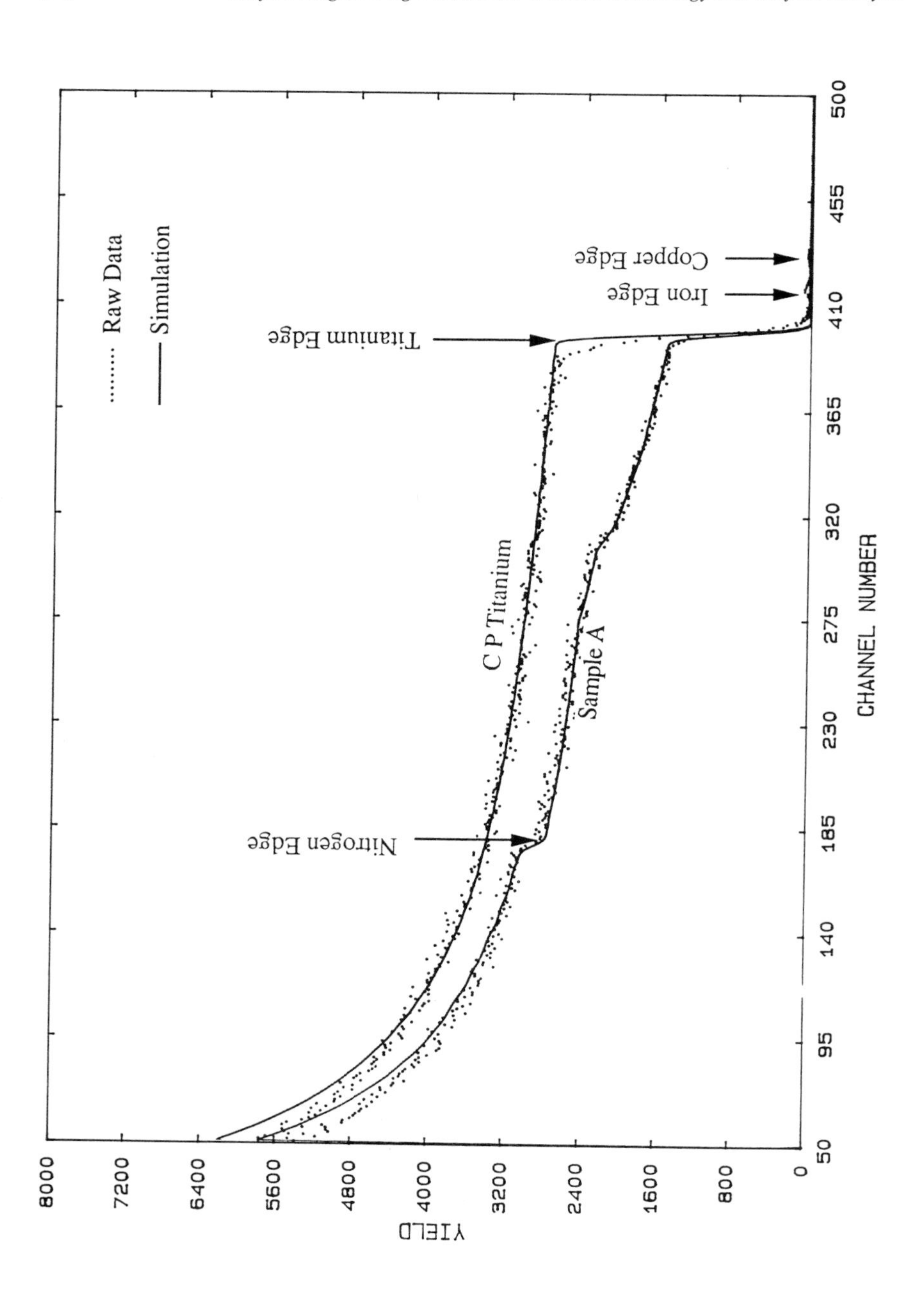

Figure 1 RBS spectra for CP titanium and sample A with their computer simulations

greater suitability of this technique for analysing heavy elements in a lighter element matrix.

Analyses of the RBS spectra were carried out using a computer simulation program. Parameters for the detector geometry and beam energy are entered into the program. A first estimate of the elemental distribution of the sample is postulated and entered into the computer as a series of layers to which a thickness and composition of elements (as a number density concentration gradient across the layer) is assigned. The computer then calculates a simulated spectrum (yield against energy) from this information, which is then compared with the measured RBS spectrum. If the shape of the simulated spectrum differs from the shape of the measured spectrum, adjustments are made to the proposed layer structure (elemental distribution) of the sample and the simulation is recalculated. This iterative process continues until a close match between the RBS spectrum and its simulation is achieved. The layer structure is then assumed to represent the elemental distribution in the sample.

The simulations for the two spectra are also shown in Figure 1. The spectrum of the CP Ti sample (which was used as a standard) was matched to the simulation using the number density of titanium ($5.68 \times 10^{22} cm^{-3}$). Figure 2 shows the nitrogen depth profile used to simulate the RBS spectrum for sample A.

The simulation method relies on knowing the number densities of the elements which make up the compounds found in the nitride coating in order to produce a depth profile. Although the number densities of elements in the bulk compounds TiN, Ti_2N and the α-Ti matrix are well known, they may not be the same as those found in a thin film of the same compound as thin films are frequently not fully dense[16]. The effect of this on the depth profile is to make the layers appear thinner than they actually are. It is also necessary to extrapolate between number densities of these compounds in order to profile the changes in nitrogen solid solution in TiN and α-Ti.

NRA complements RBS in respect to nitrogen and provides additional information on oxygen. Figure 3 shows the emitted particle energy spectrum for sample A, produced when the coating was bombarded with 1.1 MeV

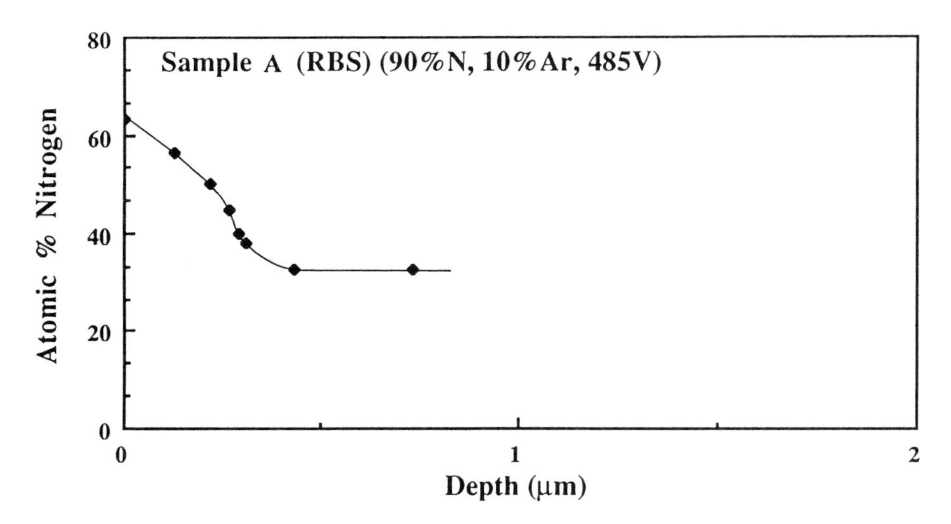

Figure 2 Depth profile of nitrogen in the first micron of sample A, used to produce the RBS simulation shown in Figure 1

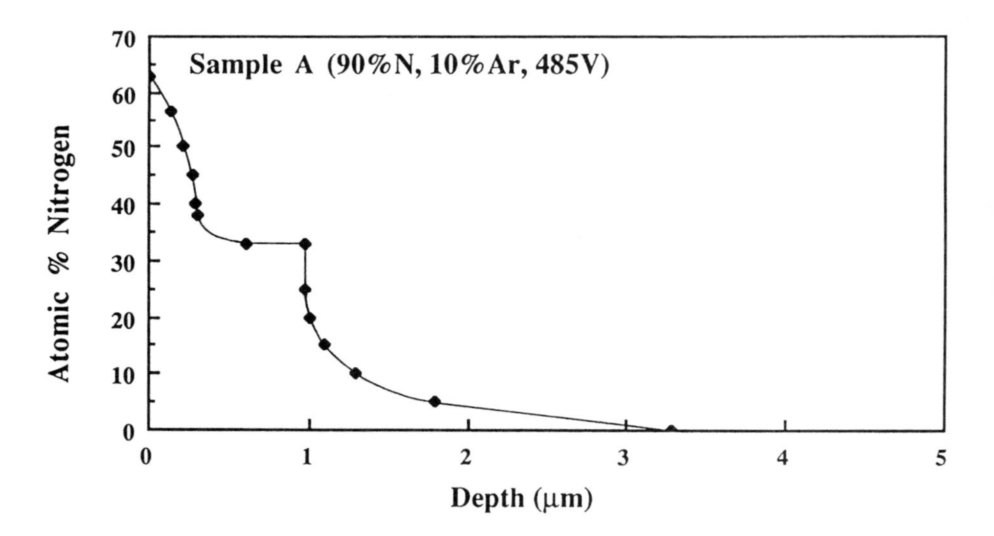

Figure 4 Nitrogen depth profile of the first 5 microns of sample A, used to produce the NRA simulation of the $Np_{1,2}$ peak (see insert Figure 3)

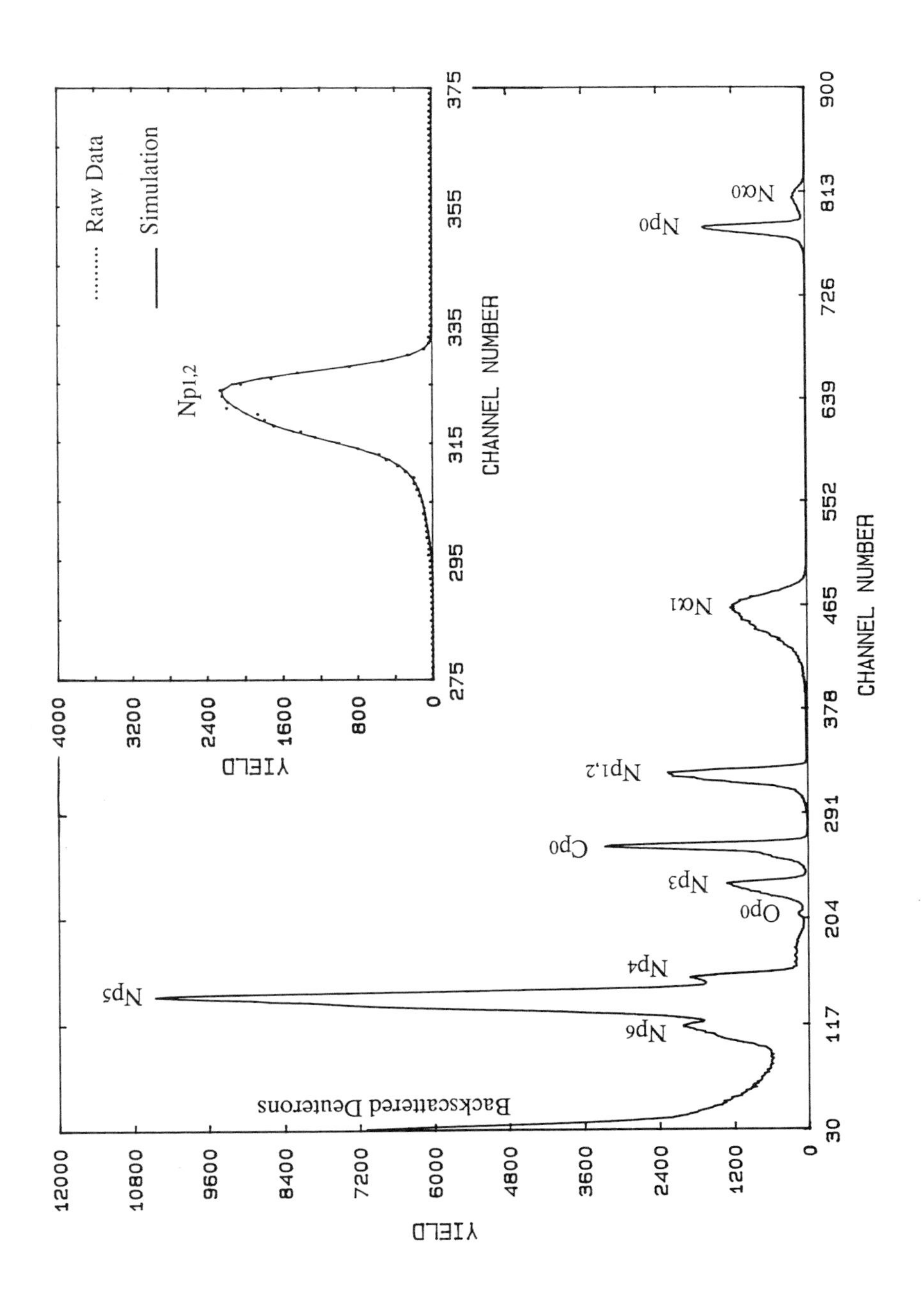

Figure 3 NRA spectrum for sample A showing the proton and alpha particle peaks from reactions with nitrogen, oxygen and carbon

deuterons, measured at a detection angle of 150^0 with an aluminium absorber foil of 12μm over the detector. The spectrum shows proton peaks from the $^{14}N(d,p)^{15}N$, $^{16}O(d,p)^{17}O$ and $^{12}C(d,p)^{13}C$ reactions and alpha peaks from the $^{14}N(d,\alpha)^{12}C$ reaction.

The large number of peaks detected when several light elements are present in a target makes the choice of beam energy and detection parameters (particularly thickness of aluminium detector foil) critical if peak overlaps and background due to backscattered deuterons are to be avoided. If the light element layer is thick then the peaks broaden towards low energy and this may also cause peak overlaps. When the beam energy is raised above 1.3 MeV titanium and aluminium undergo nuclear reactions which interfere with the nitrogen peaks.

Figure 4 shows the nitrogen depth profile used to produce a simulation of the $N(d,p_{1,2})$ peak (insert in Figure 3) to a depth of 4 to 5μm with a resolution of 0.1μm. The assumptions made about the titanium and nitrogen number densities in the coating for RBS analysis also apply for NRA. The cross-sections for two of the nitrogen peaks were available at this energy and detection angle $N(d,p_0)$ and $N(d,p_{1,2})$. Of these, the $N(d,p_{1,2})$ peak is of lower energy and therefore has better depth resolution than the $N(d,p_0)$ peak which is also overlapping with the $N(d,\alpha_0)$ peak. The $N(d,p_{1,2})$ peak was therefore used to analyse the profiles.

A small oxygen peak can be seen at channel number 207. This can be used to obtain an upper limit to the level of oxygen contamination in the coating. A carbon peak can be seen at channel number 263, which is due to surface carbon present on the coating; this is found for all materials and is usually in the form of hydrocarbons. The $C(d,p_0)$ reaction has a high cross-section at 1.1 MeV compared with the nitrogen and oxygen reactions, so the technique is very sensitive to small amounts of carbon.

EPMA on Cross-section

EPMA, unlike RBS and NRA has no inherent depth information in its spectra, so depth profiling is carried out by measuring the nitrogen concentration at intervals

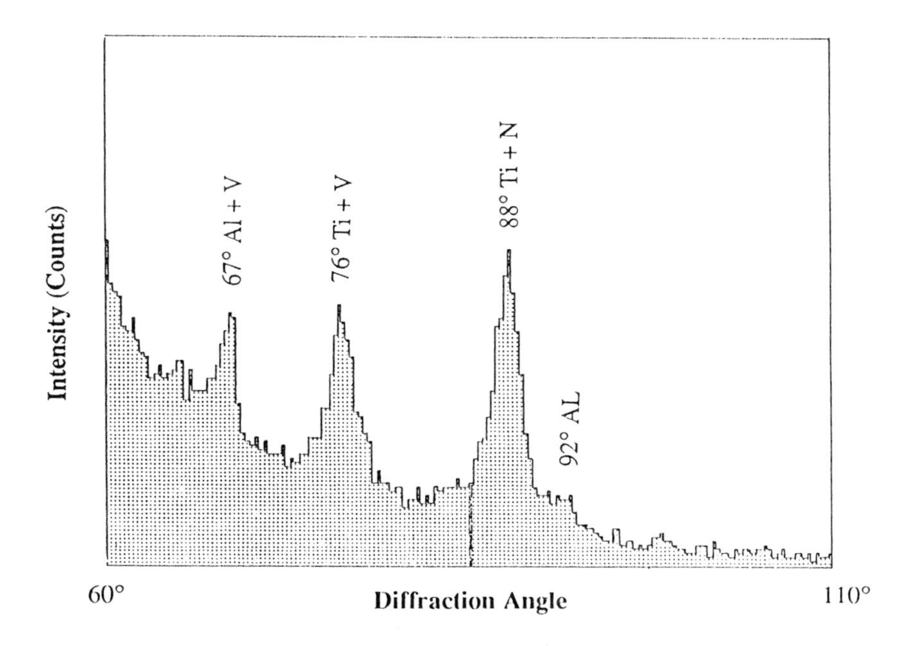

Figure 5 EPMA spectrum for the nitrided region of sample D

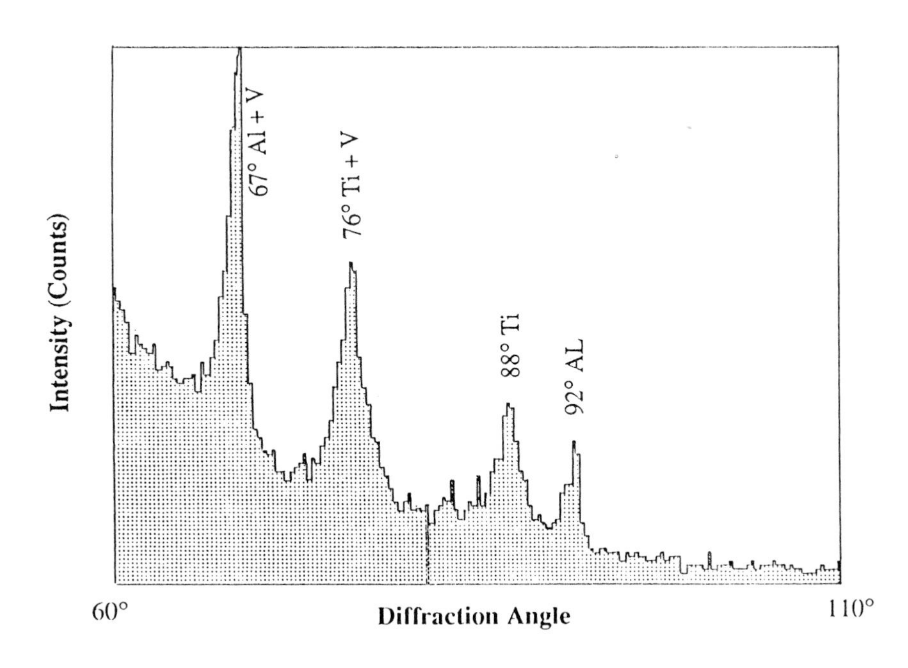

Figure 6 EPMA spectrum for IMI 318 (Ti-6Al-4V)

on a cross-section of the coating. O, Al and V concentrations can also be measured.

Figure 5 shows a wave length dispersive X-ray (WDX) scan measured on a cross-section of the nitrided region of sample D using the electron microprobe. From this it can be seen that a titanium peak overlaps the nitrogen peak at 88^0. An aluminium peak overlaps a vanadium peak at 67^0. Figure 6 shows a WDX scan of the IMI 318 matrix (Ti-6Al-4V), with titanium, aluminium and vanadium peaks. All these peak overlaps make the analysis of nitrogen in titanium difficult and it is necessary to assume that the ratio of the titanium peak heights is constant and that the overlap between the peaks remains the same.

Figure 7 shows the concentration profiles of nitrogen and titanium calculated for the first 10μm over the cross-section of the nitrided region below the sample surface. The nitrogen concentration appears to drop off very fast after 5μm and there is no clear region of Ti_2N, but it is not known whether this is an effect of the analysis method or the concentration of nitrogen.

Figure 8 shows a comparison between the depth profiles obtained by the EPMA and NRA techniques for sample D. From this it can be seen that the NRA profile does not go deep enough to allow a comparison of compound depth thickness, however the EPMA technique does suggest a higher concentration for nitrogen in the outer region. This difference is probably due to matrix effects in the EPMA technique, which cause the sensitivity of the technique to nitrogen to vary with the composition of the sample. The EPMA profile does not show a region which corresponds to Ti_2N, whcih may be because the Ti_2N layer is narrower than the beam diameter (~1μm) so is not resolved as a separate layer. The presence of a significant quantity of Ti_2N in this sample was confirmed by X-ray diffraction analysis. A sample with a thinner coating than D was not analysed with EPMA because of the poor resolution, so a complete comparison could not be made between EPMA and NRA over a whole coating thickness.

Peaks from the aluminium and vanadium present in the titanium alloy IMI 318 appear on the spectrum. Figure 9 shows the profiles calculated for the two aluminium peaks. The difference between them has been assumed to be due to

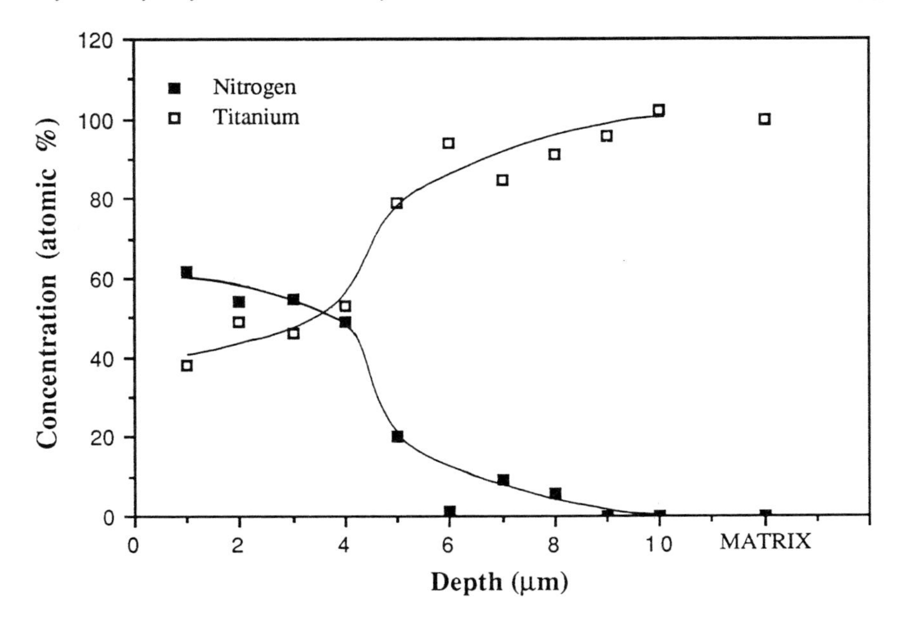

Figure 7 Concentration depth profiles for nitrogen and titanium in sample D measured by EPMA

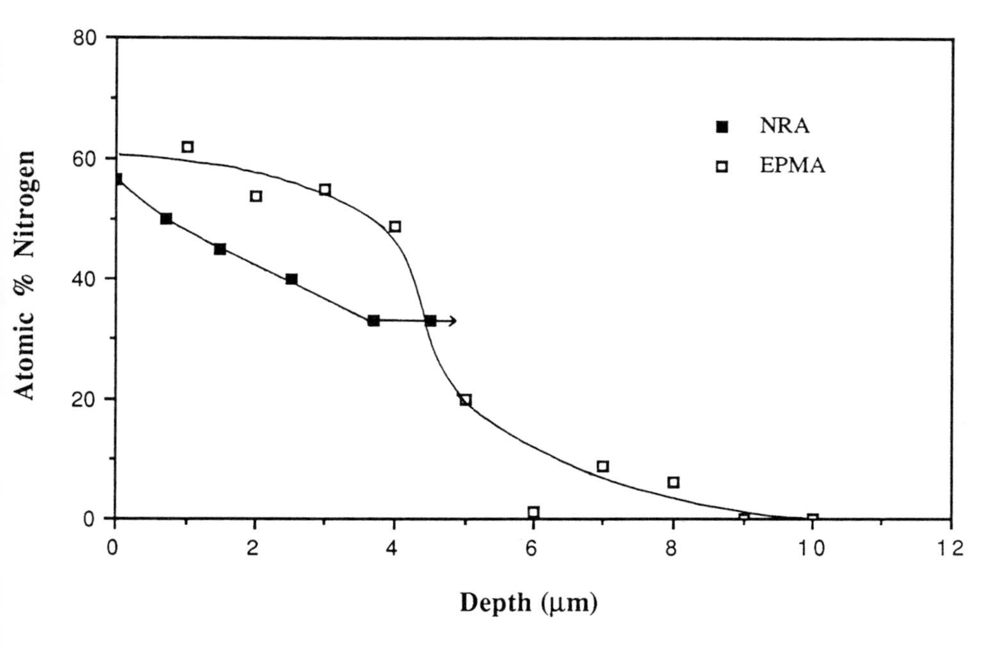

Figure 8 Comparison between the EPMA depth profile for nitrogen and the NRA depth profile for the same sample D

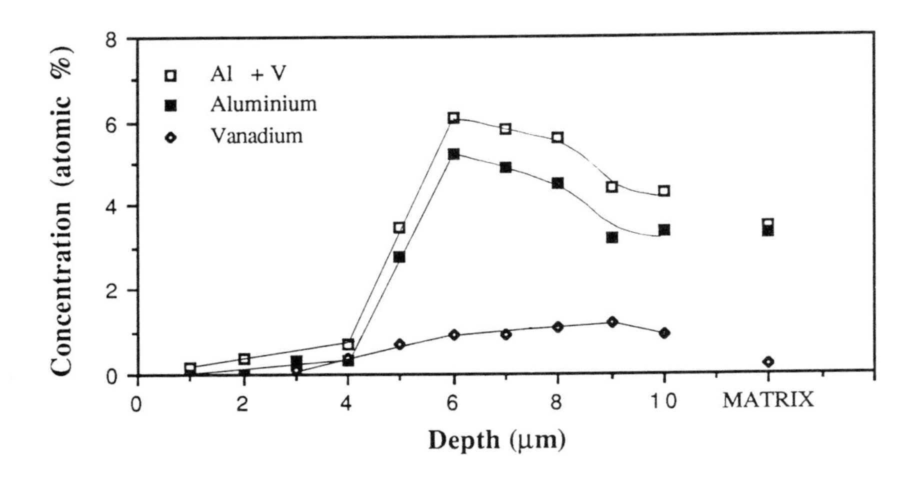

Figure 9 Concentration depth profiles for aluminium and vanadium in sample D measured by EPMA

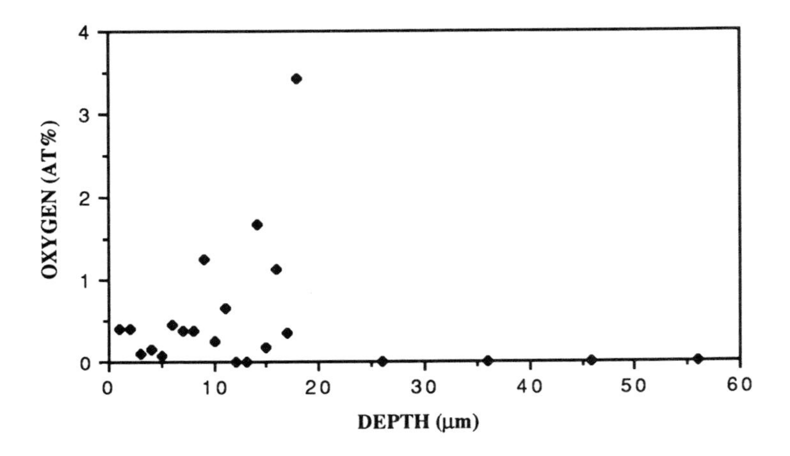

Figure 10 Oxygen depth profile in sample D measured by EPMA

vanadium, the concentration of which is small compared with the titanium concentration. Both aluminium and vanadium appear to have been pushed back from the surface by the nitriding process, thus increasing the concentration of aluminium and vanadium at a distance of about 6µm below the surface to above the normal concentration found in IMI 318.

Oxygen was found to vary in concentration between 0 to 2 atomic % in the first 20 to 25µm below the surface of sample D (Figure 10), but it was not detected in the unnitrided region of the IMI 318. This confirms the belief that oxygen is detected from the bulk of the sample (penetration depth of the beam 1µm) rather than merely on the surface exposed by producing the cross-section, where additional oxygen would be expected because of titanium's high affinity for oxygen.

XPS and In-situ Ion Etching

XPS is the only technique which shows the chemical environment of the coating constituents Ti, O, N and C. Figure 11 shows a wide scan XPS spectrum of sample B. The following lines can be seen in the direction of increasing binding energies (BE): Ti3p (together with O2s), Ti3s, Ar2p, C1s, N1s, Ti2p, O1s (Ti2s), and two sets of Auger lines of the O(KLL) and Ti(LLM) series. Lines from the copper sample holder (Cu3p, Cu3s and Cu(LMM)) and argon lines implanted in the surface during Ar^+ ion etching are also visible.

High resolution scans of the main peaks show most of these peaks to be composite peaks made up of several components. The insert in Figure 11 shows a high resolution scan of the Ti2p peak and its component peaks. The position of the component peaks is found by using a peak synthesis procedure. The various component peaks are shifted from the binding energy of pure titanium peaks because they correspond to titanium in several different chemical environments, in this case TiN, TiO_2 and TiN_xO_y, the chemical bonds of titanium with the element causing a shift in titanium's electron binding energies.

In the region of the sample analysed (~20Å into the surface), about half (50 to 60%) of the titanium is in a TiN environment, a small amount is in the TiO_2 state (14 to

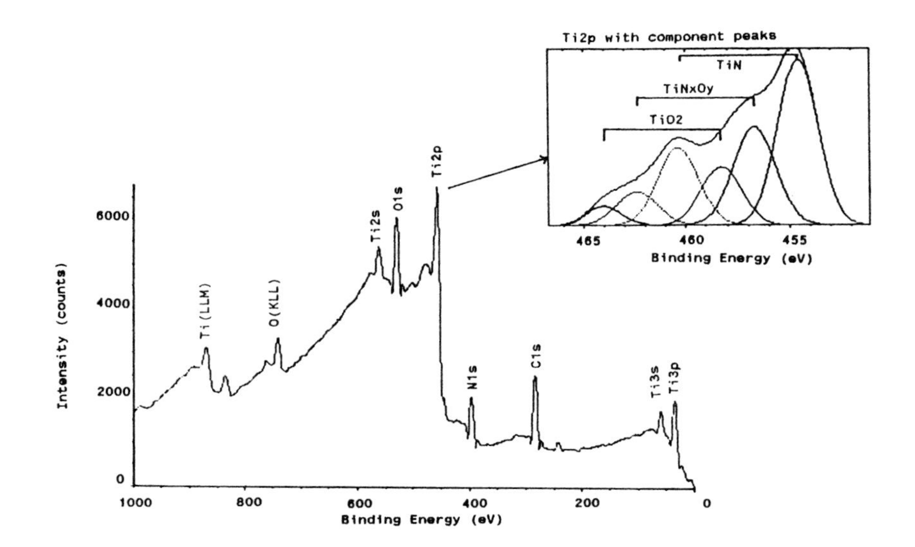

Figure 11 A wide scan XPS spectrum of sample B with a narrow (high resolution) scan of the Ti2p peak (insert), which has been split into its component peaks

18%) but a significant amount is in an environment resembling the low valency state of titanium oxides, with compositional data suggesting this to be a kind of TiN_xO_y oxynitride. A small amount of carbide type carbon was detected along with some titanium in the layers bound to hydrogen.

Information on the structure below the surface was obtained by mechanical polishing and by ion etching. Mechanical polishing down to the Ti_2N layer was carried out on sample B. However, no nitrogen deficient component corresponding to Ti_2N could be unambiguously identified by XPS.

Profiling using the ion etch (5 min. etch = removal of 4 to 5 atomic layers) on sample C (Figure 12) shows a rapid drop in carbon away from the surface with an increase in titanium and nitrogen; however the oxygen concentration remains at about the same level of 23 atomic %. This oxygen concentration is much higher than that measured with other surface analysis methods (NRA and EPMA) on plasma nitrided titanium.

Ar^+ ion etching is carried out in the spectrometer at a pressure of 5 x 10^{-6} mbar, which drops to 1.5 x 10^{-9} mbar before spectra are measured. In most circumstances this would be expected to remove surface contaminants with recontamination then kept to a minimum by the high vacuum conditions. However in the light of the unexpectedly high oxygen concentration measured by XPS compared with EPMA and NRA and also taking into account titanium's very high reactivity with oxygen and the very shallow penetration depth (of 2nm) of the XPS technique it can be concluded that reoxidation of the outer surface is taking place in the spectrometer during and after ion etching.

The fact that the oxygen concentration remains constant over four sets of 5 minutes Ar^+ ion etch, when room temperature oxidation is estimated to extend only to the first 5 or 10 atomic layers (1 to 2 nm) (5 minutes etch = removal of 4 to 5 atomic layers) backs up this conclusion of reoxidation during and after etching in the spectrometer.

The presence of surface oxide on titanium nitride, also explains why no nitrogen deficient component peak

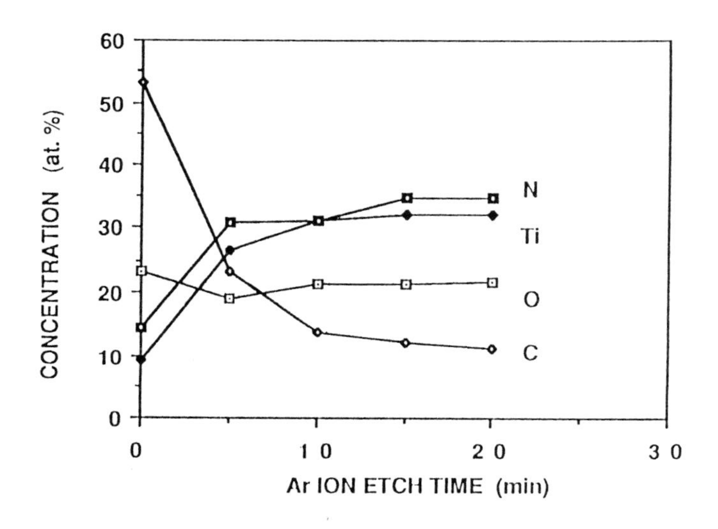

Figure 12 XPS profiles of nitrogen, titanium, oxygen and carbon in sample C

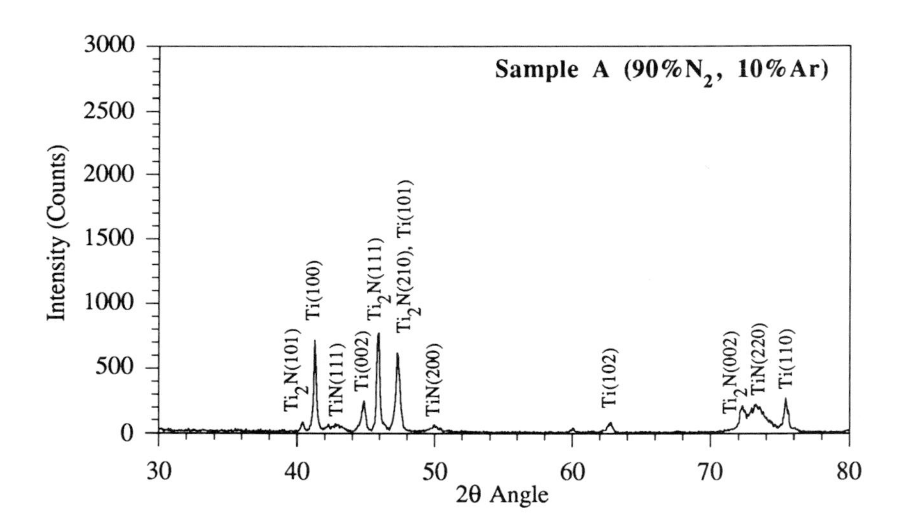

Figure 13 X-ray diffraction trace for sample A

corresponding to Ti_2N could be unambiguously detected by XPS on the sample B (polished) as with the depth analysis the material was in the form of a titanium oxynitride. The presence of surface oxide might be explained by contamination arising from moisture in the sputtering argon or from the XPS vacuum system, and it should be noted that it did occur for both instruments.

XPS thus provides strong evidence for the presence of an oxide (TiO_2) film on titanium nitride which can form at room temperature and even at very low partial pressures of oxygen such as in the XPS spectrometer.

These facts make this technique unsuitable for the analysis of the chemical environment of titanium nitride coatings, which differs significantly between the near-surface and the bulk. However, it would be of great use in corrosion type studies for analysing the very thin outermost film on the surface coatings.

X-ray Diffraction Analysis

XRD is the only technique used which provides phase information. Figure 13 shows the XRD spectrum obtained for sample A. The spectrum shows patterns from the αTi, ϵTi_2N and δTiN phases[17]. The presence of these phases fits in with the atomic percentages of nitrogen seen on the NRA/RBS profiles.

No sign of titanium oxide peaks can be seen on the XRD pattern, but the method of XRD used is not very sensitive to small amounts of phase. Some background appears on the spectra for samples nitrided at 850^0C.

The choice of radiation is important, as the penetration depth of different X-rays allows different depths of surface to be analysed. Cr Kα radiation has a low depth of penetration in TiN (2.5μm) but the pattern was found to be noisy with a high background. So Co Kα radiation has been used which has a penetration depth of 5μm.

Variations in peak width and height between the XRD spectra for different samples provide information on the preferred orientations of phases in the coatings and the variation in the amount of nitrogen in solid solution in the phases.

5 CONCLUSIONS

An overall picture of the structure and composition of plasma nitrided titanium coatings can be obtained using the surface analysis discussed.

XRD confirms the phases present in the coating to be TiN, Ti_2N and αTi. This fits in with the depth profiles obtained by RBS and NRA, which can be combined to give a nitrogen depth profile to a depth of 4 to 5μm (resolution 0.1μm) with a high resolution in the first micron (0.01μm) from the RBS. The nitrogen concentration depth profile from NRA/RBS has an initial steep region where the nitrogen concentration drops rapidly, suggesting a thin TiN layer, a flattened region containing about 33 atomic % nitrogen corresponding to the Ti_2N phase, followed by an exponential fall off in nitrogen corresponding to the region of nitrogen diffusion in αTi.

The depth of beam penetration is a limitation in the NRA technique for samples with thicker coatings. EPMA provides nitrogen profiles to a greater depth, but the resolution is poor and the TiN and Ti_2N layers are not clear. The comparison between EPMA and NRA shows the nitrogen concentration to be higher in the compound zone when measured by EPMA than by NRA. The boundary between the compound layer and the diffusion layer is clearly defined by EPMA and because it is measured on a cross-section it is not affected by coating density like NRA and RBS.

NRA, RBS and EPMA all indicate the presence of over-stoichiometric TiN on the surface of the samples[18].

The oxygen measurements by EPMA over a cross-section showed a higher concentration of oxygen in the nitrided region than in the unnitrided substrate. Oxygen measurements made on the outside surface using EPMA gave oxygen concentrations of less than 3 atomic %, the concentration being averaged over the depth of beam penetration (1μm). However, these readings were very scattered due to the small area of the beam (1μm diameter) which was comparable in size to changes in coating microstructure. Oxygen measurements obtained by NRA (beam area 2mmx2mm, penetration depth 5μm) gave similar oxygen concentrations (<3 atomic %) and also could be used to show changes in oxygen concentration with nitriding conditions[18].

XPS characterised the very near-surface of the samples (~2nm) and showed the presence of a thin oxidised layer on the outside of the coating in the form of TiO_2, an oxynitride TiN_xO_y and TiN. Etching with an Ar^+ ion beam to clean off the oxide layer proved unsuccessful as reoxidation of the surface took place in the spectrometer, probably during and after ion etching. This reoxidation also made the use of ion etching to depth profile the titanium nitride coatings misleading, as high oxygen concentrations (~23 atomic %) were measured on all the etched surfaces. NRA, RBS and EPMA all indicate low oxygen concentrations within the layers and it has therefore been concluded that after Ar^+ ion etching the titanium nitride surface region measured by XPS is not representative of the bulk of the layers.

Surface analysis can provide a large amount of detailed information on titanium nitride coatings. However, the use of several techniques is recommended, as information obtained from individual techniques can be misleading when applied to different materials.

REFERENCES

1. M. Thoma, Proc. Conf. Designing with Titanium, Institute of Metals, Bristol, July 1986.
2. T. Bell, Z.L. Zhang, J. Lanagan and A.M. Staines, "Coatings and Surface Treatment for Corrosion and Wear Resistance," Edited by K.N. Strafford, P.K. Datta and C.G. Coogan, Ellis Horwood, 1984, p. 165.
3. B. Edenhofer, Heat Treatment of Metals, 1974, 1, 23.
4. A. Von Engle, "Ionised Gases," 2nd Edition, Oxford University Press, 1965, p. 217.
5. K.T. Rie and Th. Lampe, Proc. Int. Conf. on Surface Modification of Metals by Ion Beams, Heidelberg, Sept. 1987.
6. T. Bell, H.W. Bergmann, J. Lanagan, P.H. Morton and A.M. Staines, Surface Engineering, 1986, 2(2), 133.
7. J.C. Simpson and L.G. Earwaker, Surface and Coating Technology, 1986, 27, 41.

8. J.C. Simpson and L.G. Earwaker, Nuclear Inst. & Meth. in Phys. Research, 1986, B15, 502.
9. R.A. Jarjis, Nuclear Cross Section Data for Surface Analysis V1-3, University of Manchester, Dept. of Physics Internal Report, 1979.
10. C.F. Feldman and J.W. Mayer, "Fundamentals of Surface and Thin Film Analysis," North Holland, London, 1987.
11. M.H. Lorreto, "Electron Beam Analysis of Materials," Chapman and Hall, 1984.
12. J.E. Castle, "Analysis of High Temperature Materials," Edited by O. Van der Biest, Applied Science Publishers, 1983.
13. D. Briggs and J.C. Riviere, "Particle Surface Analysis by Auger and X-ray Photoelectron Spectroscopy," Edited by D. Briggs and M.P. Seah, Ch.3, Spectral Interpretation, John Wiley & Sons, Ltd, 1983.
14. M. Wittmer, J. Noser and H. Melchior, J. Appl. Phys., 1981, 52(11), 6659.
15. B.D. Cullity, "Elements of X-ray Diffraction," Addison-Wesley Inc., 2nd Edition, 1978.
16. See, for example, Electrochem. Soc. Ext. Abstr.3, 1966 and papers therein.
17. H.A. Wreidt and J.L. Murray, "Phase Diagrams of Binary Titanium Alloys," Edited by J.L. Murray, ASM International, 1987.
18. H.J. Brading, L.G. Earwaker, P.H. Morton and T. Bell, Proc. 5th Int. Conf. on Surface Modification, 2nd-4th Sept., 1991, University of Birmingham, UK.

4.2.3
Elevated Temperature Wear Surfaces of Some Mould Materials Studied by STM and SIMS

K. A. Pischow, E. O. Ristolainen, and A. S. Korhonen

LABORATORY OF PROCESSING AND HEAT TREATMENT OF MATERIALS, HELSINKI UNIVERSITY OF TECHNOLOGY, VUORIMIEHENTIE 2 A, 02150 ESPOO, FINLAND

1 INTRODUCTION

Good thermal shock resistance, resistance to high temperature wear, resistance to oxidation and scale growth, resistance to dimensional changes, and resistance to surface breakdown are essential requirements for tools used at elevated temperatures. Good thermal diffusivity and surface finish are further requirements for mould materials used in the glass industry[1]. In both cases the overall cost effectiveness will be of vital importance.

The oxidation of TiN at elevated temperatures might have a considerable effect on the failure of the coating-substrate system. Hofmann[2] studied the formation and diffusion properties of oxide films with Auger Electron Spectroscopy (AES) and X-ray Photoelectron Spectroscopy (XPS). The thickness of a TiO_2 layer increased parabolically with time at temperatures between 400-650^{0}C, which was attributed to a dominant oxygen diffusion process. A sample oxidized for 1 h at 500^{0}C in air showed an oxidizing depth of over 100nm and no TiN at a depth of 60 nm. The oxide layer was purely TiO_2. According to the author, the diffusion of oxygen and nitrogen is mostly prevalent, leading to a relatively sharp oxide-nitride interface with practically unaltered nitride composition beneath the oxide layer. An interesting comparison was made with chromium nitride films oxidised as before at 500^{0}C. This time a chromium oxide layer (Cr_2O_3) was formed, but beneath this layer the nitride was depleted in Cr and enriched in N. This behaviour, which was also found at higher temperatures, suggests an interpretation as follows. The growth of the oxide layer occurs mainly by out-diffusion of Cr, leaving a zone depleted of Cr in the coating underneath, the depth of which increases in the same manner as the oxide layer thickness. On the other hand, the content of nitrogen increases. This can be attributed to a much reduced diffusion of nitrogen through the oxide, which appears to be stable and compact.

Selective steam oxidation of titanium and aluminium in TiN and (Ti,Al)N PVD coatings on AISI M35 HSS substrates was reported by Louw et al[3]. Oxidation was performed in an autoclave at 130^{0}C and 210 kPa steam pressure for a maximum time of 48 min. A few circular brown oxidation spots were noticed on the TiN coatings, which can be attributed to pinhole diffusion according to the authors. Using Auger, XPS and SIMS the authors were able to prove iron diffusion to the surface through grain boundary and pinhole diffusion. The steam oxidation of TiN was explained by initial titanium oxidation at the coating surface. When there is not sufficient titanium available at the coating surface for the fast oxidation rate, oxidant diffusion occurs through grain boundaries and pinholes. Iron is oxidized at the coating-HSS interface, with subsequent diffusion of oxygen or iron ions to the surface of the coating.

In our previous studies of different mould materials we showed, firstly, that there might be a correlation between the increasing chromium content of the substrate and the decreasing oxidation rate of the TiN coating, secondly, that under sliding wear conditions at high temperatures a tribofilm forms on the surface of the coated samples strongly affecting the wear rate of the sample and thirdly, that the effects of the Ti-interlayer are strongly dependent on the chemical reactions occurring between titanium and the substrate surface. From these observations three hypotheses were formed. Firstly, the reason behind the improved corrosion oxidation resistance was assumed to be the chromium diffusion through the pinholes to the coating surface. Secondly, the tribofilm was expected to reduce the oxidation rate. Thirdly, it was assumed that the titanium in the Ti-interlayer reduces the native oxides on the substrate surface[4-6].

The aim of this study is to prove the aforementioned hypotheses by analysing the wear mechanisms and oxidation behaviour of TiN coated mould materials using Scanning

Tunnelling Microscopy (STM) and Secondary Ion Mass Spectrometry (SIMS).

2 EXPERIMENTAL

STM

The scanning tunnelling microscope used was a Struers Tunnelscope 2400 with a short scanning tube for scanning within a 6 µm x 6 µm area with a lateral resolution up to 0.1 Å. The bias voltage was adjustable from 10 mV to 5 V in steps of 1 mV. The tunnel current could be varied from 0.01 nA in steps of 10 pA. During the scanning the bias voltage and the tunnelling current were low and kept constant at 0.1 V and 1 nA respectively.

PVD TiN Coatings and Pretreatments

TiN films were deposited by an industrial size reactive triode ion plating apparatus, which has been described elsewhere[7]. Substrates were argon sputter cleaned before the deposition. Ti layers of 30, 300 and 600 nm were deposited next to the substrate before adding nitrogen to the plasma. Coating parameters were chosen based on our previous experience in achieving stoichiometric TiN with good adhesive and wear resistant properties.

SIMS Analysis

Depth profiling was performed on a VG-IX70S double focusing magnetic sector SIMS, using a Xe_2^+ primary beam at 6 keV impact energy. A primary ion current of 50 nA was used with a spot size of 1000 nm and the raster scanned over an area of 250x150 µm. A mass resolution of 1000 was used in this study. Profiling with the IX70S was performed by detecting positive $^{14}N^+$, $^{28}Si^+$, $^{50}Ti^+$, $^{52}Cr^+$, and $^{56}Fe^+$ ions.

3 RESULTS AND DISCUSSION

STM Analysis

The worn surface of TiN coated mould materials after high temperature wear testing, where impact load affected the surface, was studied with STM.

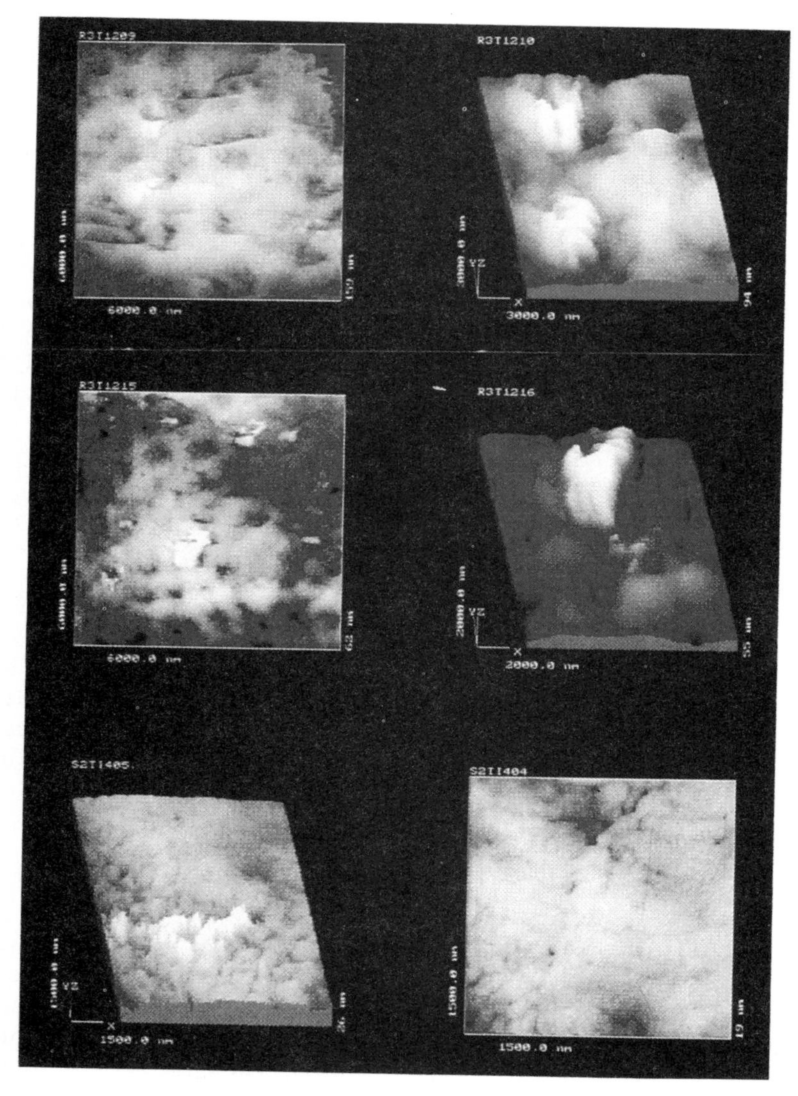

Figure 1 STM scan of TiN-coated stainless steel samples after high temperature wear tests; top row, a pit bottom showing big grains (columnar part of TiN coating) and very typically oxide perturbations in the vicinity of pin holes; middle row, a pit bottom showing small grains from the Ti-interlayer and disturbances caused by Cr_2O_3 or TiO_2; bottom row, oxidised TiN-coating

On stainless steel samples two different kinds of surfaces were detected on the bottom of the pits. Firstly, the rather smooth delaminated surface, in Figure 1 top row, showing large grains (columnar part of TiN coating) and very typically oxide perturbations in the vicinity of pin holes. Secondly, an interlayer surface, in Figure 1 middle row, showing small grains from the intermediate layer and disturbances caused by Cr_2O_3 or TiO_2. The typical oxidized TiN surface is shown in Figure 1 bottom row.

The tribofilm on the TiN coated cast irons formed at the surface during sliding against a quartz plate, maintained at a temperature of 1000^0C, was studied with the STM. The electrical properties of the thick oxides containing the tribofilm were so poor that the imaging was possible only after partial removal of the film and on places where the film was thin. On the pictured areas the STM scans gave a clear image of a solid surface formed of small particles, Figure 2(a). When a part of the tribofilm was removed, Figure 2(b), STM scans showed the cracked TiN coating surface.

Optical Microscopy of the Oxidized Surface

In order to evaluate the effect of the chromium content of the substrate on the high temperature oxidation behaviour of TiN coatings some simple oxidation tests were performed by annealing samples in an atmosphere of air.

TiN coated stainless steel and plain carbon steel sheet samples coated with PVD TiN were oxidized from 10 min to 1h at 400, 500, 600, 700 and 800^0C in air. The yellow colour of the TiN coated SS samples was unchanged at 400^0C; however, it started to change first to intense blue and red at 500 and 600^0C and then to greyish at 700^0C, being completely grey at 800^0C. The bluish colour and red spots are most probably due to the chromium and iron diffusion to the surface via pinholes and micropores. At 800^0C small brownish circular marks around pinholes were found, having the same kind of appearance as on cast iron surfaces seen by SEM-Figure 3. The colours of the plain carbon steel seem to develop mostly in the same way. However, the change to the grey colour of TiO_2 appears to be already complete at 600^0C. At 800^0C the decohesion of the coating started after 10 min. exposure. After 30

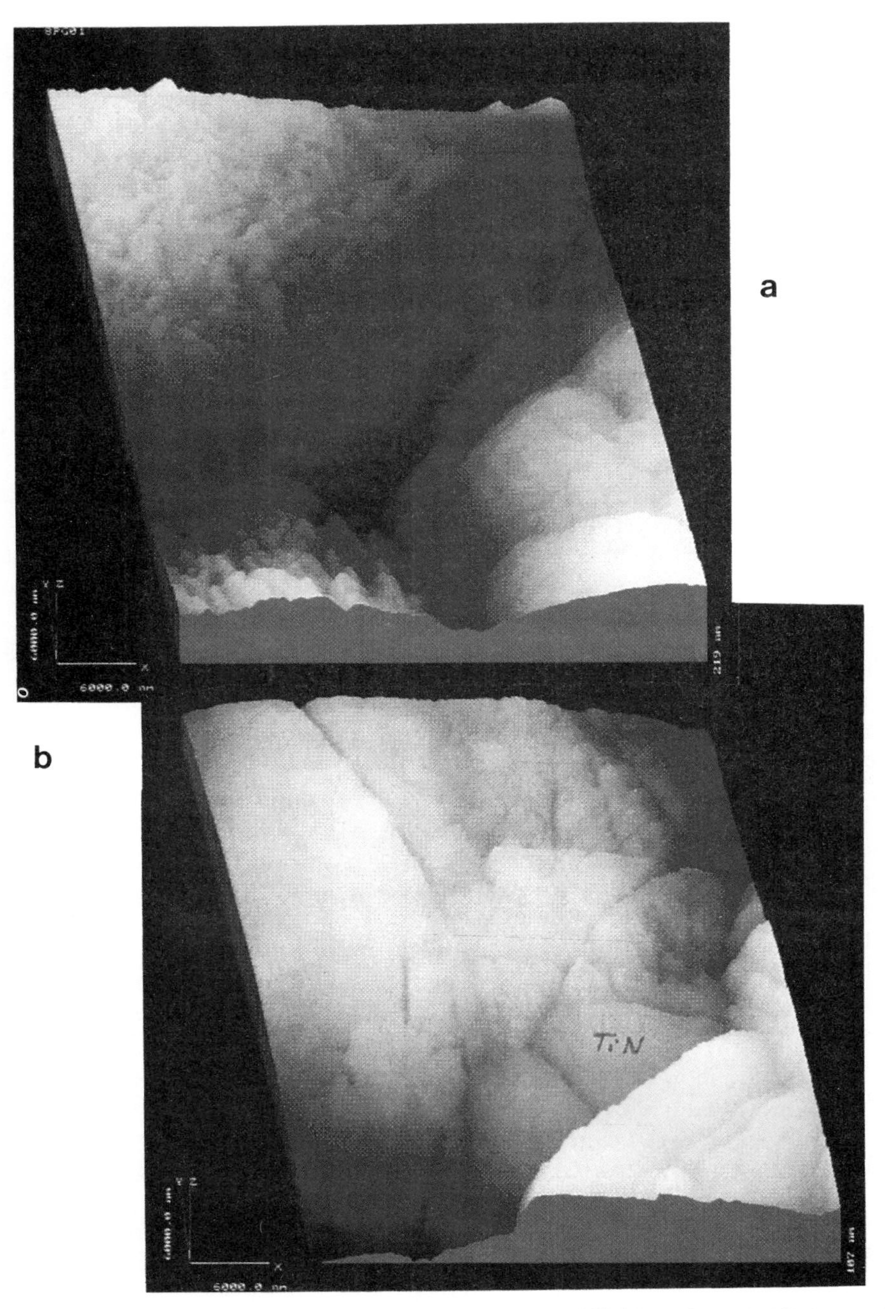

Figure 2 STM scans of the tribofilm formed on a TiN coated cast iron sample during high temperature wear tests; (a) tribofilm showing a solid surface formed of small particles; (b) when a part of the tribofilm was removed, the cracked TiN coating surface can be seen

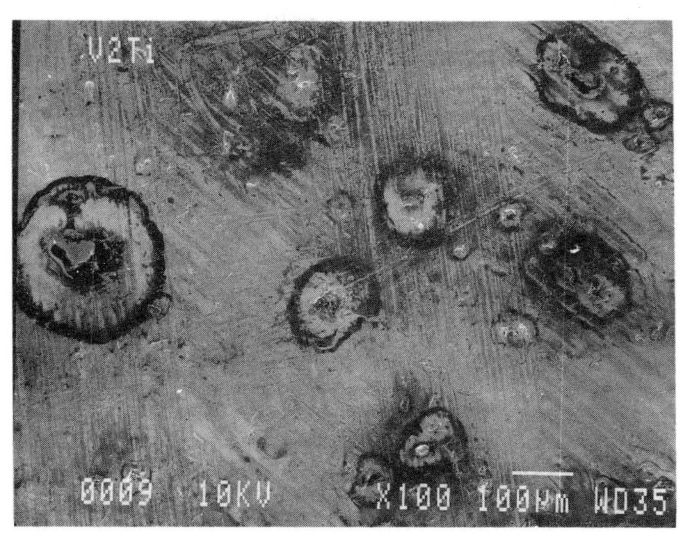

Figure 3 SEM scan of an oxidised area on the TiN coated cast iron sample surface, showing diffusion-induced circles round pinholes

minutes the decohesion was complete. In contrast, with the SS samples no signs of decohesion were noticed even after 1 hour exposure at 800^0C.

SIMS Analysis

SIMS depth profiling of the TiN coated austenitic stainless steel oxidized for 1 hour at 800^0C, Figure 4, shows high concentrates of chromium and iron on the topmost layer of the coating. It should be noticed that both nitrogen and titanium are at low levels at the beginning of the analysis, starting to increase as the chromium and iron decrease. Silicon shows also a maximum somewhere in the surface diffusion layer; the reason for and possible effects of this behaviour are not fully understood. The depth profiling of chromium shows clearly that the diffusion of Cr and Fe occurs mainly through the pinholes onto the surface.

SIMS depth profiling of the tribofilm was difficult; after 4.5 hours sputtering time the substrate surface was reached. The depth profiles of N, Si, Ti and Fe from a TiN coated cast iron sample after a high temperature wear test are shown in Figure 5. There are very strong indications of Si and Fe on the surface, derived from the worn counter piece and substrate respectively. A very high peak of nitrogen can also be seen; this must be due to the reaction of nitrogen released from the oxidized TiN

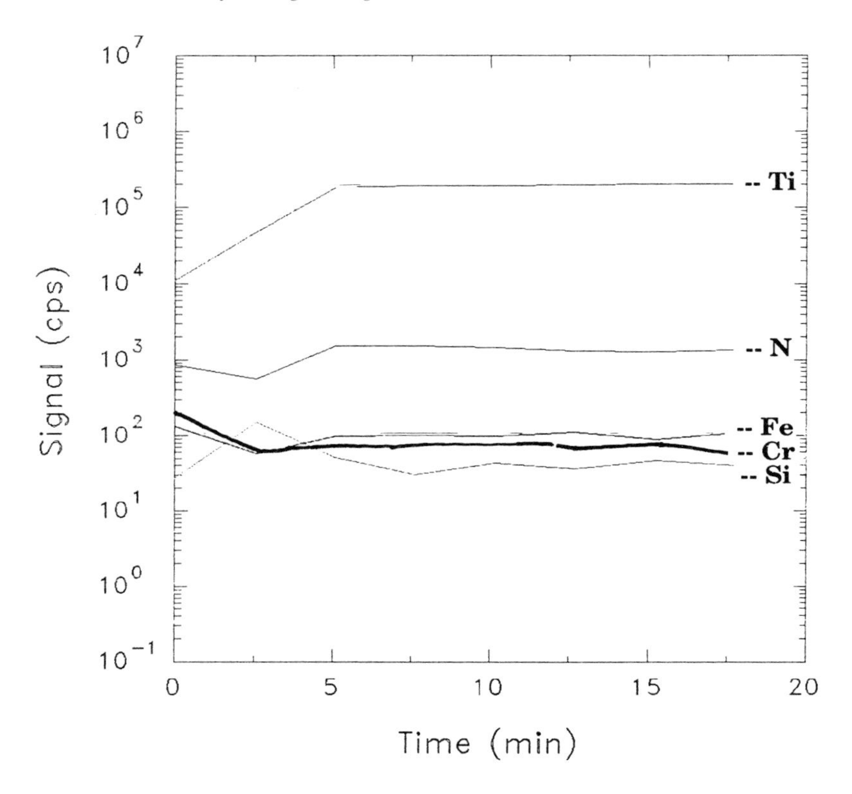

Figure 4 SIMS depth profiles of the oxidised (at 800^{0}C for one hour) TiN coated stainless steel surface

with some other components on the surface of the tribofilm. The oxidation of TiN leads to a thin but very pure layer of titanium oxide seen as a minimum in the curve of iron and nitrogen. Approaching the substrate surface the iron and silicon start to rapidly increase and the nitrogen and titanium to decrease.

Based on the aforementioned SIMS analysis and the information achieved from STM pictures it was possible to sketch the structure of the interlayers and interfaces as can be seen in Figure 6. In this figure the places of SIMS spot analyses have also been marked.

SIMS analysis of the tribofilm showed some carbon and nitrogen and considerable amounts of silicon, iron and titanium of which all seem to be in the form of oxides. The second layer is a titanium oxide layer and it seems to be very pure titanium oxide. This analysis took place on

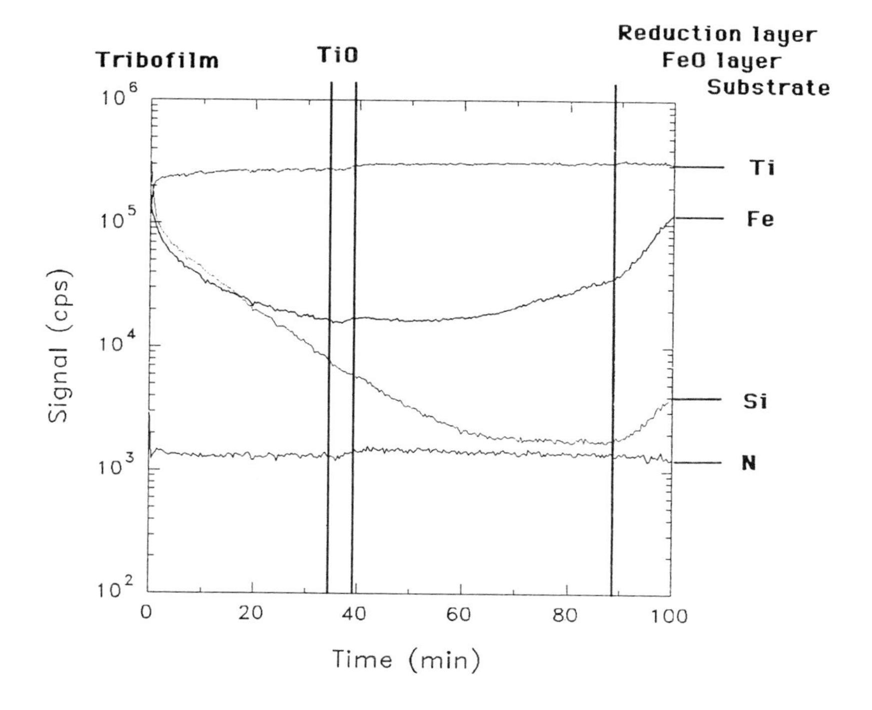

Figure 5 SIMS depth profiles of the TiN coated cast iron surface after a high temperature wear test

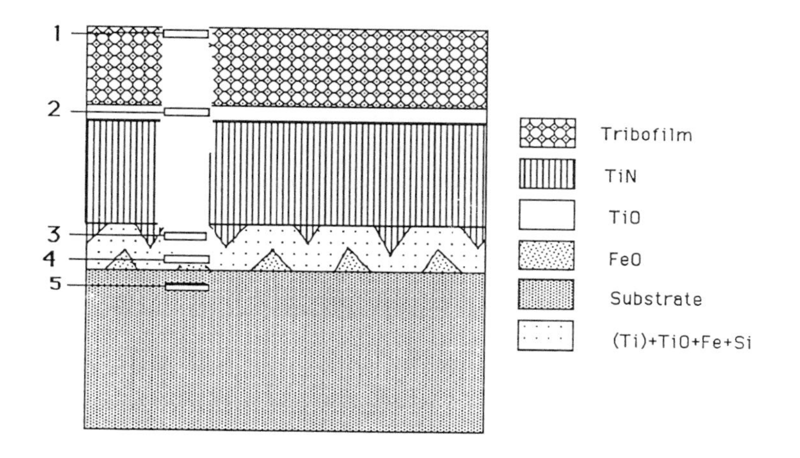

Figure 6 Schematic structure of the TiN coated cast iron surface after a high temperature wear test; SIMS analyses are numbered 1 to 5

an area which had no pinholes, since no iron can be seen. This indicates also that the diffusion of iron (and chromium) occurs through the pinholes. Under the TiN coating, SIMS analysis revealed a reduction layer, where the titanium from the 300 nm thick Ti interlayer reduces the surface oxides of the substrate. The analysis of this layer shows strong evidence of Ti and TiO and Si and Fe without the corresponding oxide peaks indicating that the reduction of silicon and iron oxides to metallic iron and silicon has occurred. Sputtering one step further gives very strong indication of iron with the peak of iron oxide showing that on the very surface of the substrate some unreacted iron oxide is left. The last scan is from the substrate, showing that some titanium has also diffused to the substrate.

4 CONCLUSIONS

The diffusion of chromium and iron through the pinholes seems to be very effective. The thin chromium oxide layer which forms on top of the coating, if the chromium content of the substrate is high enough, appears to markedly improve the oxidation resistance as well as the wear resistance of the coating-substrate system.

SIMS depth profiling of the oxidized surface showed strong evidence for the hypotheses stated in the Introduction, which was also consistent with the information achieved from the STM pictures and with the conclusion drawn from the wear test reported elsewhere[5].

STM analysis and SIMS depth profiling gave strong evidence that the high temperature sliding wear starts with the formation of a titanium oxide layer on the TiN coating. The strength and hardness of this oxide layer are low, leading to formation of a tribofilm which also includes silicon oxide particles from the counter part and iron oxides and carbon from the worn substrate.

SIMS depth profiling through the TiN/Ti-interlayer/ substrate interfaces indicated that Ti reduces iron and silicon oxides on the substrate surface which obviously contributes to coating adhesion.

The results of the STM and SIMS analyses showed that by combining the information achieved by these techniques more information about interfaces and interlayers, and processes taking place on the surface can be obtained.

ACKNOWLEDGEMENTS

The authors would like to thank R. Suominen for taking the SEM pictures and E. Harju for making the coatings.

REFERENCES

1. T.F. Ensor, Glass Technol., 1978, 19(5), 113-119.
2. S. Hofmann, Thin Solid Films, 1990, 193/194, 648-664.
3. C.W. Louw, I. LeR. Strydom, K. van den Heever and M.J. van Staden, Surface and Coatings Technology, 1991, 49, 348-352.
4. K.A. Pischow, L. Eriksson, E. Harju, A.S. Korhonen and E.O. Ristolainen, presented at ICMCTF San Diego, 6-10 April 1992.
5. K.A. Pischow, S.O. Kivivuori and A.S. Korhonen, presented at ICMPD 92 Conference, 1-3 July 1992, Siegburg, Germany.
6. K.A. Pischow, S.O. Kivivouri, J. Larkiola, P. Myllykoski, P. Ramsay, P. Terho, J. Witikkala and A.S. Korhonen, to be presented at WOM '93 San Francisco, 13-17 April 1993.
7. J.M. Molarius, Dissertation Thesis, Espoo, Finland, 1987.

Section 4.3 Surface Integrity and Properties

4.3.1
The Surface Analysis Techniques and Their Role in the Study of Coatings and Surface Modification

D. E. Sykes

INSTITUTE OF SURFACE SCIENCE AND TECHNOLOGY, UNIVERSITY OF TECHNOLOGY, LOUGHBOROUGH LE11 3TU, UK

1 INTRODUCTION

Surface analysis is concerned with the direct determination of the elemental or chemical composition of the outermost atomic layers on solid surfaces. It is precisely this part of the material, the surface, which determines many important physical and chemical properties of the material. The surface analysis techniques are well described elsewhere [1,2,3] and it is outside the scope of this paper to discuss the techniques in great detail.

The techniques can be separated into two categories:

the electron spectroscopies, Auger electron spectroscopy, Auger, and X-ray photoelectron spectroscopy, XPS, which are essentially non-destructive techniques.

the mass spectrometries, static and dynamic secondary ion mass spectrometry, SSIMS and DSIMS, laser ablation mass spectrometry, LAMS or LIMA, and their derivatives, which are inherently destructive techniques as material is physically removed from the surface for analysis.

In the electron spectroscopies the surface atoms are excited by a probe (X-rays or electrons) and electrons, with energies specific to the atom of origin, are produced as a result of the excitation. The energies of these electrons are such that they can only travel a few atomic layers in solid materials, thus the analytical signal (the electrons) is very surface specific. Electrons produced at greater depth within the sample are scattered and go to produce a background signal, thus the sensitivity of the electron spectroscopies is limited by signal to background considerations. In general detection limits are of the order of 0.1 atom % and quantitative analyses for elements from Li to U can be achieved. The local chemical environment of the atom can give rise to small chemical shifts in the electron energies, this effect is exploited to great effect in XPS.

In the mass spectrometries the surface atoms are removed; some of them are ionised in the process and these ions are detected as the analytical signal. In this case there is no background signal and consequently the mass spectrometries can be very sensitive. As the analytical signal comes from the outermost atomic layers the techniques are surface sensitive but at the same time destructive. In general all elements and isotopes can be detected to give qualitative analyses; in some cases quantitative analyses with sub ppm detection limits can be achieved. An additional feature of the mass spectrometries is that chemical information is also available through the molecular fragments produced in the material removal process. In SIMS a focused ion beam is used as the probe whilst in LIMA a pulsed laser beam is used. In a more recent development a dc glow discharge plasma has been used to remove material and subsequently ionise the sputtered particles.[4] Although this method was developed for bulk analysis it has some applications in the analysis of thick layers.

2 APPLICATIONS

The surface analysis techniques are very versatile and can be applied to a wide range of materials and processes. It is the objective of this section to provide examples of the application of the techniques to a variety of surface processes of technological importance, from the treatment of soft materials - polymers, to the production of hard coatings.

Flame Treatment of Polymers

The surfaces of polymers are relatively inert and in many materials applications good adhesion is required. In order to achieve improved adhesion some form of surface pretreatment or priming is required. The most widely used pretreatments for polyolefins are corona discharge and flame treatment. In this context XPS is a useful technique to use to study changes in the chemistry of the polymer surface following treatment as it provides a quantitative analysis of the elemental composition and, through the interpretation of chemical shifts, the presence of functional groups on the surface.

Figure 1 shows the measurement of oxygen concentration on the surface of flame treated polypropylene as a function of the air to gas ratio in the flame, together with the contact angle measurement which gives an indication of the wettability of the surface. The details of the treatment and a full discussion of the results are given elsewhere.[4] The surface oxygen level rises steadily to a maximum of 6 atom percent as the air to gas ratio is increased from 7 to 12. Over the same region the contact angle falls, indicating the effect that the increased oxygen level has on the wettability of the surface. At higher air to gas ratios the oxygen content of the surface begins to fall and the contact angle to rise, again indicating that there is an optimum air to gas ratio for the most effective treatment.

High resolution XPS spectra can be used to identify functional groups present on the sample surface; however an alternative approach is to employ the derivatization technique. A compound, containing an element not present in the sample, is reacted with the sample surface. If the compound is known to react only with specific functional groups then, by determining the concentration of the element used as a label, the concentration of the functional group can be determined. Trifluoroacetic anhydride, TFAA, is a derivatizing reagent for hydroxyl groups. Figure 2 shows the relationship between the surface oxygen concentration and the concentration of hydroxyl groups, determined by TFAA derivatization, on the flame treated polypropylene surface as a function of various process parameters. By this method it can be seen that about 20% of the oxygen present on the treated surface is in the form of hydroxyl groups.

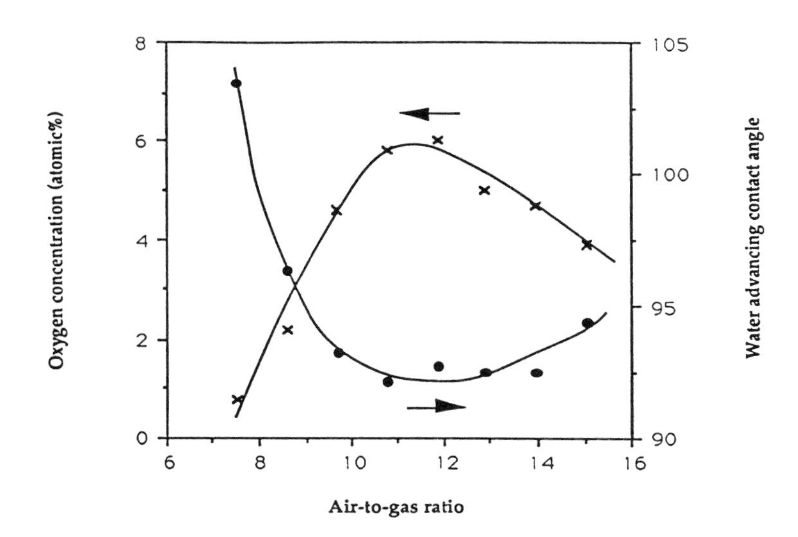

Figure 1 Surface oxygen concentration and water contact angle of flame treated polypropylene as a function of air to gas ratio in the flame

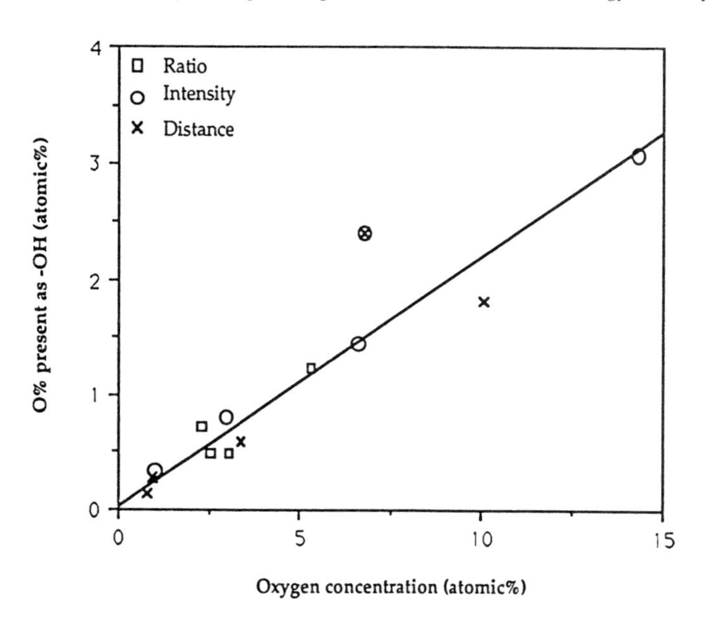

Figure 2 The relationship between surface oxygen content and surface hydroxyl concentration

Surface treatment of Carbon Fibres

In the production of carbon fibre reinforced plastics and composites the interfacial bond between the fibre and the matrix can influence the properties of the final product. In order to control the fibre-matrix interaction the surface chemistry and reactivity are modified to achieve the desired properties. One method of achieving this control is by anodic oxidation of the fibre surface. As with the polypropylene example above, XPS can be used to quantify the compositional changes of the surface.

Table 1 Surface compositions of treated and untreated polyacrylonitrile based carbon fibres

Carbon fibre type	Surface composition atom %		
	C	O	N
High modulus	96.5	3.2	0.3
High modulus oxidised	93.6	4.1	2.3
Low Modulus	92.0	4.7	3.3
Low modulus oxidised	83.0	9.2	7.8

Anodic oxidation can be seen to have increased the oxygen and nitrogen levels on both types of fibre, however the change is more pronounced on the low modulus fibre than on the high modulus material. This is consistent with the less ordered structure of the low modulus material, the high modulus material having a more graphitic, less reactive surface. More details of this study can be found elsewhere.[6]

Release Agents

An alternative method of obtaining information about the chemical groups present on a surface is to use the technique of static SIMS. In this type of analysis a very low dose of ions is used to remove molecular fragments from the sample surface. Figure 3a, for example, shows the mass spectrum from a sheet of polyethylene terephthalate, PET, that had been pressed against a sheet of PET sprayed with a polydimethylsiloxane containing release agent. Characteristic peaks at masses 73, 133, 147, 207 and 221 are seen in the spectrum. These peaks can be assigned to fragment ions from the polydimethylsiloxane, thus indicating that the release agent has transferred from the deliberately contaminated PET to the untreated sheet. Figure 3b shows the spectrum

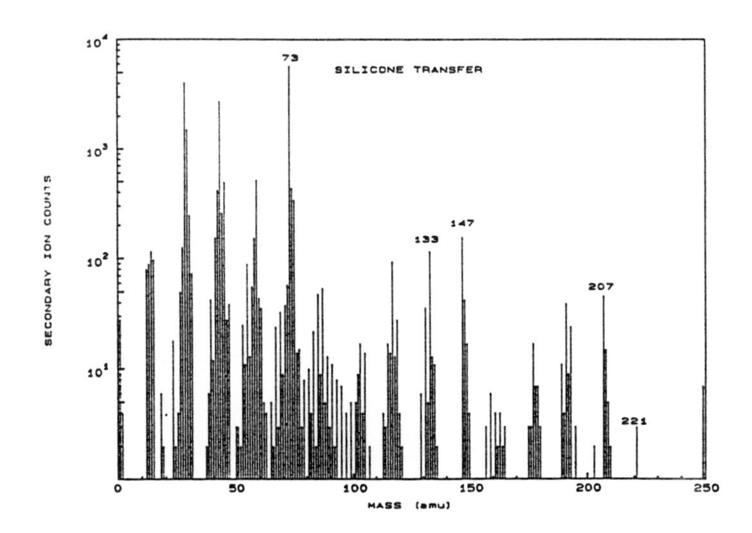

Figure 3(a) Static SIMS mass spectrum of a PET sample contaminated with polydimethylsiloxane by transfer from contact with a deliberately contaminated sample

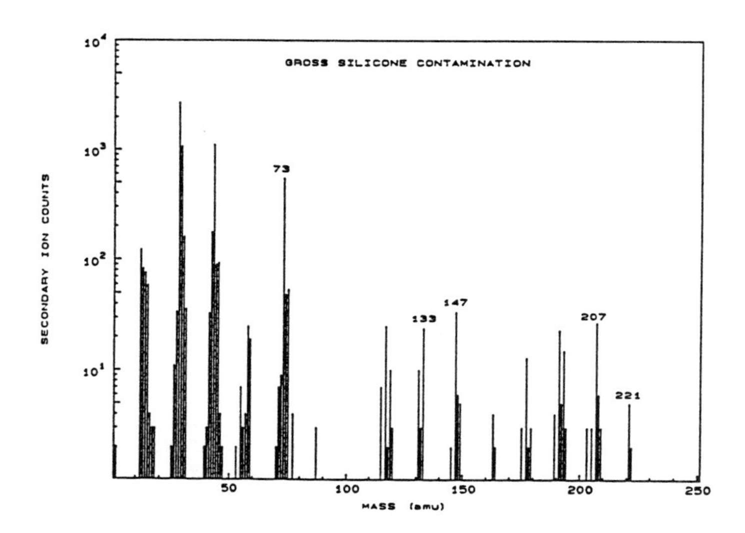

Figure 3(b) Static SIMS mass spectrum of a PET sample deliberately contaminated with polydimethylsiloxane

from the deliberately contaminated sample and again, the characteristic peaks of polydimethylsiloxane are present, as would be expected, but the signal intensity is reduced despite the fact that the concentration is higher. This illustrates the difficulty of quantification in SSIMS; the technique remains, at present, a qualitative but very powerful technique.

A drawback of the SSIMS technique, and indeed, depth profiling using ion bombardment in XPS, is that the ion bombardment process causes some modification of the sample surface; the material removal process can change the very chemistry of the surface that these two techniques are used to measure. These effects are well known to the surface analyst but may mislead the unwary. Detailed studies of the ion beam induced decomposition of transition metal oxides have been carried out by Sulivan and his co-workers.[7]

Analysis of Wear Tracks

In this example, both the spatial resolution and depth profiling capability of Auger were used to show the compositional variations between different areas of samples subjected to a sliding wear test. The samples were case hardened steel rings, the edges of which had been in contact with each other but rotating at different speeds and subjected to increasing load. In some cases a black ring was observed to develop on the rings prior to catastrophic failure. The principle objective of the analysis was to establish the composition of the black ring: to compare this with the composition of the counter face, the scuffed surfaces following failure and the original case hardened surface.

Auger identifies the major elements present as a function of position and depth. Depth profiles were performed by combining sequential argon ion bombardment of the surface with Auger analysis on samples cut from the two rings. Profiles were performed in the three regions of interest: the unworn surface, the black coloured band and the wear scar following breakdown. The major elements identified were Fe, Ni, C, O and, in the case hardened surface only, N. The depth distributions of Fe, C and O are shown in Figure 4 a to e which show the atomic concentrations in different samples as a function of argon ion etch time.

The unworn surface of the "slow" ring shows the presence of a well defined oxide layer, this layer was found to contain nitrogen at about one atom percent; no nitrogen was found in the surfaces that had been in contact during the test. The black band shows the presence of a much thicker oxide layer, with the oxygen concentration decaying slowly into the bulk material. In the wear scar region the surface is much rougher

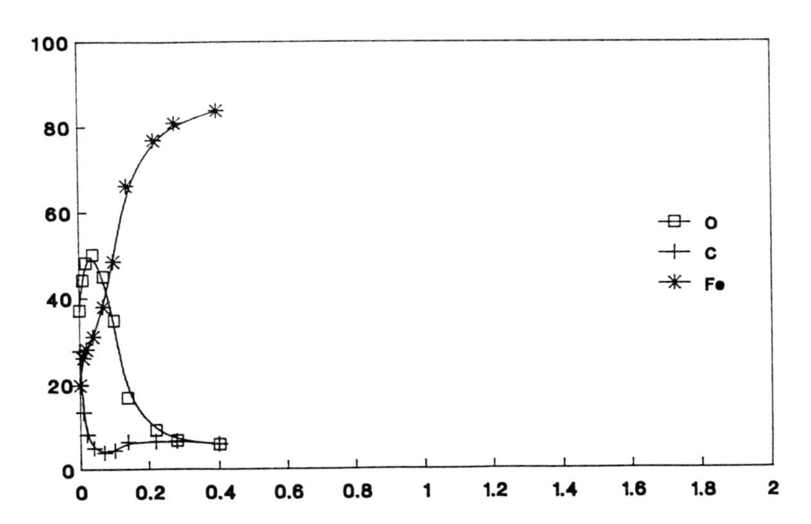

Figure 4(a) Auger depth profile through the oxide layer on an unworn region of the surface of the slow ring; atom % concentration as a function of etch time in seconds $\times 10^3$

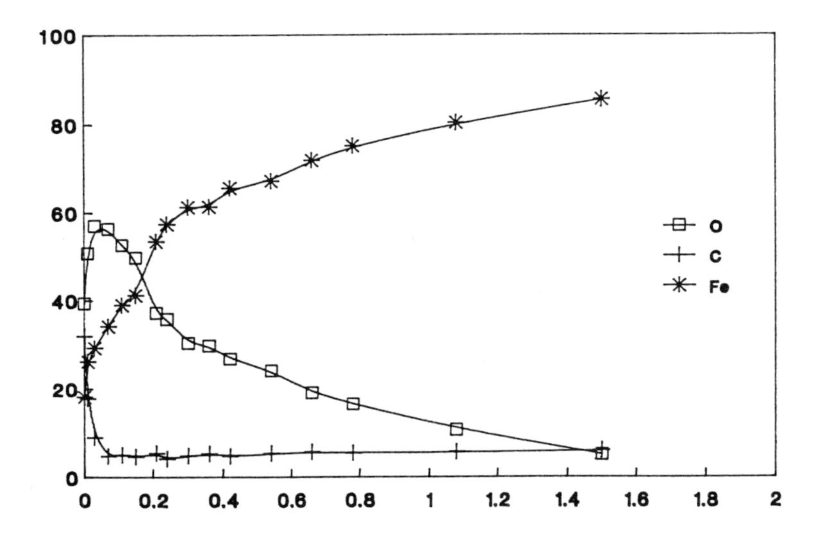

Figure 4(b) Auger depth profile through the oxide layer on the black region of the surface of the slow ring; atom % concentration as a function of etch time in seconds $x10^3$

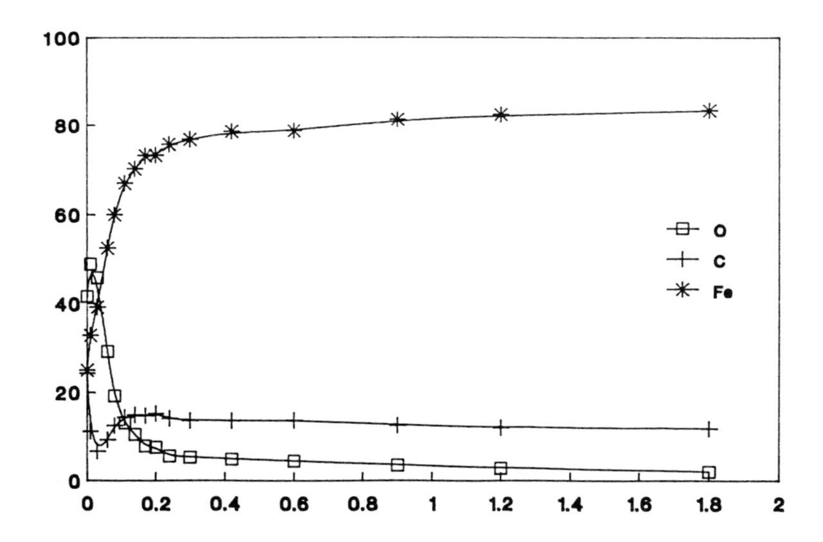

Figure 4(c) Auger depth profile through the oxide layer on the wear scar on the surface of the slow ring; atom % concentration as a function of etch time in seconds $x10^3$

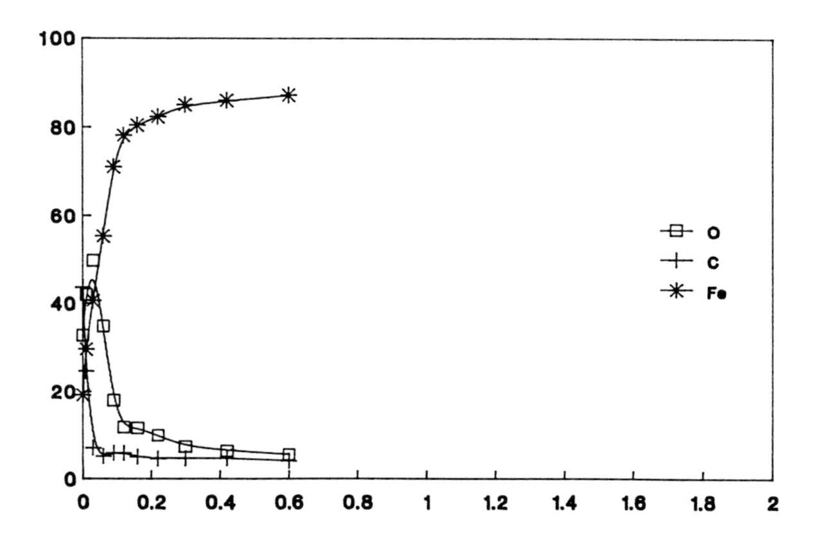

Figure 4(d) Auger depth profile through the oxide layer on the black region of the surface of the fast ring; atom % concentration as a function of etch time in seconds x10^3

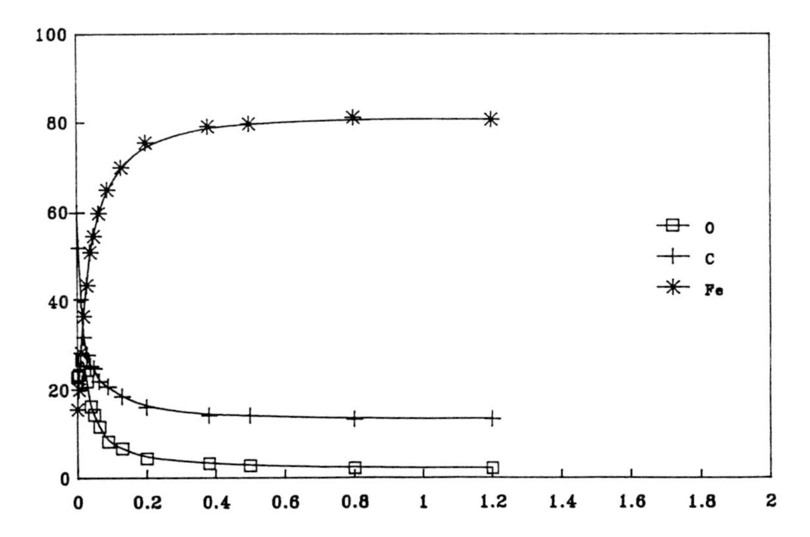

Figure 4(e) Auger depth profile through the oxide layer on the wear scar on the surface of the fast ring; atom % concentration as a function of etch time in seconds x10^3

following breakdown but the oxide layer is thinner than in the other two areas and there is a considerably higher carbon level present in the sub-surface.

The contact region on the "fast" ring shows an oxide layer that is similar in thickness to the unworn surface and not as thick as the black layer on the "slow" ring. The wear scar, however, shows the same oxide thickness and high carbon level as seen on the corresponding surface of the "slow" ring.

These measurements give an insight into the processes taking place during sliding wear and the subsequent breakdown of the surface that occurs under high load conditions.

CVD Coatings

The combination of Auger analysis with sputter profiling can be used to study surface layers up to about one micron in thickness, but beyond this the time taken to sputter profile becomes prohibitive and artefacts can be introduced into the measurement. In order to study layers of greater thickness and to probe buried interfaces, some form of mechanical sectioning can be employed and Auger used to follow compositional variations across the section. One particularly successful method developed at Loughborough is the ball cratering technique.[8] In this method an indentation is made in the surface of the sample using a spherical lap (a large ball bearing); simple geometry then relates distance across the section to depth below the surface.

Ball cratering, together with microhardness measurements and scratch testing have been used to investigate the adhesion of CVD deposited coatings on high speed steel substrates, type BT 42. Figure 5 a,b and c show the Auger profiles from ball craters in TiN, TiC and two layer (Al_2O_3 over Ti base coat) coatings. The TiN layer showed good adhesion and no flaking at the edge of the scratch whilst both the TiC and the multilayer coatings showed flaking at sub-critical loads. The Auger analysis reveals a diffuse interface for the TiN coating, a desirable feature for good layer adhesion; the TiC coating on the other hand has a much more abrupt interface. The multilayer coating however, which fails at the Al_2O_3/Ti base layer interface, also shows a diffuse interface. In order to fully explain these observations the microhardness of the layers needs to be taken into account; however the Auger analysis reveals that the layer adhesion is not simply related to interface abruptness.[9]

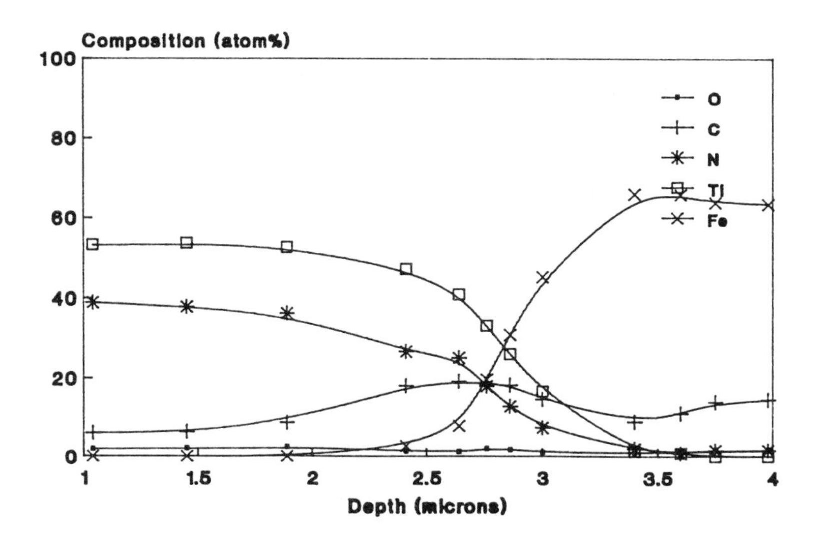

Figure 5(a) Auger ball crater profiles for the major elements through a TiN coated high speed steel sample

Plasma Nitride Samples

The plasma nitriding process can be used to produce a hard surface layer some tens of microns thick on tool steels. Although Auger analysis could be used for surface characterisation in this case as above, these samples were analysed to demonstrate the capabilities of the mass spectrometries. The samples of plasma nitride that were examined were prepared using a dc glow discharge,[10] and were analysed using a combination of imaging SIMS, LIMA and GDMS. Polished sections were analysed by SIMS and LIMA whilst GDMS was used to depth profile through the nitride layer.

In the LIMA technique a focused laser pulse is used to ablate material from an area of micron dimensions into a time of flight mass spectrometer. The technique gives a fast, qualitative small area analysis. Figure 6a shows the nitrogen to chromium ratio, in arbitrary units, as a function of position in from the edge of a polished cross section on a sample which exhibited an anomalous microhardness profile and banding in the optical image of the white layer. Because of the high sensitivity of the technique and the limited dynamic range of the detection system the nitrogen peak is off scale in the regions of high concentration. Nevertheless, an abrupt compositional change at about 60 microns is evident, in general agreement with the microhardness profile; however the data beyond the compositional change show some scatter with local high nitrogen concentrations at some positions well below the surface.

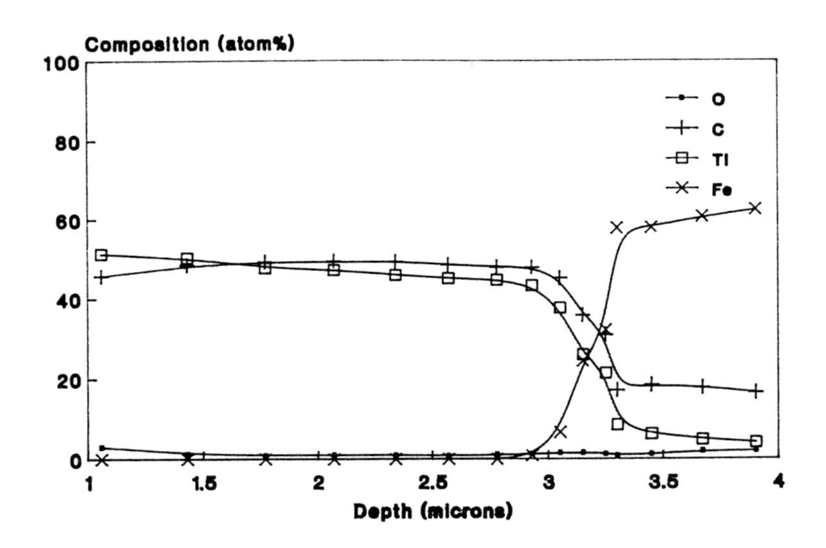

Figure 5(b) Auger ball crater profiles for the major elements through a TiC coated high speed steel sample

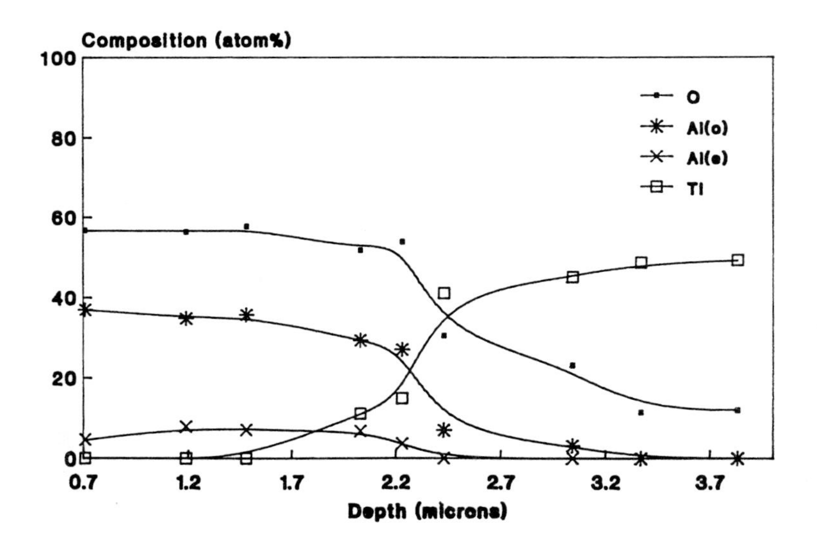

Figure 5(c) Auger ball crater profiles for the major elements through a Al_2O_3/Ti base coat multilayer on a high speed steel sample

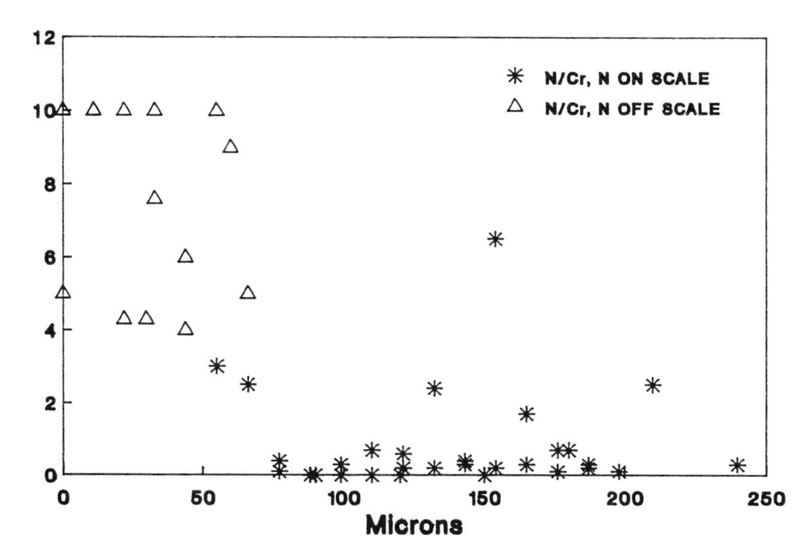

Figure 6(a) LIMA analysis: $^{14}N^+$ to $^{50}Cr^+$ ratio as a function of position on a cross section of a plasma nitride sample

Figure 6(b) and (c)
SIMS ion images of the nitrogen distribution from the plasma nitride samples; anomalous sample left (b), normal sample right, (c); images nominally 150 microns in diameter, nitrogen recorded as mass 26, CN^-

SIMS images of cross sections through the sample with the unusual microhardness profile and a sample with a conventional profile, Figures 6b and 6c, show similar nitrogen distributions to each other, with an abrupt transition in the nitrogen content at 50 to 60 microns in from the surface. Beyond the nitrogen rich layer nitrogen still appears to be present but in small regions of locally high concentration; this observation accounts for the scatter in the LIMA data and illustrates the need for large area analysis but with high spatial resolution as is achieved in imaging SIMS. No evidence for any banding in the nitrogen concentration to correspond with the banding apparent in the optical image of the white layer was found, Figure 6b. This suggests that the banding is not of a compositional nature.

The sample with the conventional microhardness profile was also depth profiled using GDMS to monitor the concentrations of major and minor elements. Figure 6d shows the nitrogen concentration as a function of depth; the other elements monitored (C, O, Si, Ti, V, Cr, Mn, Mo and W) showed no variations in concentration over the depth analysed. In agreement with the SIMS data the GDMS profile confirms the presence of a nitride layer some 50 to 60 microns in thickness.

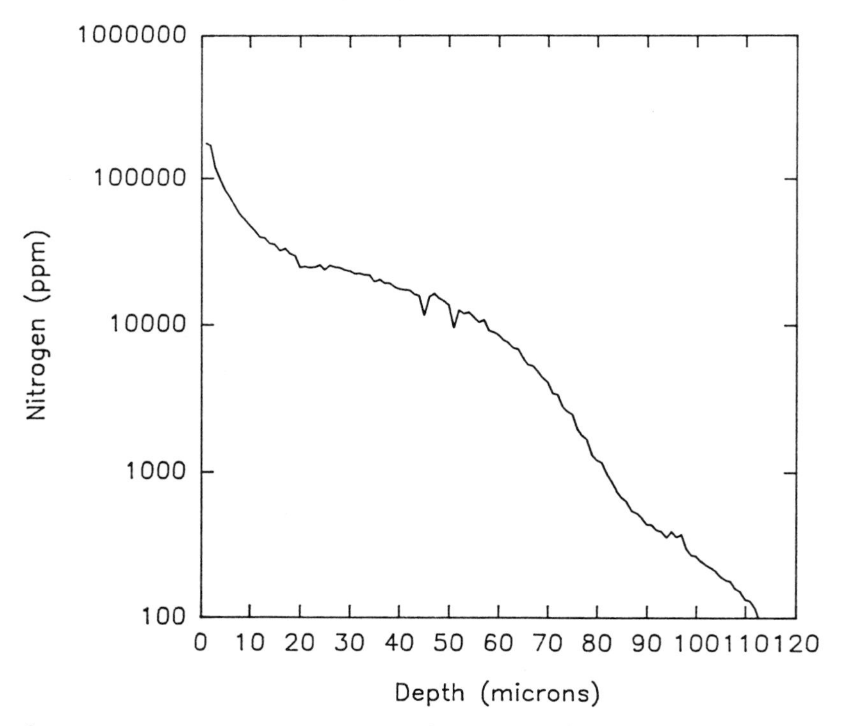

Figure 6(d) GDMS depth profile for nitrogen through the white layer on the plasma nitride sample with the conventional microhardness profile

3 CONCLUSIONS

The surface analysis techniques provide a range of analytical capabilities that can be used to study changes in surface composition deliberately produced by surface engineering or occurring as a result of surface modification. Each technique provides a different type of information and often a problem is only fully understood when several of the techniques are used in conjunction with each other. The correct choice of technique to address a particular problem quickly and efficiently requires expertise and a full understanding of the techniques and their limitations.

Despite their power as analytical tools, the information produced by the surface analysis techniques is of little value without some knowledge of the physical properties of the materials analysed. It is hoped that this paper has shown how surface analysis can be used to complement other measurements to understand processes occurring during surface engineering.

ACKNOWLEDGEMENTS

Thanks are due to my many colleagues for their contributions to this paper: Dr R H Bradley, Dr A Chew, Mrs P A Crapper, Mr G W Critchlow, Mr D D Hall, Mr E Sheng, Dr I Sutherland, Miss A E Waddilove and to Prof A Matthews (Hull), Dr D Kelly (Leicester) and Dr A B Smith (Loughborough) for supplying samples.

REFERENCES

1. Methods of Surface Analysis, ed. J M Walls, Cambridge University Press, 1989.

2. An Introduction to Surface Analysis by Electron Spectroscopy, Microscopy Handbooks 22, J F Watts, Oxford University Press, 1990.

3. Secondary Ion Mass Spectrometry, ed. J Vickerman, A Brown and N M Reed, Oxford University Press, 1989.

4. W W Harrison, Ch. 3 of Inorganic Mass Spectrometry, ed. F Adams, R Gijbels and R Van Grieken, Wiley, 1988.

5. I Sutherland, D M Brewis, R J Heath and E Sheng, Surface and Interface Analysis, 1991, 17, 507.

6. R H Bradley, X Ling and I Sutherland, Submitted to Surface and Interface Analysis.

7. J L Sulivan, S O Saied and T Choudhury, Vacuum, 1992, 43, 89.

8. J M Walls, D D Hall and D E Sykes, Surface and Interface Analysis, 1979, 1, 204.

9. Z Naeem, A B Smith, M Lamsehchi and G W Critchlow, Submitted to Surface and Interface Analysis.

10. A Leyland, K S Fancey and A Matthews, Surface Engineering, 1991, 7, 207.

4.3.2
Wetting Phenomena: Their Use and Limitations in Surface Analysis

M. E. R. Shanahan

CENTRE NATIONAL DE LA RECHERCHE SCIENTIFIQUE, ECOLE NATIONALE SUPÉRIEURE DES MINES DE PARIS, CENTRE DES MATÉRIAUX P.M. FOURT, B.P. 87, 91003 EVRY CEDEX, FRANCE

1 INTRODUCTION

Adhesion, wear and wetting properties are all interrelated in that, to a large extent, the same physical or van der Waals surface forces predominate. Although in many cases final adhesion between two phases may depend on chemical bonding, diffusion or other phenomena, it is fair to say that in the main, initial intimate contact between the two surfaces is established whilst one of the materials is in a liquid state. Without good wetting, insufficient molecular contact can rarely be obtained[1,2]. Subsequent solidification may be due simply to a temperature drop as in the case of thermoplastic materials (or rubbing metal surfaces), or to chemical reaction as in the case of thermoset polymeric resins such as epoxy adhesives. Although other mechanisms are recognised, it is thought that secondary, van der Waals bonds are often adequate to establish good mechanical strength at an adhesive interface. The relevant theory of adhesion, known as the theory of thermodynamic adsorption or wetting, was propounded by Sharpe and Schonhorn[3]. The secondary bonds referred to are those initially postulated by van der Waals in 1873 to explain deviations from the ideal gas law. These may be split into three categories : randomly oriented dipole-dipole interactions (orientation), described by Keesom[4], randomly oriented dipole-induced dipole (induction) interactions, described by Debye[5] and fluctuating dipole-induced dipole (dispersion) interactions, described by London[6]. All three types follow an inverse sixth power law of the separation distance for the energy of interaction. Theoretical modelling has been effected successfully using quantum electrodynamic considerations[7].

Since these interactions are relatively long-range (up to ca.1000 Å in the case of retarded dispersion forces), they correspond to those controlling the immediate behaviour of two phases put into contact and are thus the essential ingredients in describing wetting phenomena. Indeed, together with acid-base interactions which also contribute to adhesion and wetting phenomena [8,9], they are the cause of surface and interfacial tensions. The relationships between surface energetics and tensions and especially the fact that a surface tension acts parallel to, rather than normal to, a free surface are however relatively involved[10].

Nevertheless, accepting the existence of such surface forces, various classic equations of wetting may be derived quite simply in order to explain observed phenomena. Wetting is itself a subject of fundamental importance ; its applications lie in fields as diverse as studies of the irrigation of the human eye and the successful use of inks and paints, or tertiary oil extraction and the efficiency of sprayed insecticides. However, in addition, since wetting behaviour is sensitive to local impurities, it can be used for the characterisation of solid and/or liquid surfaces. Limitations nevertheless exist since wetting phenomena are to this day incompletelely understood, although much progress has been made in recent years.

2 FUNDAMENTAL EQUATIONS

The two fundamental equations of wetting are those attributed to Young[11] and to Laplace [12] . Young's equation considers the equilibrium configuration at the triple line where solid S, liquid L and vapour V, say, meet. Defining the equilibrium contact angle, θ_o, as the angle subtended at the triple line between the (flat) solid surface and the tangent plane to the liquid/vapour interface (in the liquid phase) we have (see Figure 1(a)) :

$$\gamma_{SV} = \gamma_{SL} + \gamma_{LV} \cos \theta_o \quad (1),$$

where γ_{ij} represents the interfacial tension between phases i and j. Equation (1) may be derived very simply by considering the force balance parallel to the solid surface. However, for those not convinced by this argument, there exist in the literature a number of other proofs invoking free energy and, in some cases, variational methods - these arrive at the same conclusion [13-16]. It is only under unusual conditions, such as when the solid is sufficiently soft to be deformed significantly by the normal component of γ_{LV} [17,18], that deviations from Young's equation, <u>at equilibrium</u>, may be expected (see Figure 1(b)).

Laplace's equation relates the pressure difference, Δp, across a liquid/fluid boundary (the fluid may be a second liquid immiscible with the first or, as assumed above, simply the vapour, V, of the initial liquid) to the corresponding interfacial tension, γ_{LV}, and the local curvature of the interface :

$$\Delta p = \gamma_{LV} \left(\frac{1}{R_1} + \frac{1}{R_2} \right) \quad (2),$$

where R_1 and R_2 are the Eulerian principal radii of curvature. Again, Laplace's equation may be derived using a force balance or an energy (work) argument [19].

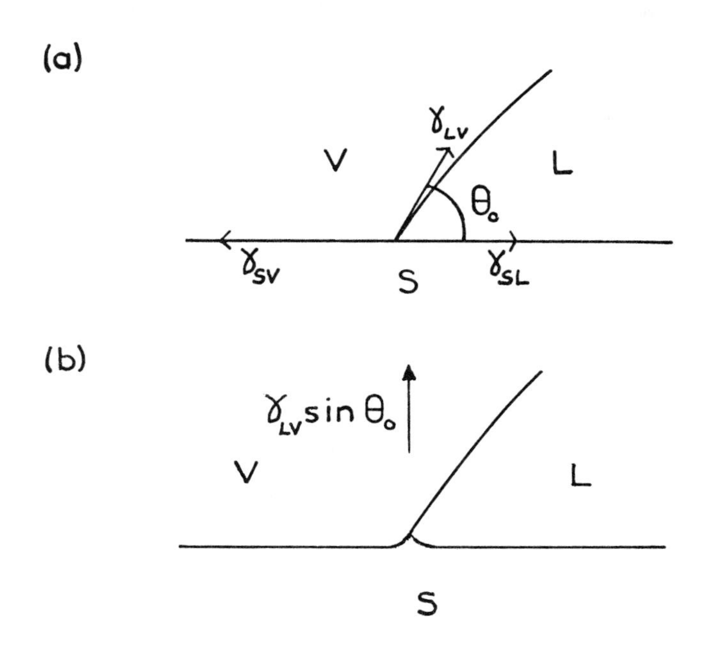

Figure 1 (a) Equilibrium of interfacial tensions at triple line
(b) Deformation of soft solid by normal component of γ_{LV}

These fundamental equations have stood the test of time and may be used in the characterisation of solid and liquid surfaces. For example, in the case of a liquid, it may be shown from Laplace's equation that a cylindrical solid plunged into a liquid bath leads to the formation of a meniscus around its perimeter, of length l, and that the apparent weight increase of the solid (neglecting buoyancy effects which can be allowed for) is equal to γ_{LV} l cos θ_o. If the liquid wets the solid completely, θ_o is zero and thus the surface tension of the liquid (interfacial tension liquid/vapour) can be ascertained. This is the principle of the Wilhelmy plate [20]. A plate of a high surface energy material, usually platinum, is used as the solid of immersion since it can usually be assumed that on such a surface, θ_o will indeed be zero. In addition, the plate is often roughened as this also aids lowering of the contact angle.

In the case of solids, surface characterisation is rather more tricky. Considering equation (1), we may obtain γ_{LV} , for example using the Wilhelmy technique, and measure the contact angle using one of a number of techniques, of which the most common is probably that of the sessile drop, but we are still left with two unknowns. The difference (γ_{SV} - γ_{SL}) may be obtained but not the value of each component. It is at this stage that additional information must be gathered in order to interpret the basic, physical wetting equations if we are to obtain more detailed knowledge of the character of the solid surface.

3 INTERPRETATION OF INTERFACIAL TENSIONS

Considering wetting using a single liquid (and its vapour), provided the solid is of relatively low surface tension, we may assimilate γ_{SV} with γ_S. The difference between the surface tension of the solid, γ_S, and its value in the presence of the vapour of the liquid, γ_{SV}, may be assumed small. (For high energy solids, this difference, known as the equilibrium spreading pressure, $\pi_e = \gamma_S - \gamma_{SV}$, may not be negligible, but is in principle obtainable from the study of adsorption isotherms [21].) Under these circumstances, the major problem is to understand the meaning of the term γ_{SL}. Several attempts have been made in the literature, of which the major ones will be briefly presented.

In the pioneering work of Zisman [22], although not stated specifically, γ_{SL} is effectively neglected. By considering the plot of the cosine of contact angle, θ_o, vs. the surface tension of a series of (homologous) liquids, a straight line is obtained. The intersection of this line with $\cos\theta_o = 1$ gives the value of the surface tension of the hypothetical liquid which will just spread on the solid. This corresponds to the critical surface tension, γ_C, of the solid.

Girifalco and Good [23] adopted a geometric rule. Considering two general contiguous phases A and B (S and L above) :

$$\gamma_{AB} = \gamma_A + \gamma_B - 2\,\Phi\,(\gamma_A\,\gamma_B)^{1/2} \qquad (3),$$

where Φ is a function involving molar volumes and is usually in the range 0.5 to 1.2.

Fowkes [24] later considered the specificity of certain types of interaction and, in particular, when only dispersive forces are involved, he suggested the formula (with certain restrictions) :

$$\gamma_{AB} = \gamma_A + \gamma_B - 2\left(\gamma_A^D\,\gamma_B^D\right)^{1/2} \qquad (4).$$

It is implicitly assumed that one (or both) of A and B is (are) capable of exchanging only dispersive interactions and thus at least one of the superfixes D in the geometric mean will disappear.

Since equation (4) will only be valid provided that one phase, at least, is uniquely dispersive in nature, workers have attempted to allow for supplementary (polar) interactions, I_{AB}^P :

$$\gamma_{AB} = \gamma_A + \gamma_B - 2(\gamma_A^D \gamma_B^D)^{1/2} - I_{AB}^P \qquad (5).$$

Whereas Fowkes made theoretical assumptions concerning the potential function of interaction between molecules in his geometric mean relation, the functional form for I_{AB}^P is less clear. Several suggestions have been made including the geometric mean as above[25,26] (with polar components of surface tension, γ^P, replacing dispersive components, γ^D), and the harmonic mean [27].

Neumann et al[28] observed a systematic relation between $\gamma_{L(V)} \cos\theta$ and $\gamma_{L(V)}$ for various liquids on polymer surfaces and thus inferred a linear variation of Φ in equation (3). This led to the expression :

$$\gamma_{SL} = \frac{[\gamma_S^{1/2} - \gamma_L^{1/2}]^2}{1 - 0.015(\gamma_S \gamma_L)^{1/2}} \qquad (6).$$

Interest has been shown for some years now in the possibility of acid-base interactions (of the Lewis type) intervening in interfacial tensions. Fowkes[8] made early predictions based on the work of Drago[29] and more recent work by van Oss et al [9] suggests that surface tensions, and by extension interfacial tensions, may be usefully characterised by putting such components as those due to dispersive and dipole-dipole interactions together in one term referred to as van der Waals/Lifshitz interactions (see above) and treating the electron donor/acceptor capabilities separately. Again, geometric mean relations have been found suitable for the mathematical treatment. Although the acid-base aspect of interfacial tensions has met with success, a full understanding of the situation can still not be claimed. The nature of interfacial tensions in general is a field in which much progress has been made, but much still remains to be done, especially concerning high energy solids, such as metals.

4 SOLID SURFACE HETEROGENEITY AND WETTING HYSTERESIS

From the above, it can be seen how solid surfaces may be characterised energetically from wetting measurements. In simple terms, for a given liquid, the higher the surface tension of the solid, the more the liquid is likely to spread i.e. the lower the contact angle. Nevertheless, there are often experimental problems due to surface heterogeneity and to wetting hysteresis. The above sections presuppose a flat, homogeneous, isotropic, smooth, non-deformable solid surface - rarely obtained in practice ! If the solid is composite, say for simplicity consisting of a fraction, f, of material 1 of intrinsic contact angle θ_1 and a fraction (1 - f) of material 2 of angle θ_2

(for a given liquid), Cassie and Baxter[30] showed that the overall measured contact angle, θ_M, should be given by :

$$\cos \theta_M = f \cos \theta_1 + (1 - f) \cos \theta_2 \qquad (7).$$

θ_M is in fact an 'artificial' angle corresponding to a mean value if the regions of A and B are small compared to liquid drop dimensions. A somewhat similar relation was developed by Wenzel[31] to allow for surface rugosity.

Wetting hysteresis is the variability of contact angle between a maximum, advancing value, θ_A, obtained when the wetting front is moving forward to cover more solid and a minimum, receding value, θ_R, observed during, for example, draining (see Figure 2). The equilibrium value, θ_0, is clearly between the two, but where ? Unfortunately the range can easily exceed 60° thus rendering experimental evaluation difficult. Possible causes of hysteresis include local specific adsorption, straightforward chemical heterogeneity of the solid, chemical interaction between the solid and liquid leading to solid swelling or dissolution, restructuration of the solid due to the proximity of a liquid (reorientation), roughness of the surface and local strain of the solid due to capillary forces. Space will not permit a review here but the reader may consult, for example, references 16, 32-38.

Whatever be the primary cause of contact angle hysteresis, it is generally regarded as a 'nuisance' hindering evaluation of the 'true' equilibrium contact angle. However, recent work suggests that it may potentially be used constructively in order to characterise the overall state of a solid surface more completely [39, 40].

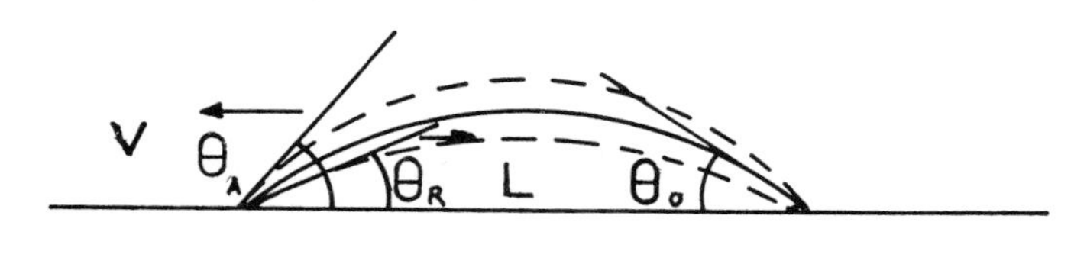

Figure 2 Wetting hysteresis as shown with a sessile drop

Assuming for simplicity that wetting hysteresis is caused by chemical heterogeneity alone, we may consider an isolated, small, circular flaw of diameter w and effective surface tension (or surface free energy) $\gamma_{SV} + \varepsilon$ on an otherwise homogeneous substrate of surface tension γ_{SV}. Let us suppose that the distance between the (undisturbed) triple line and the flaw is δ_0. If the contact line is advancing, it does not 'know' of the presence of the flaw and remains (locally) straight until it contacts the heterogeneity at which point it engulfs it. However a receding contact line, with the flaw initially within the liquid, becomes deformed as contact occurs (see Figure 3). For a while, the extra energy associated with the deformed liquid surface is less than that corresponding to 'uncovering' the heterogeneity. The deformation of the line takes on a logarithmic form [16, 41,42] :

$$\delta(x) \sim \frac{w\,\varepsilon}{\pi\,\gamma_{LV}\,\theta_0^2}\left[\ln\frac{r_0}{|x|} - \frac{1}{2}\right] \; ; \; |x| \gtrsim \frac{w}{2} \qquad (8),$$

where $\delta(x)$ is the perpendicular distance between perturbed and unperturbed contact lines, x is the distance from the flaw following the direction of the undisturbed wetting front and r_0 is a macroscopic cut-off ; either the drop contact radius or a distance comparable to the capillary length, whichever is smaller. The actual length of the triple line disturbance is given by a similar expression :

$$\delta_0 \sim \frac{w\,\varepsilon}{\pi\,\gamma_{LV}\,\theta_0^2}\left[\ln\frac{2\,r_0}{w} + \frac{1}{2}\right] \qquad (9).$$

A force increasing linearly with distance between the flaw and the undisturbed position of the triple line results from separation in the receding mode (but not the advancing mode, for the example given) until a threshold is reached at which 'unhooking' occurs - it becomes cheaper energetically to 'uncover' the flaw than to extend the liquid surface yet further.

The model may be extended to a solid surface covered by a population of average density, ν, of such small flaws and the overall average excess force, F, adding to γ_{SV} in the receding mode [cf. equation (1)] can be shown to be :

$$F \sim \frac{\nu\,(w\,\varepsilon)^2\,L}{2\,\pi\,\gamma_{LV}\,\theta_0^2} \qquad (10),$$

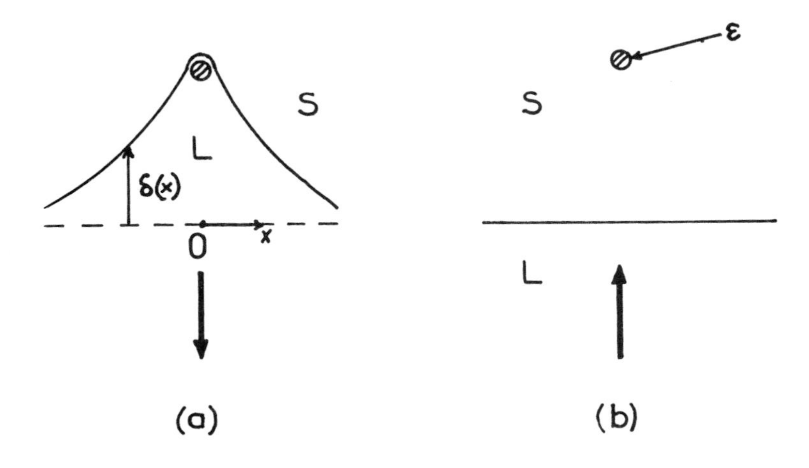

Figure 3 (a) Local deformation of wetting front near (positive) surface tension heterogeneity during receding mode
(b) No modification to front occurs during advancing mode

where L represents the terms in brackets in equation (9) (see Figure 4). This modifies the average contact angle to a lower value, θ_R, whilst for an advancing triple line, F is absent and a higher average angle, θ_A , pertains.

Provided contact angles are fairly small, the difference $\Delta\theta = \theta_A - \theta_R$, or hysteresis, may be expressed as [40]:

$$\Delta\theta \sim \frac{\nu\,(w\,\varepsilon)^2\,L}{2\,\pi\,\gamma_{LV}^2\,\theta_0^3} \qquad (11).$$

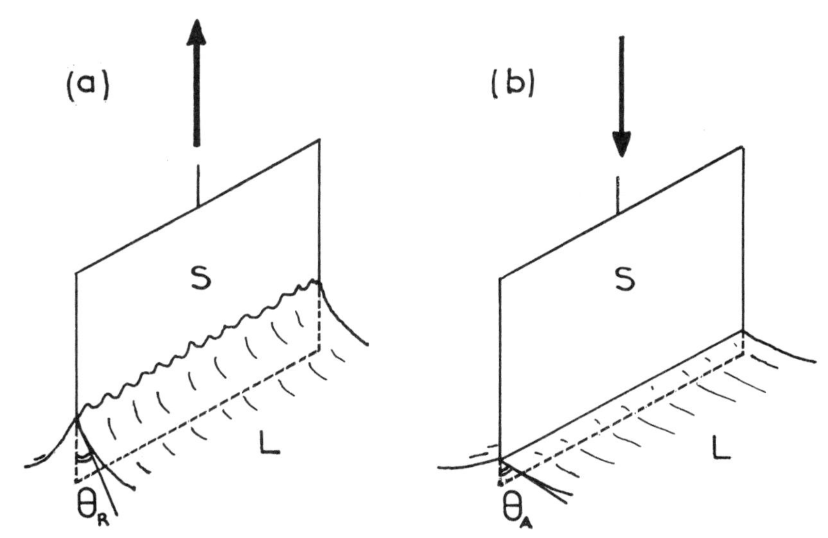

Figure 4 Wilhelmy plate in (a) receding and (b) advancing modes. Additional force (and thus lower average contact angle) in (a) is due to triple line 'hooking' onto heterogeneities

Thus by using the Wilhelmy technique[20] in immersion (advancing) and emersion (receding) modes, it is possible in principle to relate hysteresis to the flaw population. In addition, an individual force-fluctuation, or 'ripple' obtained during emersion is of the order of $\pi \gamma_{LV} \theta_0^2 L^{-1} \delta_0$. From equation (9), $w\varepsilon$ may thus be estimated and using equation (11), a value of ν obtained. Alternatively, ν may be estimated by comparing the occurrence of ripples to a stochastic phenomenon such as the frequency of births or car accidents in a given town or similar and then invoking the Poisson process for analysis [40].

The model treated is simple and further development is required, but it does give guidelines as to how it may be possible not only to characterise solid surfaces as far as their average energetic properties are concerned, but also to gain information about the spatial and energetic distributions of inhomogeneities.

5 CONCLUSIONS

Wetting phenomena have fascinated scientists for centuries and the first quantitative appraisals date from the beginning of the nineteenth century with the work of Young and Laplace (and others). The qualitative assessment of solid surface characteristics can be made from the propensity of given liquids to wet, or spread - the higher the solid surface tension, the more a liquid is likely to spread. Nevertheless, some limitations are evident from Young's equation which includes both solid surface tension and solid/liquid interfacial tension. Suitable evaluation of the latter is required.

Although a complete understanding is not yet available, much progress has been made over recent years and various explanations of solid/liquid interactions have been suggested.

Another problem recognised in wetting experiments is the existence of wetting hysteresis. Many, but perhaps not all, causes have been isolated. Whilst wetting hysteresis has been largely regarded as a nuisance until quite recently, there is now reason to believe that it may in fact be exploited as a means to obtaining further information about the solid. It would seem that knowledge of energetic and spatial distributions of surface flaws could be accessible from suitable analyses of experimental data.

REFERENCES

1. J.R. Huntsberger, 'Treatise on Adhesion and Adhesives', ed. R.L. Patrick, Marcel Dekker, 1967, 1, 119.
2. H. Schonhorn, 'Adhesion : Fundamentals and Practice', McLaren, 1969, 4, 29.
3. L.H. Sharpe and H. Schonhorn, 'Symposium on Contact Angles', Chem. Eng. News, 1963, 15, 67.
4. W.H. Keesom, Phys. Z., 1921, 22, 129 and 643.
5. P. Debye, Phys. Z., 1921, 22, 302.
6. F. London, Z. Phys., 1930, 63, 245.
7. I.E. Dzyaloshinskii, E.M. Lifshitz and L.P. Pitaevskii, Adv. Phys., 1961, 10, 165.
8. F.M. Fowkes, J. Adhesion Sci. Tech., 1987, 1, 7.
9. C.J. van Oss, M.K Chaudhury and R.J. Good, Chem. Rev., 1988, 88, 927.
10. G. Bakker, 'Wien Harms Handbuch der Experimental Physik, VI', Akad. Verlag, Leipzig, 1928.
11. T. Young, Phil. Trans. Roy. Soc., 1805, 95, 65.
12. P.S. Laplace, 'Mécanique Céleste, Suppl. au X livre', Coureier, Paris, 1805.
13. R.E. Collins and C.E. Cooke, Trans. Faraday Soc., 1959, 55, 1602.
14. R.E. Johnson, J. Phys. Chem., 1959, 63, 1655.
15. M.E.R. Shanahan, 'Adhesion 6', ed. K.W. Allen, Appl. Sci. Pub., London, 1982, 75.
16. P.G. de Gennes, Rev. Mod. Phys., 1985, 57, 827.
17. M.E.R. Shanahan and P.G de Gennes, 'Adhesion 11', ed. K.W. Allen, Elsevier Appl. Sci. Pub., London, 1987, 71.
18. M.E.R. Shanahan, J. Phys.D : Appl. Phys., 1987, 20, 945.
19. A.W. Adamson, 'Physical Chemistry of Surfaces, 4th Edition', Wiley and Sons, New York, 1982, 6.
20. L. Wilhelmy, Ann. Phys., 1863, 119, 177.
21. Reference 19, 354.
22. W.A. Zisman, Adv. Chem. Ser, 1964, 43, 1.
23. L.A. Girifalco and R.J. Good, J. Phys. Chem., 1957, 61, 904.
24. F.M. Fowkes, Adv. Chem., Ser., 1964, 43, 99.
25. D.K. Owens and R.C. Wendt, J. Appl. Polym. Sci., 1969, 13, 1741.
26. D.H. Kaelble and K.C. Uy, J. Adhesion, 1970, 2, 50.
27. S. Wu, J. Adhesion, 1973, 5, 39.

28. A.W. Neumann, R.J. Good, C.J. Hope and M. Sejpal, J. Colloid Interf. Sci., 1974, 49, 291.
29. R.S. Drago, G.C. Vogel and T.E. Needham, J. Am. Chem. Soc., 1971, 93, 6014.
30. A.B.D. Cassie and S. Baxter, Trans. Faraday Soc., 1944, 40, 546.
31. R.N. Wenzel, Ind. Eng. Chem., 1936, 28, 988.
32. R.S. Hansen and M. Miotto, J. Am. Chem. Soc., 1957, 79, 1765.
33. R.E. Johnson and R.H. Dettre, Adv. Chem. Ser., 1964, 43, 112.
34. J.D. Eick, R.J. Good and A.W. Neumann, J.Colloid Interface Sci., 1975, 53, 235.
35. R.J. Good, J. Colloid Interface Sci., 1977, 59, 398.
36. H. Yasuda, A.K. Sharma and T. Yasuda, J. Polym. Sci., 1981, 19, 1285.
37. A. Carré, S. Moll, J. Schultz and M.E.R. Shanahan, 'Adhesion 11', ed. K.W. Allen, Elsevier Appl. Sci. Pub., London, 1987, 82.
38. M.E.R. Shanahan, J. Phys. D : Appl. Phys., 1988, 21, 1981.
39. M.E.R. Shanahan, Surface Interface Sci., 1991, 17, 489.
40. M.E.R. Shanahan, J. Adhesion Sci. Techn., 1992, 6, 489.
41. J.F. Joanny and P.G. de Gennes, J. Chem. Phys., 1984, 81, 552.
42. M.E.R. Shanahan, J. Phys. D : Appl. Phys., 1989, 22, 1128.

4.3.3
Surface Analysis of Some Conducting Polymer Coatings

J. R. Bates,[1] R. W. Miles,[1] P. Kathirgamanathan,[2] R. Hill,[1] and D. Adebimpe[2]

[1] DEPARTMENT OF ELECTRICAL AND ELECTRONIC ENGINEERING AND PHYSICS, UNIVERSITY OF NORTHUMBRIA AT NEWCASTLE, NEWCASTLE UPON TYNE, UK

[2] UNIT FOR SPECIALITY ELECTRONICS POLYMERS, UNIVERSITY COLLEGE LONDON, 20 GORDON STREET, LONDON, UK

1 INTRODUCTION

Following the discovery in 1977 that trans-polyacetylene could be prepared in the form of free standing films[1] and could be doped to electrical conductivities exceeding 10^3 S/cm it was soon discovered that either chemical or electrochemical doping could produce high electrical conductivities in other polymers, such as polypyrrole[3], polyparaphenylene[4], polythiophene[5-8], polyfuran[9] and polyaniline[10]. As this work progressed it was realised conducting polymers have a wide range of applications including potential uses in corrosion protection[11], RFI/EMI shielding[11], gas sensors devices[12,13] and rechargeable batteries[14].

In order to realise the potential applications of conducting polymers it is necessary to understand how the preparation conditions influence the conductivity, microstructure and surface morphology. These properties depend upon a number of factors including the polymerisation temperature; the deposition potential, whether the growth was under potentiostatic, galvanostatic or potential cycling conditions; the nature of the anion (size, charge, and polarisability); the monomer/anion ratio; the viscosity of the reaction medium; the dielectric constant of the solvent(s); the nature of the electrode etc. The influence of some of these factors on poly(pyrrole) and co-poly(3-methylthiophene-pyrrole) are examined in this paper.

2 EXPERIMENTAL DETAILS

Films of poly(pyrrole) and co-poly(3-methylthiophene-pyrrole) were grown on either tin oxide coated glass slides or platinum foil electrodes in the presence of the tetrabutylammonium salt of either hexafluorophosphate, perchlorate, p-toluene sulphonate or trifluoromethane sulphonate in either acetonitrile or nitrobenzene. The films were grown either potentiodynamically, cycling between -0.5 and 1.5 volts (vs SCE) or potentiostatically at 1.1, 1.2, 1.3 and 1.4 volts (vs SCE). Scanning electron micrographs were taken of the free standing films and the coated tin oxide electrodes. Conductivity measurements were made using the four point probe technique.

3 RESULTS AND DISCUSSION

The conductivities and comments regarding the observed morphologies of the poly(pyrrole) hexafluorphosphate are presented in Table 1. The microstructure of the film grown under potentiodynamic conditions was very similar to that of the film grown at a constant potential of 1.1 volts (vs SCE). As the potential was increased to

Table 1 Electropolymerisation of poly(pyrrole) hexafluorophosphate in acetonitrile.

TEMPERATURE (C)	POTENTIAL (V vs SCE)	CONDUCTIVITY (S/cm)	COMMENTS
20	-0.5 to 1.5	too thin to measure	granular deposits 4-40 microns see Fig 1(a)
20	1.1	too thin to measure	as above see Fig 1 (b)
20	1.2	7.8	granular deposits with peaks see Fig 1(c)
20	1.3	4.5	granular deposits with large peaks (60-80 microns) see Fig 1(d)
10	1.2	55	globular nodules (10 microns) & secondary growths see Fig 2(b)
0	1.2	86	globular nodules & dense secondary growths see Fig 2(c)

Pyrrole 2mM; tetrabutylammonium hexafluorophosphate 10mM

Electrode: Tin oxide coated glass.

1.2 volts (vs SCE) the nodular growths increased in size and films with thicknesses from 40 to 100μm could easily be obtained. The conductivity of the film was 7.8 S/cm. Increasing the deposition potential to 1.3 volts changed the microstructure significantly (see Figure 1) with the nodular deposits being more closely packed. The conductivity was slightly lower (4.5 S/cm) than that of the film grown at 1.2 volts. This observation is consistent with the fact that the potential at which a film is grown controls the oxidation state and therefore the doping level and conductivity of the polymer. It is well recorded that the conductivity of poly(pyrrole) has an optimum potential regarding conductivity versus deposition potential.

As the deposition temperature was reduced from 20C to 10C (at a constant potential of 1.2 volts (vs SCE)) the peaks became globular and secondary growths appeared (see Figure 2). The conductivity increased from 7.8 S/cm at 20C to 55.0 S/cm at 10C. On reducing the deposition temperature to 0C the microstructure became more "mountainous" with secondary microgranules filling the gaps between peaks. This change in microstructure was accompanied by a consequent increase in the conductivity to 86.0 S/cm. Such an increase in conductivity with decreasing deposition temperature is to be anticipated as fewer structural defects such as α-β mis-linkages or cross-linkages between polymer chains occur[14].

Poly(pyrrole) trifluoromethanesulphonate on tin oxide coated glass had a significantly different microstructure and morphology to the poly(pyrrole) hexafluorophosphate (see Figure 3). The film grown at 1.2 volts (vs SCE) was again similar to the film grown potentiodynamically (σ=27.5 S/cm) with a nodular

Figure 1 Dependence of the microstructure of poly(pyrrole) hexafluorophosphate on deposition potential

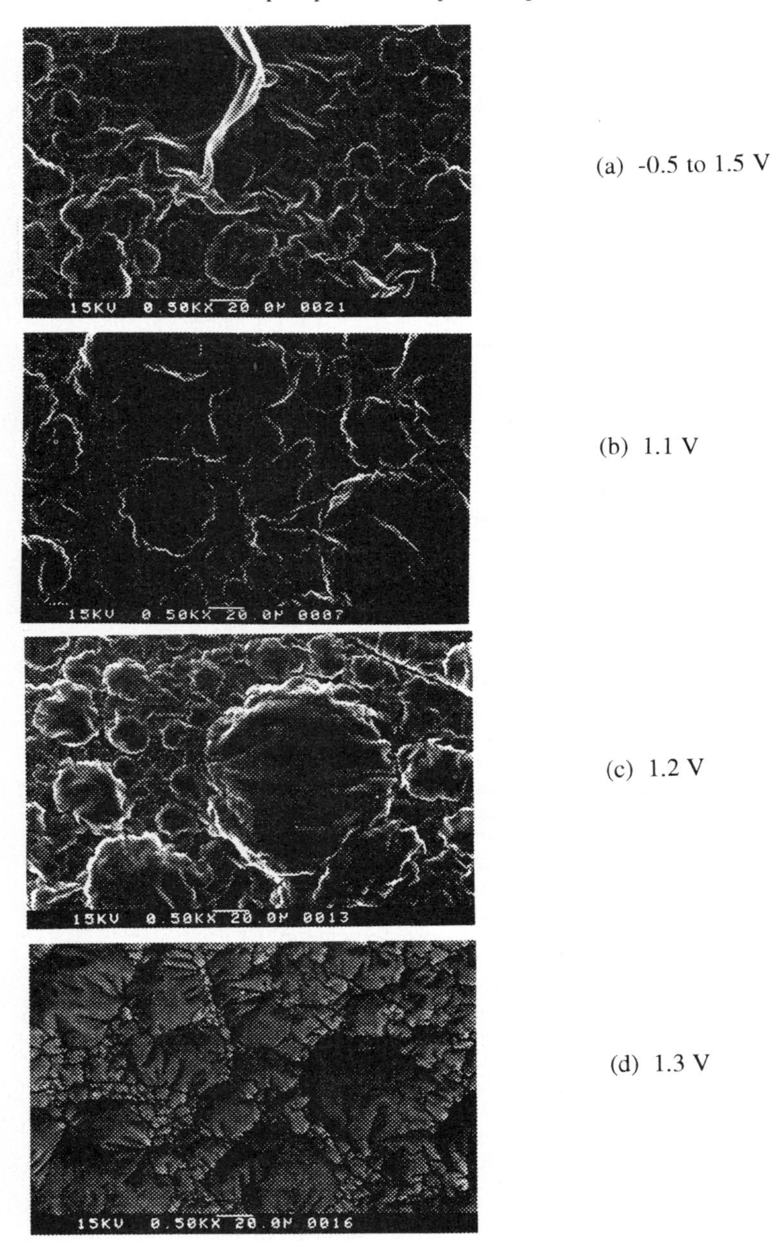

Conditions: Pyrrole 2 mM; $Bu_4N\,PF_6$ 10 mM; Electrode SnO_2 ;Temperature 20C; Solvent acetonitrile.

Figure 2 Dependence of the microstructure of poly(pyrrole) hexafluorophosphate on deposition temperature

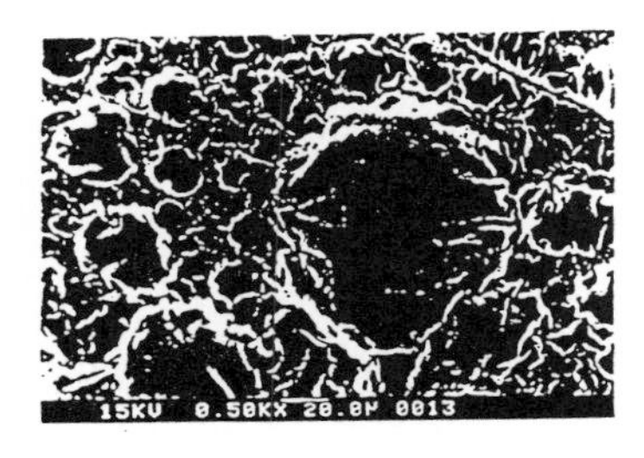

(a) 20C
7 S/cm

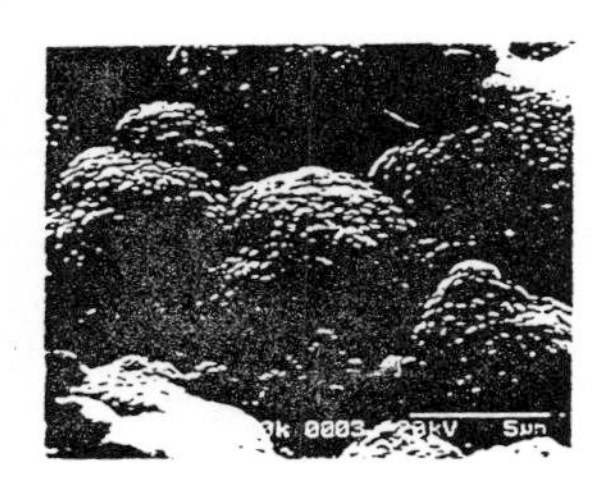

(b) 10C
55 S/cm

(c) 0C
86 S/cm

Conditions: Pyrrole 2 mM; $Bu_4N\ PF_6$ 10 mM; Electrode SnO_2 ;Deposition potential 1.2 V (vs SCE); Solvent acetonitrile.

Figure 3 Dependence of the microstructure of poly(pyrrole) trifluoromethane sulphonate on deposition potential

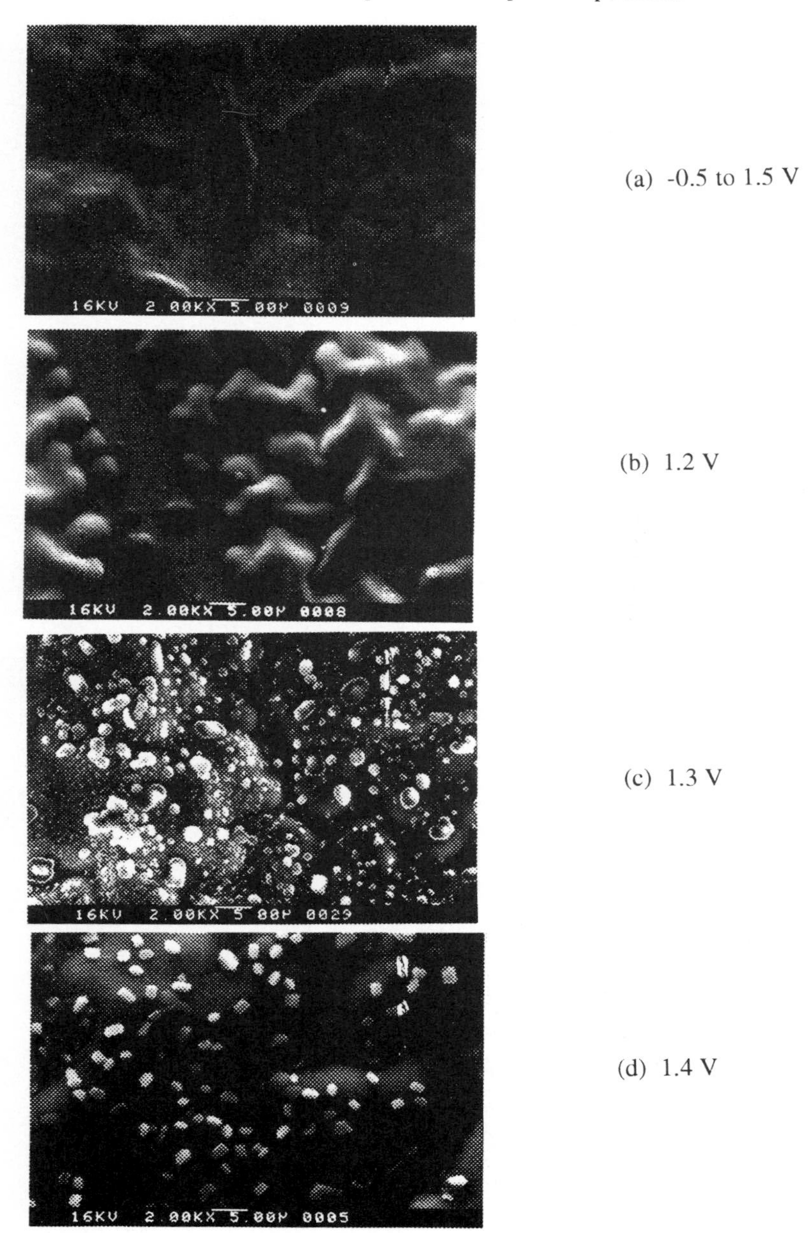

Conditions: Pyrrole 2 mM; $Bu_4N\ CF_3SO_3$ 10 mM; Electrode SnO_2 ;Temperature 20C; Solvent acetonitrile.

Table 2 Electropolymerisation of poly(pyrrole) trifluoromethane sulphonate in acetonitrile.

TEMPERATURE (C)	POTENTIAL (V vs SCE)	CONDUCTIVITY (S/cm)	COMMENTS
20	-0.5 to 1.5	film too thin to measure	long nodules see Fig 3(a)
20	1.2	27.5	small nodular growths (10 microns) see Fig 3(b)
20	1.3	film too fragile to measure	ill-defined crystallites on amorphous globular film see Fig 3(c)
20	1.4	film too fragile to measure	well defined crystals on globular film see Fig 3(d)

Pyrrole 2mM; tetrabutylammonium trifluoromethane sulphonate 10mM.

Electrode: Tin oxide coated glass.

morphology; as the deposition potential was increased to 1.3 volts small crystallites appeared. These crystallites became more highly defined as the deposition potential was increased to 1.4 volts. The conditions for the electropolymerisation of poly(pyrrole) trifluoromethane sulphonate are summarised in Table 2.

The influence of anion on film conductivity and microstructure is compared in Table 3 and Figures 5(a), (b) and (c). The polymers were grown under identical conditions. The films prepared with the larger anions (trifluoromethane sulphonate and p-toluene sulphonate) had higher conductivities than those prepared with the smaller anions (hexafluorophosphate and perchlorate).

The effect of the solvent on the morphology of poly(pyrrole) hexafluorophosphate is compared in Figures 4(a) and (b). The microstructure of the film grown in acetonitrile was significantly different from that grown in nitrobenzene. The film grown in acetonitrile had the "mountainous" structure shown Figure 1 whereas the film grown in nitrobenzene consisted of rod-like deposits on an amorphous matrix. The conductivity of the film grown in nitrobenzene had a conductivity of 480 S/cm compared to 4.5 S/cm for the film grown in acetonitrile.

Co-poly(3-methylthiophene-pyrrole) perchlorate was grown in acetonitrile at 1.15, 1.25 and 1.35 volts (vs SCE). It was decided to co-polymerise pyrrole and 3-methyl thiophene as the oxidation potential of the monomers is comparable (ca. 0.8 volts (vs SCE)). The morphology of these films was significantly different from that of the poly(pyrrole) perchlorate. This would suggest that a co-polymer has been formed although further work is necessary to confirm this. The conductivity of the co-poly(3-methylthiophene-pyrrole) perchlorate films is given in Table 4 and was in the 5-10 S/cm range for all three films. The scanning electron micrographs of the films are shown in Figure 6.

Figure 4 Dependence of the microstructure of poly(pyrrole) hexafluorophosphate on deposition solvent

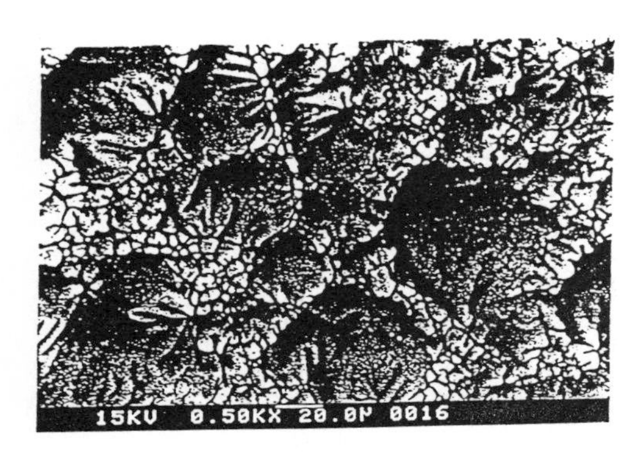

(a) acetonitrile

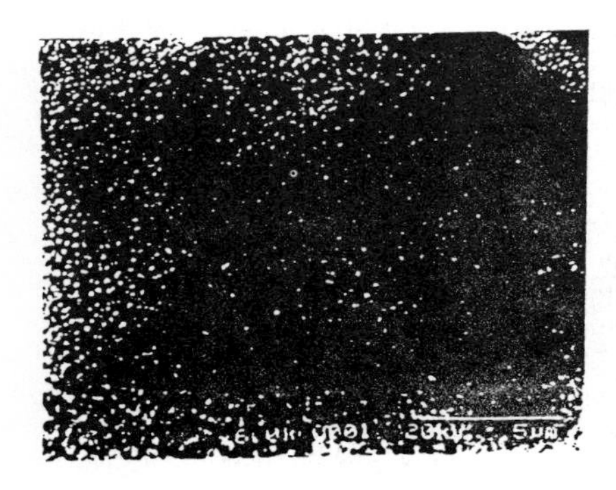

(b) nitrobenzene

Conditions: Pyrrole 2 mM; $Bu_4N\ PF_6$ 10 mM; Electrode SnO_2 ;Deposition potential 1.3V (vs SCE); Deposition temperature 20C

Figure 5 Dependence of the microstructure of poly(pyrrole) films on counter ion

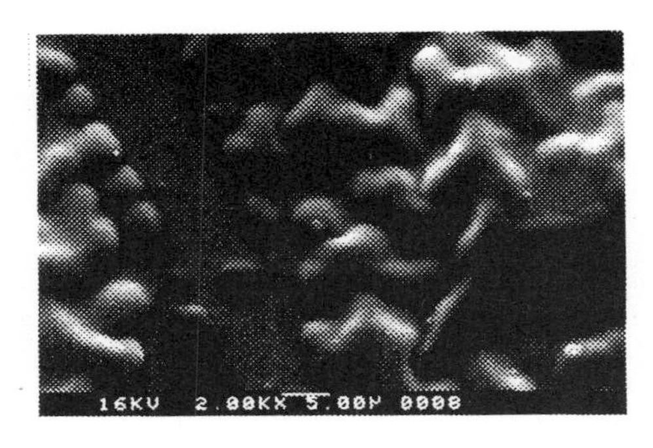

(a) $CF_3SO_3^-$
27.5 S/cm

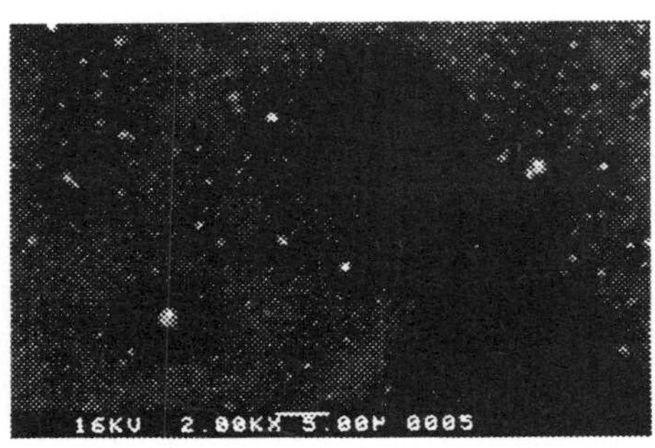

(b) pTS^-
30 S/cm

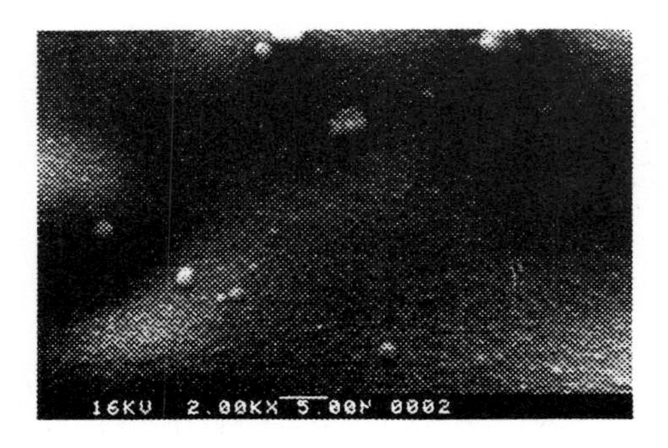

(c) ClO_4^-
6 S/cm

Conditions: Pyrrole 2 mM; Bu_4N X 10 mM; Electrode SnO_2 ;Deposition potential 1.2V (vs SCE); Deposition temperature 20C; Solvent acetonitrile

Figure 6 Dependence of the microstructure of co-poly(3-methylthiophene-pyrrole) perchlorate on deposition potential

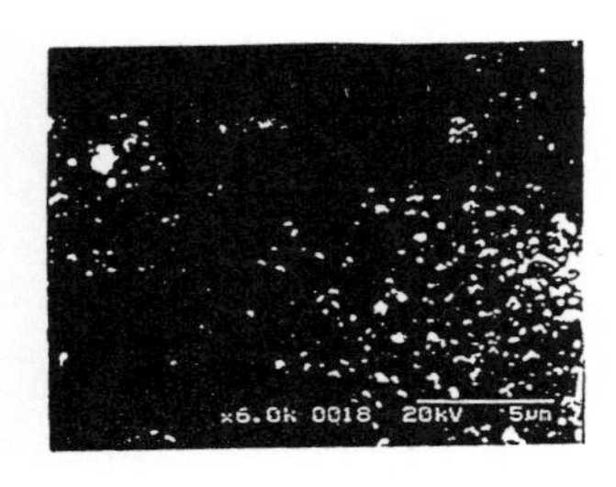

(a) 1.15 V
9 S/cm

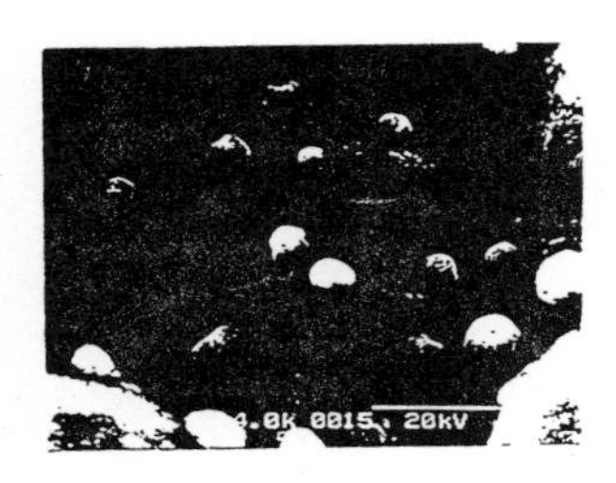

(b) 1.25 V
6 S/cm

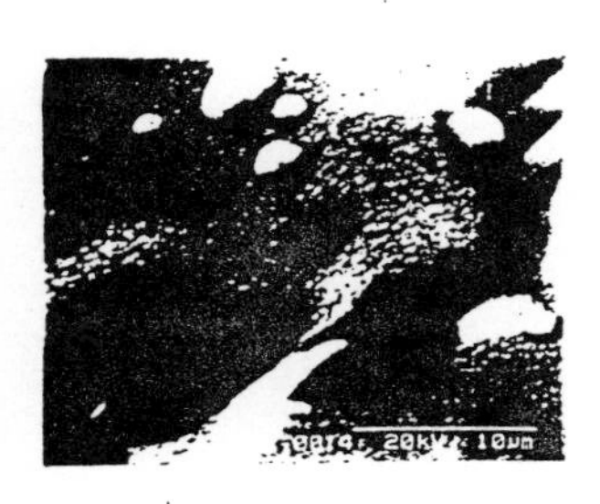

(c) 1.35 V
8 S/cm

Conditions: Pyrrole 2 mM; 3-MeT 2 mM; $Bu_4N ClO_4$ 10 mM; Electrode SnO_2 ;Temperature 20C; Solvent acetonitrile.

Table 3 Influence of anion on the conductivity of poly(pyrrole)

TEMPERATURE (C)	ANION	CONDUCTIVITY (S/cm)	COMMENTS
20	$CF_3SO_3^-$	27.5	globular deposit (2-20 microns) see Fig 5(a)
20	pTS^-	30.0	fine particulate deposit on compact amorphous structure see Fig 5(b)
20	ClO_4^-	6.0	compact amorphous structure with particulate deposit see Fig 5(c)

Table 4 Electropolymerisation of co-poly(3-methylthiophene-pyrrole) perchlorate on tin oxide coated glass electrodes

TEMPERATURE	POTENTIAL (V vs SCE)	CONDUCTIVITY (S/cm)	COMMENTS
20	1.15	9.0	granular deposits (0.1-1 micron) see Fig 6(a)
20	1.25	6.0	globular structure with secondary spherical growth (2-3 micron) see Fig 6(b)
20	1.35	8.0	globular structure with secondary growths see Fig 6(c)

4 CONCLUSION

The effects of various parameters on the microstructure, morphology and conductivity of films of poly(pyrrole) have been investigated. It has been demonstrated that changes in deposition temperature lead to an increase in the film conductivity and to significantly different surface morphologies. The effects of the supporting electrolyte and solvent have been shown to have significant effects on the conductivity, microstructure and surface morphology of poly(pyrrole) and co-poly(3-methylthiophene-pyrrole) films.

Although the potential applications of conducting polymer films are vast, this work has demonstrated that great care is needed in the preparation of these films as small changes in deposition potential and deposition temperature can have significant effects on the conductivity and morphology of the film. Great care is also needed in the choice of counter ion and solvent as these also have considerable effects on the film morphology.

REFERENCES

1. C.K.Chaing, C.R.Fincher, Y.W.Park, A.J.Heeger, H.Schirakawa, E.J.Louis, S.C.Gau, and A.G.MacDiarmid, Phys. Rev. Lett., 1977, 39, 1098.
2. H.Schirakawa and S.Ikeda, Polymer, 1971, 2, 231.
3. K.K.Kanazawa, A.F.Diaz, R.H.Geiss, W.D.Gill, J.F.Kwak, J.A.Logan, J.F.Rabolt and G.B.Street, J.Chem. Soc., Chem. Commun., 1979, 854.
4. L.W.Schacklette, R.R.Chance, D.M.Ivory, G.G.Miller and J.Baughman, Synth. Met., 1979, 1, 307.
5. K.Kaneto, K.Yoshino and Y.Inuishi, Jap. J. Appl. Phys., 1982, 21, L567.
6. K.Kaneto, Y.Kohno, K.Yoshino and Y.Inuishi, J.Chem. Soc., Chem. Commun., 1983, 382.
7. K.Kaneto, K. Yoshino and Y.Inuishi, Solid State Commun., 1983, 46, (5), 389.
8. K.Kaneto, S.Ura, K.Yoshino and Y.Inuishi, Jap. J. Appl. Phys., 1984, 23, (3), L189.
9. Yoshino, Jap. J. Appl. Phys., 1984, 23, (9), L663.
10. A.G.MacDiarmid, J.C.Chaing, M.Halpern, S.L.Mu, N.L.D.Somasiri, W.Wu and S.I.Yaniger, Mol. Cryst. Liq. Cryst., 1985, 121, 173.
11. P. Kathirgamanathan, "Surface Engineering Practice", Ellis Horwood Ltd., 1990.
12. T.Hanawa, S.Kuwabata, H.Hashimoto and H.Yoneyama, Synth. Met., 1989, 30, 173.
13. K.Yoshino, H.S.Nalwa, J.G.Rabe and W.F.Schmidt, Polym. Commun., 1985, 26, 103.
14. P.J.Nigrey, D.MacInnes, D.P.Nairns, A.G.MacDiarmid and A.J.Hegger, J. Electrochem. Soc., 1981, 128, 1651.
15. K.Tanaka, T.Shichiri and T.Yamabe, Synth. Met., 1986, 16, 207.

4.3.4
Measurement of the Depth of the Work Hardening Layer on a Machined Surface by an *X*-Ray Diffraction Method

Z. Y. Zhu and J. F. L. Chan

DEPARTMENT OF MECHANICAL, MATERIALS, AND MANUFACTURING ENGINEERING, THE UNIVERSITY, NEWCASTLE UPON TYNE NE1 7RU, UK

1 INTRODUCTION

One of the purposes of metal machining is to obtain an appropriate surface integrity on a machined component. The Surface integrity on a machined surface includes mainly the surface finish, residual stress on the surface, and work hardening on the surface[1]. Therefore there is no doubt that the degree of work hardening on a machined surface is one of the most important aspects in judging the surface integrity of a machined component. To judge the degree of work hardening, both surface work hardening and the depth of work hardened layer have to be considered[2]. Although it is generally not difficult, for example, to use a hardness meter, to measure the surface hardness, it is to date still extremely difficult to obtain accurately the depth of the work hardened layer of a machined surface[3]. There is of course no shortage of techniques which have been developed and can be used for determining this depth, as they are shown in Table 1[4]; however, many of these employ different approaches, some of which are rather complicated. These would lead to different and sometimes confusing results, which can be illustrated by Table 2[4]. There is therefore an urgent need for the development of a technique, which is both reliable and convenient to use for measuring the depth of the work hardened surface layer.

This paper presents an X-ray diffraction method developed by the authors. Based on the X-ray metallography theory and X-ray diffraction technique, the changes in the distribution curve of diffraction intensity were taken as a criteria for determining whether or not work hardening, or plastic deformation exists in a certain subsurface layer of a component. An electrolytic corrosion technique was employed to strip the surface layers step by step, and thus to establish how deep these layers were from the original surface datum by using a "weight analysis" technique.

Table 1 Different Methods for Determining the Depth of Work hardening Layer

Methods	Description
Erosion	Erosion from the top surface, measuring the speed of erosion.
Microscope	Observing the microstructure of the cross section.
Hardness	Measuring the distribution of the hardness of the cross section.
Recrystallization	Heating the observed material, observing the depth of recrystallization by microscope.

Table 2 Different Values of Depth of Work hardening Layer from Different Methods

		Depth of Work hardened Layer (μm)			
Workpiece Materials	Depth of Cut	Erosion	Microscope	Hardness	Recrystallization
0.2 % C	0.1	15 ~ 20	30 ~ 40	50 ~ 90	50 ~ 80
	0.5	30 ~ 40	50 ~ 60	130 ~ 150	100 ~ 130
0.63% C	0.1	25 ~ 35	30 ~ 50	50 ~ 130	90 ~ 130
	0.5	40 ~ 70	60 ~ 70	55 ~ 65	100 ~ 180

2 THE PRINCIPLES OF THE TECHNIQUE

It is well known that the work hardening of metallic materials is a result of the plastic deformation which exists in the materials. It follows that if the depth of the plastic deformation on a machined component can be determined, this depth can be taken as that of the work hardened layer.

To determine the depth of plastic deformation, the X-ray diffraction technique can be employed.

X-ray scatters when it passes through crystalline materials, for example, metals. Due to the regular arrangement of the electrons in these materials, when scattered waves interfere with each other, their intensity will be enhanced in certain directions, while counteracted in some other directions. Thus the intensity of the synthesis waves changes in different directions and tends to lead to diffraction. However, the conditions for forming the diffraction must satisfy the Bularge Equation[5], that is:

$$2\, d\, \sin \Theta = n\, \lambda \tag{1}$$

where, Θ - the angle between the incoming X-ray and a certain crystal plane
d - the distance between crystal planes
λ - the length of the incoming X-ray waves
n - an integer

This equation illustrates that when the length of the incoming X-ray wave is constant, corresponding to an unchanged distance between crystal planes, there should be a fixed value of the angle between incoming X-ray and crystal plane. On the contrary, if the distance between crystal planes changes, angle Θ should also change correspondingly.

When plastic deformation exists in the crystal material, the distance between crystal planes **d** changes to some extent. If **d** is bigger than a normal value, Θ will decrease. On the other hand, if **d** becomes smaller than the normal value, Θ will increase. Therefore, the irregular changes of **d** resulting from the plastic deformation will lead to the decrease of the diffraction intensity, and the widening of the distribution curve of the diffraction intensity. The wideness of the distribution curve can be represented by B, the width at the half-height of the curve, which is shown in Fig. 1.

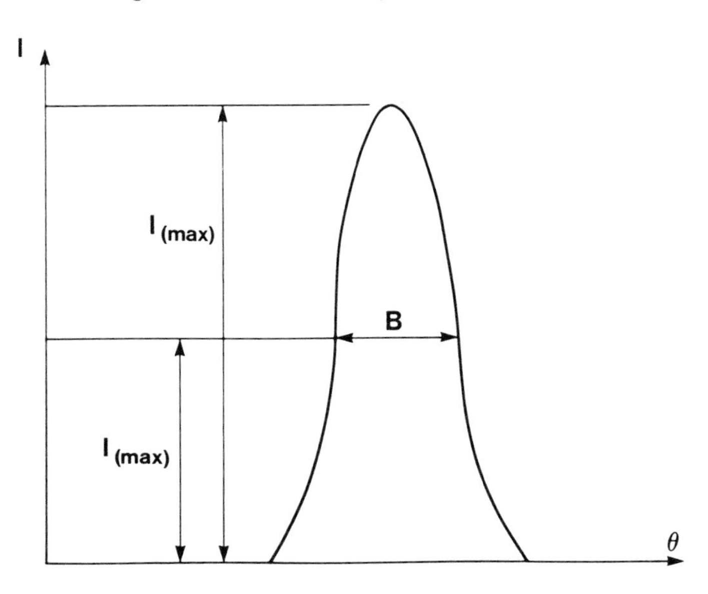

Figure 1 The width at the half-height of the distribution curve of the diffraction intensity

According to the above principle, the depth of work hardened or plastic deformation layer can be measured by judging the change of the distribution curve of the diffraction intensity, or the change of value B. If B is normal on a certain subsurface layer of the materials, it can be said that there is no work hardening on this layer. But if B is bigger than the normal value, it can therefore be determined that work hardening took place up to the depth of this layer.

3 METHOD OF MEASUREMENT

Based on the above procedure, a D/max - **γβ** X - r a y diffractometer was used, combined by a electrolytic corrosion strip technique and a "weight analysis" method, to measure the depth of the work hardening layer. The technique includes the following steps:

(a) A distribution curve of the diffraction intensity of a standard specimen was obtained. The standard specimen should possess the same property as that of the machined component to be measured, and have no plastic deformation.

(b) To expose the subsurface of the machined component, which was going to be measured. The surface of the component was stripped step by step by the electrolytic corrosion method. The depth of the stripped layer was measured by the "weight analysis" method. Both the electrolytic corrosion and weight analysis techniques will be introduced in Section 4.

(c) After each thin surface layer had been stripped, the diffractometer was employed to obtain the distribution curve of the diffraction intensity on each newly exposed surface. Then the B value of the distribution curve was compared with that of the standard specimen. If the B value on the newly exposed surface was not the same as that of the standard specimen, the procedure should continue until the B value was exactly the same as that of the standard specimen.

(d) When the B value of a certain newly exposed surface reached the value of that of the standard specimen, it can be concluded that the stripped surface layer was the layer of the work hardening. The depth of this layer, which had been recorded with the weight analysis method step by step, was that of the work hardening layer.

Figure 2 is a distribution curve of diffraction intensity taken from the standard specimen, on which the value B is 0.179 deg.

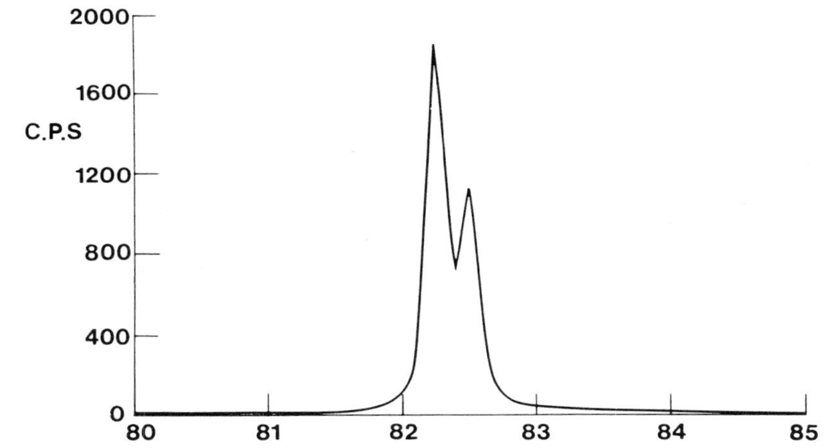

Figure 2 A distribution curve of diffraction intensity taken from the standard specimen

Figure 3 is a distribution curve of diffraction intensity taken from the original surface of the machined component, which possesses exactly the same property as that of the specimen. From Fig. 4 it can been seen that the curve is much more shorter and wider than that of the specimen, and the value B is of course much bigger than that of the specimen. This means clearly that the diffraction intensity on the original machined component surface is quite weak, and plastic deformation, or work hardening is therefore extremely severe.

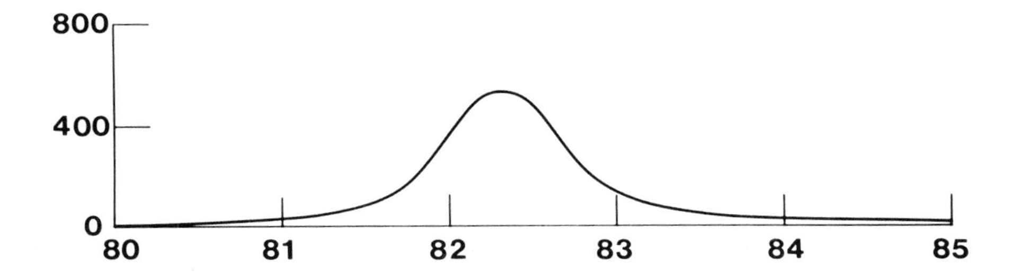

Figure 3 A distribution curve of diffraction intensity taken from an original surface of the machined component

Figure 4 shows the changing process of the distribution curves of diffraction intensity from the original surface to a certain depth of subsurface layer on the machined component. The numbers pointing to each curves represent the order of the stripping time. Curve 1 was obtained from the original surface, and Curve 2 obtained after the first stripping, then Curve 3 the second stripping .... It is quite clear from Figure 4 that the deeper a subsurface layer is from the original surface, the closer the value B is to that of the standard specimen.

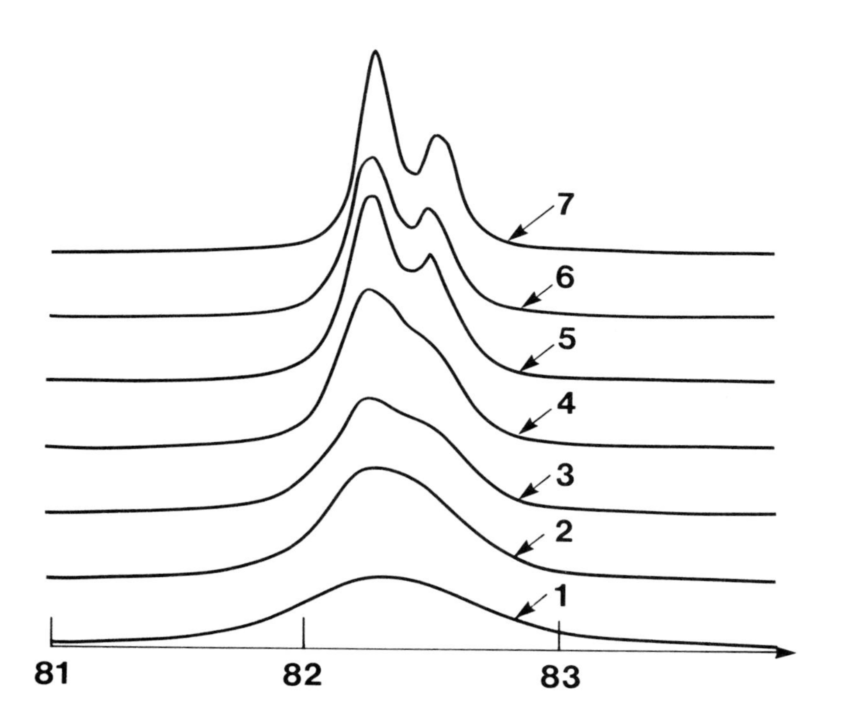

Figure 4 Diffraction intensity from the original surface changing to a subsurface layer of the machined component

Figure 5 shows the relationship between B and the depth of the subsurface layer, from which it can be easily seen that the deeper the subsurface layer, the closer B value is to reaching that of the standard specimen, which means that plastic deformation, on work hardening tends to disappear on a certain subsurface layer.

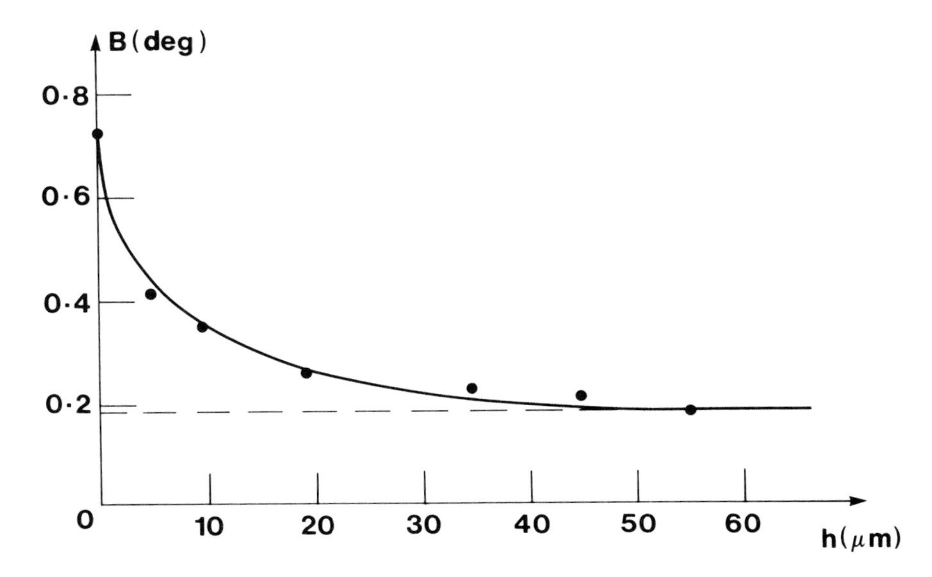

Figure 5 The relationship between B and the depth of the subsurface layer

4 STRIPPING METHOD AND MEASUREMENT OF THE DEPTH BY A WEIGHT ANALYSIS TECHNIQUE

The depth of work hardened layer on a machined surface is usually between 20 μm ~ 100 μm, which is very difficult to measure by common methods. Moreover, if the X-ray diffraction method is used to determine whether or not plastic deformation exists on a certain subsurface layer, a series of stripping has to be done, and the depth of each stripped layer should be extremely thin, say, no more than 5 μm. In order to strip and measure the depth of each stripped layer carefully and accurately, the electrolytic corrosion and a weight analysis method was employed.

Electrolytic corrosion is a widely used technique. In order to adapt it to strip a machined surface for determining the depth of work hardening, a lot of experiment has to be carried out to select a suitable corrosion condition, which depends on the workpiece material properties and the corrosion rate wanted. Whenever a desirable corrosion condition is found, the stripping can be carried out without difficulties.

When the machined component is stripped by

electrolytic corrosion, its weight reduces slightly. By weighing the stripped specimen with an analytical balance, the depth of each stripped layer can be measured.

Figure 6 illustrates the experimental results about the relationship between electrolytic corrosion stripping time and the change of the weight, as well as the depth of the stripped surface layer, of the specimen. It can also be seen that there is an excellent proportional relationship between the stripping time and the change of the weight or the depth.

In fact, there is no need to weigh the specimen all the time, because of the proportional relationship between the stripping time and the stripped layer depth. What should be done is simply control the time of electrolytic corrosion; thus the stripped depth can be judged exactly according to the line as shown in Fig. 6. Both the stripping and the measurement of the depth of stripped surface layer can be done easily, conveniently and accurately.

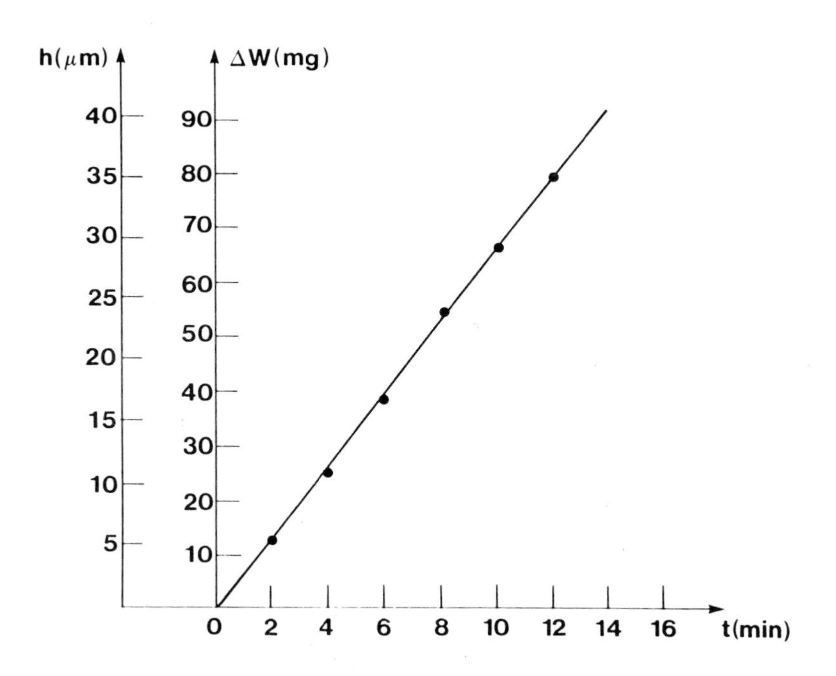

Figure 6 The relationship between stripping time, weight change, and depth of specimen

5 DISCUSSION

There have always been arguments about how thick the work hardened layer is on a machined component surface under a certain machining condition.[1, 2, 3, 4] As is shown in Table 2, different approaches lead to different results, which are sometimes rather confusing. This shows that a reliable and accurate technique has not been developed. X-ray diffraction is entirely sensitive to any disorder of electrons in crystalline materials. Whenever plastic deformation happens, which certainly leads to work hardening in a metallic material, X-ray can sense it without any error, so that it could be said that X-ray diffraction should be one of the best methods to tell whether work hardening exists on a subsurface layer of a machined component; and the width of the distribution curve, which can be represented by B, of the diffraction intensity, can be taken as a criterion for determining whether or not work hardening happens.

If X-ray diffraction is taken to sense the existence of work hardening on a subsurface layer, the only problem which remains is how to strip the surface layer and how to measure the depth of the stripped layer accurately.

A commonly used stripping method is abrading, which is taken to strip the component surface when hardness method is employed to measure the depth of work hardening. Although abrading is very convenient to strip the surface, it leads to extra plastic deformation, or extra work hardening on the newly exposed surface, which will certainly result in some errors in the measurement.

To avoid this type of error, electrolytic corrosion should provide the best access because there is no extra work hardening resulting from this technique and the stripped depth can be easily controlled with all the appropriate corrosion conditions.

The measurement of the depth of stripped layer is another key problem. Because each stripped layer should be extremely thin, sometimes within 5 μm, it is really difficult to get accurate results. But with the weight analysis technique, the measurement becomes quite easy. Whether an accurate measurement can be obtained depends entirely on the accuracy of the analytical balance, which has been highly developed in recent years.

To sum up the above discussion, a combination of X-ray diffraction, electrolytic corrosion and weight analysis techniques can lead to an accurate experiment for the determination of the depth of a work hardened layer.

The measurement with this technique is a time consuming task, although it is very reliable, convenient and accurate as well. It really takes a great deal of time to prepare the standard specimen, find an appropriate electrolytic corrosion condition, strip and weigh the machined component step by step, etc. Therefore, further study should be undertaken to find some more convenient and quicker methods measuring depth of work hardening on a machined surface.

6 CONCLUSIONS

It has been proved by experiments that by judging the changes of the distribution curves of the X-ray diffraction intensity, as well as the electrolytic corrosion and weight analysis, the depth of the work hardened layer can be reliably, conveniently and accurately measured. This X-ray diffraction technique has been taken to examine the work hardening layer of carbon steel components machined by a metal cutting process[7]. In fact, it is not process-dependent, and it can therefore be used to measure the depth of work hardening layer of the components machined by various machining processes such as grinding, rolling, pressure working, and forging.

REFERENCES

1. Z.Y. Zhu, 'The Integrity of Machined Metal Surface', Nanjing University of Aeronautics Press, 1984.
2. Z.Y. Zhu and Y.B. Yen, 'The Effect of the Roundness of a Tool Edge on the Surface Integrity of Broached Components', Mechanical Engineer (in Chinese), 1986, 6, 32-35.
3. Z.Y. Zhu, 'A Theoretical and Experimental Investigation of the Depth of Work Hardening on the Fine Machined Metal Surface', Proceedings of the 4th Chinese University Conference of Metal Cutting, August 1989.
4. X.J. Gao, 'Metal Cutting Theory', Harzhong University of Technology Press, 1985, China.
5. X. Fan, 'X-ray Metallography', Harzhong University of Technology Press, 1986, China.
6. J.H. Zhang, 'The Theory and Technique of Fine Metal Cutting', Shanghai Science and Technology Press, 1986, China.
7. Z.Y. Zhu, 'The Influence of Cutting Tool Radius on the Depth of Work Hardening Layer', Proceedings of the 2nd Sino-Japanese Symposium of Super Fine Machining, March 1988.

Conference Photographs

The following plates show speakers, guests and delegates at the 3rd International Conference on Advances in Coatings and Surface Engineering for Corrosion and Wear Resistance, and the 1st European Workshop on Surface Engineering Technologies and Applications for SMEs. Both events were organized by the Surface Engineering Research Group of the University of Northumbria and took place between 11-15th May 1992 at Newcastle upon Tyne, UK.

City of
Newcastle upon Tyne

TECVAC

Sheffield
City Polytechnic

Contributor Index

This is a combined index for all three volumes. The volume number is given in roman numerals, followed by the page number.

Subject Index

This is a combined index for all three volumes. The volume number is given in roman numerals, followed by the page number.